高等学校规划教材

半微量有机化学实验

第二版

（中英文对照）

李英俊　孙淑琴　编著

化学工业出版社
·北京·

本书在第一版基础上修订，增加了多步合成实验内容，补充了新研制的半微量实验装置，对合成实验增加了英文对照的内容。全书包括有机化学实验的一般知识、有机化学实验基本操作、半微量有机化合物的合成实验三个部分，其中半微量有机化合物的合成实验部分列入了 65 个实验。书末附有实验参考数据等。

本书可作为高等师范院校、综合性大学、工、农、医等院校化学、化工及相关专业的本科生教材，也可供从事相关专业的科技人员参考。

图书在版编目（CIP）数据

半微量有机化学实验/李英俊，孙淑琴编著．—2 版．
北京：化学工业出版社，2009.8（2022.2 重印）
高等学校规划教材
ISBN 978-7-122-05732-7

Ⅰ．半⋯　Ⅱ．①李⋯②孙⋯　Ⅲ．半微量有机合成-化学实验-高等学校-教材　Ⅳ．O621.3-33

中国版本图书馆 CIP 数据核字（2009）第 100673 号

责任编辑：何　丽　徐雅妮　　　　文字编辑：向　东
责任校对：顾淑云　　　　　　　　装帧设计：史利平

出版发行：化学工业出版社（北京市东城区青年湖南街 13 号　邮政编码 100011）
印　　装：北京盛通数码印刷有限公司
787mm×1092mm　1/16　印张 19¼　字数 509 千字　2022 年 2 月北京第 2 版第 6 次印刷

购书咨询：010-64518888　　　　售后服务：010-64518899
网　　址：http://www.cip.com.cn
凡购买本书，如有缺损质量问题，本社销售中心负责调换。

定　　价：58.00 元　　　　　　　　　　　　　　　版权所有　违者必究

第二版前言

《半微量有机化学实验》自第一版出版以来，已经使用五年。本书是根据近年来有机化学实验教学内容的更新及教学改革实践，并结合广大读者使用该书第一版所反馈的意见进行修订的。

1. 本版保持了第一版的体系和特色，全书仍分为有机化学实验的一般知识、有机化学实验基本操作、半微量有机化合物的合成实验三大部分。半微量有机化合物的合成实验所用试剂仍采用半微量，符合"绿色化学"的思想，满足环境保护的要求。

2. 对个别实验内容做了调整和补充，对一些欠准确之处做了修正。与第一版相比，第三章的实验数目由 58 个增加至 65 个。其中甲烷的制备与性质、乙烯的制备与性质 2 个实验，是为了满足师范院校学生教学实践的需要所设置的，同时对实验装置也进行了相应的改进，以适应课堂演示实验的需要。为了满足扩大招生和开放实验室的要求，又增加了 4 个合成实验和 1 个多步合成实验。

3. 为了推动有机化学实验双语教学课程建设，使学生在掌握有机化学实验技能的同时，提高专业英语水平，本书在编写过程中融入双语教学内容。对第三章的 65 个实验增加了相应的英文表述。这为学生今后直接使用英语从事科学研究、阅读英文文献及撰写科技论文打下基础。

此外，在附录 7 中还列出了实验室仪器名称英文对照，以便查阅。

本书为方便教学，另配有实验指导（电子文件），有需要的教师可与作者联系，chem-lab.lnnu@163.com

本书可作为高等师范院校、综合性大学、工农医等院校化学系本科生教材，也可供从事相关专业的科技人员参考。

限于编者水平，书中不妥之处在所难免，恳切期望专家和读者批评指正。

编　者

2009 年 4 月于大连

第一版前言

"绿色"已经成为21世纪的一个重要词语,其含义不再仅仅是对一种颜色的表述,而是具有了社会、经济、生产、文化等各方面的内涵,"绿色化学"、"绿色食品"等词语已经频繁出现在各种媒体上,"绿色化学"尤其引起了社会各界的瞩目。

化学实验是大学化学教育的重要组成部分,也是产生污染的源头,因为实验室中使用的试剂种类、数量较多,污染物成分较复杂,并且废物处理一般费用较高。同时随着扩大招生,学生人数的增多也导致实验室的排污量增大,所以实施化学实验的"绿色化"势在必行。

有机化学实验用的多半是易燃、毒性较大的药品,这些药品极易产生对人体和环境有害的物质,因而减少有机实验中"三废"排放量、提高经济效益和环保效益、保证实验室安全等,已成为一个十分重要的问题。半微量有机化学实验正体现了近年来国际和国内化学界以"绿色化学"理念预防化学污染的新思想。近年来,国内外已有许多高等院校都相继开设了半微量有机化学实验。作为"绿色化学"的一项实验方法与技术,半微量有机化学实验具有节约药品、节省经费、缩短实验时间、安全、便捷、减轻环境污染、节省能源等几大优点。

鉴于此,为了加强和提高学生的操作技能、培养学生的创新思维和环保意识,我们在多年半微量有机化学实验的教学经验基础上,参考国内外有关资料,编写了这本半微量有机化学实验教材,以满足教学需要。本书有如下几个特点。

1. 所有实验使用的试剂均为半微量(本书固体、液体主要反应原料一般为1~5g),能够满足基础有机化学实验课程教学的需要,符合"绿色化学"的思想,满足环境保护的要求。

2. 增加了自行设计的水蒸气蒸馏与重结晶单元半微量化的实验装置。

3. 引入了超声波辐射法合成、光化学合成、电化学合成、相转移催化合成等新的有机合成技术。

4. 增加了多个适合于专业基础课教学的多步合成实验内容,以使学生的实验设计能力、操作能力等得到较大的提高。

本书包含有机化学实验的一般知识、有机化学实验基本操作、半微量有机化合物的合成实验三大部分。半微量有机化合物的合成实验中列入了58个实验。书末附有各类实验参考数据,以便查阅。

本书可作为高等师范院校、综合性大学、工农医等院校化学化工及相关专业的本科生教材,也可供从事相关专业的科技人员参考。

限于编者水平,书中不妥之处在所难免,恳切期望专家和读者批评指正。

编 者

2004年11月于大连

目 录

第1章 有机化学实验的一般知识 1
1.1 有机化学实验室规则 1
1.2 有机化学实验室的安全知识 1
1.2.1 实验室的安全守则 2
1.2.2 实验室事故的预防、处理和急救 2
1.3 有机化学实验预习、实验记录和实验报告的基本要求 5
1.3.1 实验记录本 5
1.3.2 实验预习 5
1.3.3 实验记录 5
1.3.4 实验报告的基本要求 8
1.3.5 总结讨论 8
1.4 有机化学实验常用仪器、实验装置及设备 9
1.4.1 玻璃仪器 9
1.4.2 金属用具 12
1.4.3 常用实验装置 12
1.4.4 仪器的选择 14
1.4.5 仪器的装配与拆卸 14
1.4.6 电器设备 15
1.4.7 其他设备 17
1.5 常用半微量有机化学实验制备仪器 19
1.6 常用玻璃器皿的洗涤和保养 20
1.6.1 玻璃器皿的洗涤 20
1.6.2 玻璃仪器的干燥 20
1.6.3 常用仪器的保养和清洗 21

第2章 有机化学实验基本操作 22
2.1 加热与冷却 22
2.1.1 加热与热源 22
2.1.2 冷却 23
2.2 干燥与干燥剂 24
2.2.1 基本原理 24
2.2.2 液态有机化合物的干燥 24
2.2.3 固体有机化合物的干燥 27
2.2.4 气体的干燥 28
2.3 塞子的钻孔和简单玻璃工操作 28
2.3.1 塞子的钻孔 28
2.3.2 简单玻璃工操作 29
2.4 熔点的测定 31
2.4.1 基本原理 32
2.4.2 测定熔点的方法 32
2.4.3 注意事项 36

2.5 沸点的测定 37
2.5.1 基本原理 37
2.5.2 测定沸点的方法 37
2.5.3 注意事项 37
2.6 蒸馏 38
2.6.1 基本原理 38
2.6.2 蒸馏装置 39
2.6.3 蒸馏操作 39
2.6.4 注意事项 40
2.6.5 蒸馏操作练习 41
2.7 分馏 41
2.7.1 分馏原理 41
2.7.2 影响分馏效率的因素 43
2.7.3 分馏装置 43
2.7.4 操作方法 44
2.7.5 分馏操作练习 44
2.7.6 注意事项 44
2.8 共沸蒸馏 45
2.8.1 基本原理 45
2.8.2 共沸蒸馏装置 46
2.9 减压蒸馏 46
2.9.1 基本原理 46
2.9.2 减压蒸馏装置 47
2.9.3 减压蒸馏操作要点 50
2.9.4 注意事项 50
2.10 水蒸气蒸馏 51
2.10.1 基本原理 51
2.10.2 馏出液组成的计算 51
2.10.3 水蒸气蒸馏装置 52
2.10.4 操作方法 54
2.10.5 注意事项 54
2.11 萃取 54
2.11.1 基本原理 55
2.11.2 操作方法 56
2.12 重结晶 59
2.12.1 基本原理 59
2.12.2 操作方法 59
2.12.3 重结晶提纯的操作练习 62
2.12.4 注意事项 63
2.13 升华 65
2.13.1 基本原理 65
2.13.2 实验操作 65

2.14 旋光度的测定 …………………………… 66
  2.14.1 旋光仪的结构 …………………… 67
  2.14.2 旋光度的测定方法 ……………… 68
  2.14.3 注意事项 ………………………… 69
2.15 折射率的测定 …………………………… 69
  2.15.1 基本原理 ………………………… 69
  2.15.2 阿贝（Abbé）折射仪 …………… 70
2.16 色谱法 …………………………………… 72
  2.16.1 柱色谱 …………………………… 73
  2.16.2 薄层色谱 ………………………… 77
  2.16.3 纸色谱 …………………………… 81
2.17 谱学分析技术 …………………………… 81
  2.17.1 红外光谱 ………………………… 82
  2.17.2 核磁共振谱 ……………………… 87

第 3 章 半微量有机化合物的合成实验 …… 91

3.1 烷烃和烯烃的制备 ……………………… 91
  实验 1 甲烷的制备与性质 ……………… 91
  实验 2 乙烯的制备与性质 ……………… 92
  实验 3 环己烯的合成 …………………… 94
3.2 卤代烃的制备 …………………………… 94
  实验 4 溴乙烷的合成 …………………… 95
  实验 5 正溴丁烷的合成 ………………… 96
  实验 6 2-氯丁烷的合成 ………………… 97
  实验 7 2-甲基-2-氯丙烷的合成 ………… 98
3.3 醚的制备 ………………………………… 98
  实验 8 乙醚的合成 ……………………… 99
  实验 9 正丁醚的合成 …………………… 100
  实验 10 苯乙醚的合成 ………………… 102
  实验 11 β-萘乙醚的合成 ……………… 103
3.4 傅-克反应 ……………………………… 104
  实验 12 苯乙酮的合成 ………………… 106
  实验 13 二苯甲酮（酰基法）的合成 … 107
  实验 14 二苯甲酮（烷基法）的合成 … 109
  实验 15 2-叔丁基对苯二酚的合成 …… 110
3.5 酯化反应 ………………………………… 110
  实验 16 乙酸乙酯的合成 ……………… 111
  实验 17 乙酸正丁酯的合成 …………… 112
  实验 18 乙酸正戊酯的合成 …………… 113
  实验 19 乙酸异戊酯（香蕉油）的合成 … 113
  实验 20 苯甲酸乙酯的合成 …………… 114
  实验 21 乙酰水杨酸（阿司匹林）的合成 …………………………… 115
  实验 22 水杨酸甲酯（冬青油）的合成 …………………………… 116

实验 23 五乙酸葡萄糖酯的合成 ……… 118
3.6 坎尼扎罗反应 …………………………… 119
  实验 24 呋喃甲酸和呋喃甲醇的合成 … 119
  实验 25 苯甲酸和苯甲醇的合成 ……… 121
3.7 缩合反应 ………………………………… 121
  实验 26 肉桂酸的合成 ………………… 121
  实验 27 乙酰乙酸乙酯的合成 ………… 123
3.8 氧化还原反应 …………………………… 125
  实验 28 环己酮的合成 ………………… 125
  实验 29 己二酸的合成 ………………… 126
  实验 30 苯甲酸的合成 ………………… 127
  实验 31 对甲苯胺的合成 ……………… 127
  实验 32 二苯甲醇的合成 ……………… 128
3.9 重氮化反应 ……………………………… 129
  实验 33 对氯甲苯的合成 ……………… 130
  实验 34 甲基橙的合成 ………………… 132
3.10 Diels-Alder 反应 ……………………… 133
  实验 35 环戊二烯与马来酸酐的反应 … 133
  实验 36 环戊二烯与对苯醌的反应 …… 135
  实验 37 蒽与马来酸酐的反应 ………… 136
3.11 相转移催化反应 ……………………… 136
  实验 38 7,7-二氯二环[4.1.0]庚烷的合成 ………………………… 137
  实验 39 对甲苯硫代乙酸的合成 ……… 138
3.12 超声波辐射反应 ……………………… 138
  3.12.1 格氏反应 ………………………… 138
  实验 40 三苯甲醇的合成 ……………… 139
  实验 41 2-甲基-1-苯基-2-丙醇的合成 … 141
  3.12.2 羟醛缩合反应 …………………… 142
  实验 42 苯亚甲基苯乙酮的合成 ……… 142
3.13 鲁卡特反应 …………………………… 143
  实验 43 （±）-α-苯乙胺的合成 ……… 143
3.14 外消旋体的拆分 ……………………… 144
  实验 44 （±）-α-苯乙胺的拆分 ……… 145
3.15 光化学反应 …………………………… 146
  实验 45 苯频哪醇的合成 ……………… 147
  实验 46 苯与马来酸酐的反应 ………… 148
3.16 有机电化学反应 ……………………… 149
  实验 47 碘仿的合成 …………………… 149
3.17 天然产物的提取 ……………………… 150
  实验 48 从茶叶中提取咖啡因 ………… 150
3.18 金属有机化合物的制备 ……………… 152
  实验 49 二茂铁的合成 ………………… 152
3.19 维悌希（Wittig）反应 ……………… 153
  实验 50 （E）-1,2-二苯乙烯的合成 … 154
3.20 霍夫曼酰胺降解反应 ………………… 156
  实验 51 邻氨基苯甲酸的合成 ………… 156

3.21 多步合成 ·················· 157
　3.21.1 以苯酚为原料的多步合成 ······ 157
　　实验52 苯氧乙酸的合成 ············ 157
　　实验53 4-碘苯氧乙酸的合成 ······· 159
　3.21.2 以甲苯为原料的多步合成 ······ 160
　　实验54 间硝基苯甲酸乙酯的合成 ······ 160
　3.21.3 以苯胺为原料的多步合成 ······ 161
　　实验55 乙酰苯胺的合成 ············ 162
　　实验56 对溴乙酰苯胺的合成 ······· 163
　　实验57 对溴苯胺的合成 ············ 164
　　实验58 对硝基乙酰苯胺的合成 ····· 165
　　实验59 对硝基苯胺的合成 ·········· 166
　3.21.4 以对甲苯胺为原料的多步
　　　　 合成 ······················ 166
　　实验60 对甲基-N-乙酰苯胺的合成 ···· 167
　　实验61 对氨基苯甲酸的合成 ········ 167
　3.21.5 以苯甲醛为原料的多步合成 ····· 168
　　实验62 安息香的合成 ·············· 168
　　实验63 二苯基乙二酮的合成 ········ 170
　　实验64 5,5-二苯基乙内酰脲的合成 ···· 171
　　实验65 赤-1,2-二苯基-1,2-乙二醇的
　　　　　合成 ······················ 172

Chapter 3 Semimicro Synthetic Experiment of Typical Organic Compounds ·········· 173

3.1 Preparation of alkanes and alkenes ······ 173
　Experiment 1　Preparation and properties of methane ·············· 173
　Experiment 2　Preparation and properties of ethylene ·············· 175
　Experiment 3　Preparation of cyclohexene ·························· 176
3.2 Preparation of alkyl halides ············ 177
　Experiment 4　Preparation of ethyl bromide ······················ 178
　Experiment 5　Preparation of n-butyl bromide ······················ 179
　Experiment 6　Preparation of 2-chlorobutane (sec-butyl chloride) ······ 181
　Experiment 7　Preparation of 2-methyl-2-chloropropane (t-butyl chloride) ······················ 183
3.3 Preparation of ethers ···················· 184
　Experiment 8　Preparation of diethyl ether ·························· 184
　Experiment 9　Preparation of n-butyl ether ·························· 186
　Experiment 10　Preparation of phenetole (ethyl phenolate) ············ 188
　Experiment 11　Preparation of β-naphthyl ethyl ether ··············· 189
3.4 The Friedel-Crafts reaction ·············· 191
　Experiment 12　Preparation of acetophenone ·························· 193
　Experiment 13　Preparation of benzophenone (Friedel-Crafts acylation method) ······················ 195
　Experiment 14　Preparation of benzophenone (Friedel-Crafts alkylation method) ······················ 196
　Experiment 15　Preparation of 2-t-butyl hydroquinone ················ 198
3.5 Esterification reaction ·················· 199
　Experiment 16　Preparation of ethyl acetate ·························· 200
　Experiment 17　Preparation of n-butyl acetate ························ 201
　Experiment 18　Preparation of n-pentyl acetate (n-amyl acetate) ······ 202
　Experiment 19　preparation of isoamyl acetate (banana oil) ·········· 203
　Experiment 20　Preparation of ethyl benzoate ······················ 204
　Experiment 21　Preparation of acetylsalicylic acid (aspirin) ·········· 205
　Experiment 22　Preparation of methyl salicylate (oil of wintergreen) ················ 207
　Experiment 23　Preparation of glucose pentaacetate ················ 209
3.6 The Cannizzaro reaction ················ 210
　Experiment 24　Preparation of 2-furoic acid and 2-furancarbinol ······ 211
　Experiment 25　Preparation of benzoic acid and benzyl alcohol ········ 213
3.7 Condensation reaction ················ 214
　Experiment 26　Preparation of cinnamic acid ······················ 214
　Experiment 27　Preparation of ethyl acetoacetate ················ 216
3.8 Oxidation and reduction reaction ······· 219
　Experiment 28　Preparation of cyclohexanone ···················· 219

Experiment 29	Preparation of adipic acid	220	Experiment 45	Preparation of benzo pinacol ... 251

- Experiment 29 Preparation of adipic acid 220
- Experiment 30 Preparation of benzoic acid 222
- Experiment 31 Preparation of p-toluidine 223
- Experiment 32 Preparation of benzhydrol (diphenyl carbinol) 224
- 3.9 Diazo reaction 225
- Experiment 33 Preparation of p-chlorotoluene 226
- Experiment 34 Preparation of methyl orange 229
- 3.10 The Diels-Alder reaction 232
- Experiment 35 The Diels-Alder reaction of cyclopentadiene with maleic anhydride 232
- Experiment 36 The Diels-Alder reaction of cyclopentadiene with 1,4-benzoquinone 233
- Experiment 37 The Diels-Alder reaction of anthracene with maleic anhydride 235
- 3.11 Phase-transfer catalytic reaction 236
- Experiment 38 Preparation of 7,7-dichlorobicyclo [4.1.0] heptane (7,7-dichloronocarane) 236
- Experiment 39 Preparation of p-methyl phenylthioacetic acid 238
- 3.12 Ultrasonic radiation reaction 239
- 3.12.1 The Grignard reaction 239
- Experiment 40 Preparation of triphenylmethanol (triphenylcarbinol) 239
- Experiment 41 Preparation of 2-methyl-1-phenyl-2-propanol 243
- 3.12.2 Aldol condensation reaction 244
- Experiment 42 Preparation of benzalacetophenone 244
- 3.13 Leuckart Reaction 245
- Experiment 43 Preparation of (±)-α-phenylethylamine 246
- 3.14 Resolution of racemic compound 248
- Experiment 44 Resolution of (±)-α-phenylethylamine 248
- 3.15 Photochemical reaction 250
- Experiment 45 Preparation of benzo pinacol 251
- Experiment 46 Addition of benzene to butenedioic anhydride (maleic anhydride) 253
- 3.16 Organic electrochemical reaction 254
- Experiment 47 Preparation of iodoform 254
- 3.17 Isolation of natural product 255
- Experiment 48 Isolation of caffeine from tea 256
- 3.18 Preparation of organometallic compound 259
- Experiment 49 Preparation of ferrocene 259
- 3.19 The Wittig reaction 261
- Experiment 50 Preparation of (E)-1,2-diphenylethene (stilbene) 262
- 3.20 Hofmann degradation reaction 264
- Experiment 51 Preparation of anthranilic acid (2-aminobenzoic acid) 264
- 3.21 Multistep synthesis 266
- 3.21.1 The multistep synthesis of 4-iodophenoxyacetic acid from phenol 266
- Experiment 52 Preparation of phenoxyacetic acid 267
- Experiment 53 Preparation of 4-iodo phenoxyacetic acid 268
- 3.21.2 The multostep synthesis of ethyl 3-nitrobenzoate from toluene 269
- Experiment 54 Preparation of ethyl 3-nitrobenzoate 269
- 3.21.3 The multistep synthesis of p-bromoaniline and p-nitroaniline from aniline 270
- Experiment 55 Preparation of acetanilide 271
- Experiment 56 Preparation of p-bromoacetanilide 272
- Experiment 57 Preparation of p-bromoaniline 273
- Experiment 58 Preparation of p-nitroacetanilide 274
- Experiment 59 Preparation of p-

nitroaniline 276
3.21.4 The multostep synthesis of *p*-aminobenzoic acid from *p*-toluidine 276
Experiment 60 Preparation of *N*-acetyl-*p*-toluidine 277
Experiment 61 Preparation of *p*-aminobenzoic acid 278
3.21.5 The multistep synthesis of 5,5-diphenylhydantion and *erythro*-1,2-diphenyl-1,2-ethandiol from benzaldehyde 279
Experiment 62 Preparation of benzoin 279
Experiment 63 Preparation of benzil 281
Experiment 64 Preparation of 5,5-diphenylhydantion 281
Experiment 65 Preparation of *erythro*-1,2-diphenyl-1,2-ethandiol 282

附录 284

附录1 常用元素的相对原子质量（2004） 284
附录2 常用有机溶剂的纯化 284
附录3 常用酸碱溶液的质量分数、相对密度和溶解度 288
附录4 水的饱和蒸气压 290
附录5 部分共沸混合物的性质 290
附录6 常用酸碱的相对分子质量及浓度 292
附录7 实验室仪器（Laboratory Equipment） 292

参考文献 296

第 1 章　有机化学实验的一般知识

化学是一门以实验为基础的学科，而有机化学实验在整个化学实验中占有非常重要的地位，是化学专业教学计划中的必修基础课程，重视和学好这门课程非常重要。首先，介绍有机化学实验的一般知识，学生在进行有机实验之前必须认真学习这部分知识，了解进入实验室后应该注意的事项及有关规定。

1.1　有机化学实验室规则

为了保证有机化学实验正常、有效、安全地进行，保证实验课的教学质量，学生必须遵守有机化学实验室的规则。

① 做好实验前的准备工作，包括预习有关实验的内容、查找实验中所需的有关试剂与药品的物理常数、查阅相关的理论和参考资料、找全所需要的器材，以免临时慌乱，预习方法见 1.3。写好实验预习报告，方可进入实验室；没达到预习要求者，不得进行实验。准备工作做得好，不仅能保证实验顺利进行，而且可以从实验中获得更多的知识。

② 进入实验室时，应熟悉实验室及其周围的环境，熟悉灭火器材、急救药箱的使用及放置的地方。严格遵守实验室的安全守则和每个实验操作中的安全注意事项。如有意外事故发生应立即报请老师处理。进入实验室必须穿着实验服，实验室内不得吃东西和吸烟。

③ 每次实验时，将实验装置组装好后，要经实验指导教师检查合格后，方可进行下一步操作。在操作前，要想好每一步操作的目的、意义、实验中的关键步骤、难点及注意事项，了解所用药品的性质及应注意的事项。

④ 实验时应保持安静，遵守纪律。要求精神集中、认真操作、细致观察、积极思考、忠实记录。不得擅自离开。

⑤ 实验中要遵从教师的指导，严格按照实验指导书所规定的步骤、试剂的规格和用量进行实验。若要更改，必须经指导教师同意后，才可改变。

⑥ 应经常保持实验室的整洁。暂时不用的器材，不要放在桌面上，以免碰倒损坏。污水、污物、残渣、火柴梗、废纸、塞芯和玻璃碎片等分别放在指定的地点，不得乱丢，更不得丢入水槽，废酸和废碱应分别倒入指定的回收缸中。

⑦ 爱护公共仪器和工具，在指定的地点使用，并保持整洁。要节约用水、电、煤气和药品。仪器损坏应如实填写破损单，办理赔偿手续。

⑧ 实验完毕后，要将个人实验台面打扫干净，仪器洗净并放好，拔掉电源插头，关闭水、电和煤气开关。实验结果由指导教师登记，实验产品回收统一管理。请指导教师检查、签字后方可离开实验室。课后，按时写出符合要求的实验报告。

⑨ 学生要轮流值日，值日生应负责整理好公用器材，打扫实验室，倒净废物缸，检查水、电、煤气开关是否关闭，关好门窗。离开实验室前，再请指导教师检查、签字。

1.2　有机化学实验室的安全知识

有机化学实验所用的药品多数是有毒、可燃、有腐蚀性或爆炸性的，所用的仪器大部分是玻璃制品，所以在有机实验室中工作，若粗心大意，就易发生事故，如割伤、烧伤，乃至

1

火灾、中毒、爆炸等，必须认识到化学实验室是潜在危险的场所。虽然在选择实验时，尽量选用低毒性的溶剂和试剂，并在教学中采用了半微量有机实验及利用有机实验CAI课件进行微机辅助教学，减少了实验室的污染，改善了师生的实验环境，使燃烧、爆炸、中毒等隐患事故相应减少，但是大量使用时对人体也会造成伤害。因此，一定要重视安全问题，思想上提高警惕，实验时严格遵守操作规程，加强安全措施，这样大多数事故就可以避免。下面介绍实验室的安全守则和实验室事故的预防、处理和急救。

1.2.1 实验室的安全守则

① 实验开始前应检查仪器是否完整无损，装置是否正确稳妥，在征求指导教师同意后，方可进行实验。

② 实验进行时，不得离开岗位，要经常注意反应进行的情况和装置有无漏气、破裂等现象。

③ 当进行有可能发生危险的实验时，要根据实验情况采取必要的安全措施，如戴防护眼镜、面罩、橡皮手套等。

④ 使用易燃、易爆药品时，应远离火源。实验试剂不得入口。严禁在实验室内吸烟、吃食物。实验结束后要认真洗手。

⑤ 熟悉安全用具，如灭火器材、砂箱、石棉布、急救药箱的放置地点和使用方法，并妥善保管。安全用具和急救药品不准移作它用。

1.2.2 实验室事故的预防、处理和急救

1.2.2.1 火灾的预防、处理和急救

实验室中使用的有机溶剂大多数是易燃的，着火是有机实验室常见的事故之一，应尽可能避免使用明火。防火的基本原则如下。

(1) 在操作易燃的溶剂时要特别注意：①应远离火源；②勿将易燃、易挥发液体放在敞口容器中，如烧杯中用直接火加热；③加热必须在水浴中进行，切勿使容器密闭，否则会造成爆炸。当附近有暴露放置的易燃溶剂时，切勿点火。

(2) 在进行易燃物质实验时，应先将酒精一类易燃的物质搬开。

(3) 蒸馏易燃的有机物时，装置不能漏气，如发现漏气，应立即停止加热，检查原因，若因塞子被腐蚀，则待冷却后，才能换掉塞子。接收瓶不宜用敞口容器，如广口瓶、烧杯等，而应用窄口容器，如三角瓶等。蒸馏装置接收瓶的尾气出口应远离火源，最好用橡皮管引到下水道口或室外。

(4) 回流或蒸馏低沸点易燃液体时应注意以下几点。

① 应放入数粒沸石、素烧瓷片或一端封口的毛细管，以防止暴沸。如在反应后才发觉未放入沸石这一类物质时，绝不能急躁，不能立即揭开瓶塞补放，而应先停止加热，待被回流或蒸馏的液体冷却后才能加入，否则会因暴沸而发生事故。

② 严禁直接加热。

③ 瓶内液量最多只能装至半满。

④ 加热速度宜慢、不能快，避免局部过热。总之，回流或蒸馏低沸点易燃液体时，一定要谨慎。

(5) 油浴加热或回流时，必须注意避免由于冷凝用水溅入热油浴中致使油溅到热源上而引起火灾的危险。通常发生危险的主要原因是橡皮管在冷凝管上套得不紧密，开动水阀过快，水流过猛把橡皮管冲掉，或者由于套不紧而漏水。所以，要求橡皮管套入时要很紧密，开动水阀时要慢动作，使水流慢慢通入冷凝管中。

(6) 当处理大量可燃性液体时，应在通风橱或指定地方进行，室内应无火源。

(7) 不得把燃着的或带有火星的火柴梗、纸条等乱抛乱掷，也不得丢入废液缸中，否则

会发生危险。

（8）实验室不得存放易燃、易挥发物质。

（9）有煤气的实验室，应经常检查管道和阀门是否漏气。

（10）一旦发生火灾，应沉着镇静地及时采取正确措施，以免事故的扩大。首先，立即切断电源，移走易燃物。然后，根据易燃物的性质和火势采取适当的方法进行扑救。有机物着火通常不用水进行扑救，因为一般有机物不溶于水或遇水可能发生更强烈的反应而引起更大的事故。小火可用石棉布盖熄，火势较大时，应用灭火器扑救。

常用的灭火器有二氧化碳、四氯化碳、干粉、泡沫等灭火器。目前，实验室中常用的是干粉灭火器。使用时，拔出销钉，将出口对准火点，将上手柄压下，干粉即可喷出。二氧化碳灭火器也是有机实验室中常用的灭火器。灭火器内存放着压缩的二氧化碳气体，适用于油脂、电器、较贵重的仪器着火时使用。虽然四氯化碳和泡沫灭火器都具有较好的灭火性能，但四氯化碳在高温下能生成剧毒的光气，而且与金属钠接触会发生爆炸；泡沫灭火器会喷出大量的泡沫而造成严重污染，给后处理带来麻烦。因此，这两种灭火器一般不用。不管采用哪一种灭火器，都是从火的周围开始向中心扑灭。

地面或桌面着火时，还可用沙子扑救，但容器内着火时不易用沙子扑救。身上着火时，应就近在地上打滚（速度不要太快）将火焰扑灭。千万不要在实验室内乱跑，以免造成更大的火灾。

1.2.2.2 爆炸的预防

在有机化学实验里预防爆炸的一般措施如下。

（1）蒸馏装置必须正确。常压蒸馏不能造成密闭体系，应使装置与大气相连通。减压蒸馏时，要用圆底烧瓶作为接收器，不能用三角烧瓶、平底烧瓶等不耐压容器作为接收器，否则会发生爆炸。

（2）无论是常压蒸馏还是减压蒸馏，均不能将液体蒸干，以免局部过热或产生过氧化物而发生爆炸。

（3）切勿使易燃易爆的物体接近火源，有机溶剂，如乙醚和汽油一类的蒸气与空气相混合时极为危险，可能会由一个热的表面或者一个火花、电火花引起爆炸。

（4）使用乙醚时，必须检查有无过氧化物存在，如果有过氧化物存在时，应立即用硫酸亚铁除去过氧化物，才能使用，除去乙醚中过氧化物的方法可参考有关书籍。同时，使用乙醚时应注意在通风较好的地方或在通风橱内进行。

（5）对易爆炸的固体，如重金属乙炔化物、苦味酸金属盐、三硝基甲苯等都不能重压或撞击，以免引起爆炸。对于这些危险的残渣，必须小心销毁。例如，重金属乙炔化物可用浓盐酸或浓硝酸使其分解，重氮化合物可加水煮沸使其分解等。

（6）卤代烷勿与金属钠接触，因反应太猛往往会发生爆炸。

（7）在用玻璃仪器组装实验装置之前，要检查玻璃仪器是否有破损。

1.2.2.3 中毒的预防

（1）剧毒药品要妥善保管，不许乱放，实验中所用的剧毒物质应有专人负责收发，并向使用毒物者提出必须遵守的操作规程。实验后对有毒残渣必须进行妥善而有效的处理，不准乱丢。

（2）有些剧毒物质会渗入皮肤，因此接触这些物质时必须戴橡皮手套，操作后立即洗手，切勿让剧毒药品沾及五官或伤口。例如，氰化钠沾及伤口后就会随血液循环至全身，严重者会造成中毒死亡事故。

（3）称量药品时应使用工具，不得直接用手接触，尤其是有毒药品。任何药品不得用嘴尝。

（4）在反应过程中可能生成有毒或有腐蚀性气体的实验应在通风橱内进行，使用后的器皿应及时清洗。在使用通风橱时，实验开始后不要把头部伸入橱内。

（5）如发生中毒现象，应让中毒者及时离开现场，到通风好的地方，严重者应及时送医院。

（6）当发现实验室漏煤气时，应立即关闭煤气开关，打开窗户，并通知实验室工作人员进行检查和修理。

1.2.2.4　触电的预防

进入实验室后，首先应了解水、电、气的开关位置，而且要掌握它们的使用方法。在实验中，应先将电器设备上的插头与插座连接好后，再打开电源开关。使用电器时，应防止人体与电器导电部分直接接触，不能用湿手或用手握湿的物体接触电源插头。为了防止触电，装置和设备的金属外壳等都应连接地线。实验结束后应切断电源，再将连接电源插头拔下。

1.2.2.5　玻璃割伤的预防、处理和急救

有机化学实验中主要使用的是玻璃仪器，玻璃割伤是常见的事故之一。使用玻璃仪器时最基本的原则是：不得对玻璃仪器的任何部位施加过度的压力。

（1）需要用玻璃管和塞子连接装置时，用力处不要离塞子太远，尤其是插入温度计时，要特别小心。

（2）新割断的玻璃管或玻璃棒的断口处特别锋利，使用时要将断口处用火烧至熔化，使其成圆滑状。

玻璃割伤后，要仔细观察伤口有没有玻璃碎片，如有应用消毒过的镊子取出。若为一般轻伤，应及时挤出污血，用生理盐水洗净伤口，在伤口处涂上碘酒，再用绷带包扎；若伤口严重，流血不止，应立即用绷带扎紧伤口上部，使伤口停止流血，急送医疗所。

实验室应备有急救药箱，内有急救药品，如：生理盐水、紫药水、碘酒、双氧水、饱和硼酸溶液、1%醋酸溶液、5%碳酸氢钠溶液、70%酒精、玉树油、烫伤油膏、万花油、药用蓖麻油、硼酸膏或凡士林、磺胺药粉等。还应备有洗眼杯、消毒棉花、纱布、胶布、绷带、剪刀、镊子、橡皮管等急救用具。

1.2.2.6　灼伤的预防、处理和急救

皮肤接触了高温、低温或腐蚀性物质之后均能被灼伤。为避免灼伤，在接触这些物质时，最好戴上橡胶手套和防护眼镜。发生灼伤时应按下列要求处理。

（1）酸灼伤

皮肤上——立即用大量水冲洗，然后用5%碳酸氢钠溶液清洗，涂上油膏，并将伤口扎好。

眼睛上——抹去溅在眼睛外面的酸，立即用水冲洗，用洗眼杯清洗，或将橡皮管套在水龙头上，用慢水对准眼睛冲洗后，即到医院就诊，或者再用稀碳酸氢钠溶液洗涤，最后滴入少许蓖麻油。

衣服上——依次用水、稀氨水和水冲洗。

地板上——撒上石灰粉，再用水冲洗。

（2）碱灼伤

皮肤上——先用水冲洗，然后用饱和硼酸溶液或1%醋酸溶液清洗，再用水冲洗，最后涂上油膏，并包扎好。

眼睛上——抹去溅在眼睛外面的碱，用水冲洗，再用饱和硼酸溶液清洗后，滴入蓖麻油。

衣服上——先用水洗，然后用10%醋酸溶液洗涤，再用氢氧化铵中和多余的醋酸，最

后用水冲洗。

(3) 溴灼伤　如溴弄到皮肤上，应立即用水冲洗，涂上甘油，敷上烫伤膏，将伤处包好。如眼睛受到溴的蒸气刺激，暂时不能睁开时，可对着盛有酒精的瓶口注视片刻。

(4) 水灼伤　被热水烫伤后一般在患处涂上红花油，然后擦烫伤膏。

上述各种急救方法，仅为暂时减轻疼痛的措施。若伤势较重，在急救之后，应速送医院诊治。

1.3　有机化学实验预习、实验记录和实验报告的基本要求

有机化学实验是一门综合性较强的理论联系实际的课程，是培养学生独立工作能力的重要环节。学生在进行每个实验时，必须做好实验预习、实验记录和实验报告。

1.3.1　实验记录本

每个学生都必须准备一个实验记录本，并编上页码，不能用活页本或零星纸张代替。不能撕下记录本的任何一页。如果写错了可以用笔勾掉，但不能涂抹或用橡皮擦掉。文字要简练明确，书写整齐，字迹清楚。写好实验记录本是从事科研实验的一项重要训练。

在实验记录本上做预习笔记、实验记录及课后实验总结。

1.3.2　实验预习

为了使实验能够达到预期的效果，在实验之前要做好充分的预习和准备。预习时除了要求反复阅读实验内容，领会实验原理、有关实验步骤和注意事项外，还需在实验记录本上准备好预习提纲。以合成实验为例，预习提纲包括下列内容：

① 实验目的；
② 主反应和重要副反应方程式；
③ 原料、产物和副产物的物理常数；
④ 原料用量（g、mL、mol），计算过量试剂的过量百分数，计算理论产率；
⑤ 正确而清楚地画出主要反应装置图，并注明仪器名称；
⑥ 用图表形式表示整个实验步骤的流程。

预习时，应该清楚每一步操作的目的是什么，为什么这么做，要清楚本次实验的关键步骤和难点，实验中有哪些安全问题。预习是做好实验的关键，只有预习充分，实验时才能做到又快又好。

1.3.3　实验记录

实验记录是科学研究的第一手资料，实验记录的好坏直接影响对实验结果的分析。因此，学会做好实验记录也是培养学生科学作风及实事求是精神的一个重要环节。

在实验过程中，实验者必须养成一边实验一边直接在记录本上做记录的习惯，不许事后凭记忆补写，或以零星纸条暂记再转抄。记录的内容包括：实验的日期，试剂的规格和用量，仪器的名称、规格、牌号，实验的全部过程，如加入药品的数量，仪器装置，每一步操作的时间、内容和所观察到的现象（包括温度、颜色、体积或质量的数据等）。记录要实事求是，准确反映真实的情况，特别是当观察到的现象与预期的不同，以及操作步骤与教材规定的不一致时，要按实际情况记录清楚，以便作为总结讨论的依据。应该牢记，实验记录是原始资料，科学工作者必须重视。

判断记录本内容的标准是记录必须完整，且组织得好和清楚，不仅自己现在看得懂，甚至几年后也能看懂，而且还能使他人看懂。记录是宁可多记录一些，也不要漏记，在写实验

报告时从中精选。如漏记了主要内容,将难以补救。

实验记录示例

实验××× 溴乙烷的制备

【实验目的】
- 学习从醇制备溴代烷的原理和方法。
- 学习蒸馏装置和分液漏斗的使用方法。

【反应式】
主反应
$$NaBr + H_2SO_4 \longrightarrow NaHSO_4 + HBr$$
$$HBr + C_2H_5OH \rightleftharpoons C_2H_5Br + H_2O$$

副反应
$$2C_2H_5OH \xrightarrow{H_2SO_4} C_2H_5OC_2H_5 + H_2O$$
$$C_2H_5OH \xrightarrow{H_2SO_4} C_2H_4 + H_2O$$

【物理常数】

名称	相对分子质量	相对密度	熔点/℃	沸点/℃	溶解度/g·100g⁻¹(溶剂)	
					水中	醇中
乙醇	46	0.79	−117.3	78.4	∞	
溴化钠	103				79.5 (0℃)	
硫酸	98	1.83	10.38	340(分解)	∞	
溴乙烷	109	1.46	−118.6	38.4	1.06 (0℃)	∞
硫酸氢钠	120				50 (0℃) 100 (100℃)	
乙醚	74	0.71	−116	34.6	7.5 (20℃)	∞
乙烯	28		−169	−103.7		

【计算】

名称	实际用量/g	理论量/mmol	过量/%	理论产量/g
95%乙醇	2.6 (3.3mL、56.5mmol)	48.6	16.3	
NaBr	5 (48.6mmol)			
浓硫酸(98%)	11.6 (6.3mL、118.3mmol)	48.6	143.4	
C_2H_5Br		48.6		5.30

【仪器装置图】

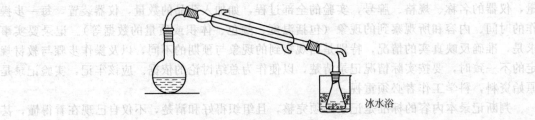

(a) 反应装置

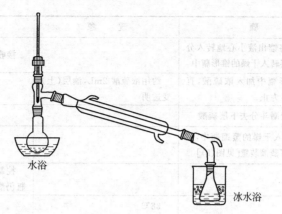

(b) 蒸馏装置

【实验步骤流程图】

```
         C₂H₅OH、NaBr、H₂SO₄、H₂O
                    │ 加热
         ┌──────────┴──────────┐
       残留物                 馏出物
     H₂SO₄、NaHSO₄      C₂H₅Br、C₂H₅OC₂H₅、
                        C₂H₅OH、H₂O、HBr
         ┌──────────────────┴─┐
      油层（下）              水层（上）
  C₂H₅Br、C₂H₅OH、C₂H₅OC₂H₅   H₂O、HBr
         │ 浓 H₂SO₄ 洗，分离
         ├──────────────────┐
      油层（上）           硫酸层（下）
  C₂H₅Br、H₂SO₄（微量）   H₂SO₄、C₂H₅OH、
         │ 蒸馏（水浴）    C₂H₅OC₂H₅
    C₂H₅Br（37~40℃）
```

【实验记录】

日期： 年 月 日

时间	步 骤	现 象	备 注
8:30	安装反应装置[见图(a)]		接收瓶内放入少量冰水，并将其置于冰水浴中。接引管的支口用橡皮管导入下水道或室外
8:45	在烧瓶中加入 3.3mL 95%乙醇和3mL 水,混合均匀		
8:55	在不断振摇和冷却下慢慢加入 6.3mL 浓硫酸		
9:00	混合物冷却至室温后,在冷却条件下加入 5g 溴化钠及两粒沸石	固体成碎粒状,未全溶,溶液淡黄色	
9:10	小火加热使反应平稳进行	液体微沸有泡沫产生,在冷凝管中有无色液体馏出,乳白色油状物沉在接收瓶的水底	
9:50		固体消失	
10:00	停止加热	馏出液中已无油滴,瓶中残留物冷却成无色晶体	

续表

时 间	步 骤	现 象	备 注
10:15	待烧瓶冷却后,将馏出液小心地转入分液漏斗中,将有机层转入干燥的锥形瓶中		接收瓶仍然浸在冰水浴中
10:25	边振摇边向锥形瓶中加入浓硫酸,直至溶液有明显分层为止	约用浓硫酸2mL,油层(上)变透明	
10:30	再用干燥的分液漏斗分去下层硫酸		
10:35	油层(粗产品)转入干燥的蒸馏瓶中,加入两粒沸石,安装好蒸馏装置[见图(b)]		
10:50	水浴加热蒸馏		控制水浴温度在50～55℃,接收瓶仍然浸在冰水浴中
10:55	开始有馏出液	38℃	
11:00	蒸完	39.5℃	得到3.2g

【产物】

溴乙烷,无色透明液体,沸程38～39.5℃,产量3.2g,产率60.4%。

【讨论】

本次实验产物产量和质量基本合格。产率较低的原因可能是反应时加热太猛,使副反应增加。另外,溴乙烷沸点很低,后处理纯化时一部分产物因挥发而损失。

1.3.4 实验报告的基本要求

实验报告应包括实验目的,反应式,主要试剂的规格、用量,实验步骤和现象,产率计算,注意事项,讨论等。要如实记录填写报告,文字精练,图要准确,讨论要认真。关于实验步骤的描写,不可照抄书上的实验步骤,应该对所做的内容作概要的描述。下面介绍的实验报告的格式仅供参考。

① 实验目的
② 反应式
　主反应
　副反应
③ 主要试剂及产物的物理常数

名 称	相对分子质量	性 状	密 度	熔点/℃	沸点/℃	溶解度		
						水	醇	醚

④ 仪器装置图
⑤ 实验步骤和现象
⑥ 产品外观、质量、产率、物理常数测试结果
⑦ 注意事项
⑧ 总结与讨论

1.3.5 总结讨论

做完实验后,除了整理实验报告,写出产物的产量、产率、状态和实际测得的物性,如沸点、熔程等数据,以及回答指定的问题外,还要根据实际情况对产物的质量和数量、实验

过程中出现的问题等进行讨论，以总结经验和教训，写出做实验的体会。通过讨论来总结、提高和巩固实验中所学到的理论知识和实验技术，这是把直接的感性认识提高到理性思维的必要步骤，也是科学实验中不可缺少的一个环节。

一份完整的实验报告可以充分体现学生对实验理解的深度、综合解决问题的能力以及文字表达的能力。

1.4　有机化学实验常用仪器、实验装置及设备

在有机化学实验中经常会用到一些玻璃仪器、实验装置及有关设备，了解实验中所使用的这些仪器、装置、设备及其维护方法是十分必要的，也是最基本的要求。

1.4.1　玻璃仪器

玻璃仪器一般是由软质或硬质玻璃制作而成的。软质玻璃耐温、耐腐蚀性较差，但价格便宜，因此一般用它制作的仪器均不耐温，如普通漏斗、量筒、吸滤瓶、干燥器等。硬质玻璃具有较好的耐温和耐腐蚀性，制成的仪器可在温度变化较大的情况下使用，如烧瓶、烧杯、冷凝管等。

常用的玻璃仪器一般分为两类，普通玻璃仪器和标准磨口仪器。在有机化学实验室，常用的普通玻璃仪器有非标准口的锥形瓶、烧杯、吸滤瓶、布氏漏斗、热水漏斗、玻璃钉漏斗、普通漏斗、分液漏斗等，如图1-1所示。

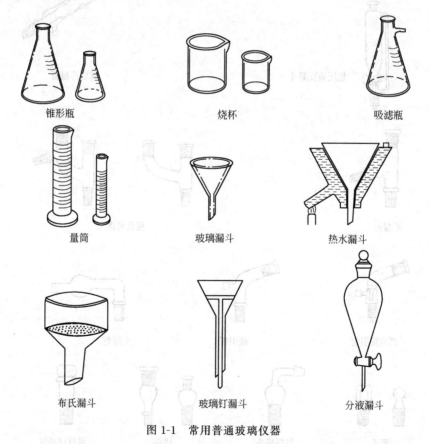

图1-1　常用普通玻璃仪器

常用的标准磨口仪器有圆底烧瓶、三口烧瓶、蒸馏头、冷凝器、接引管等，如图1-2所示。

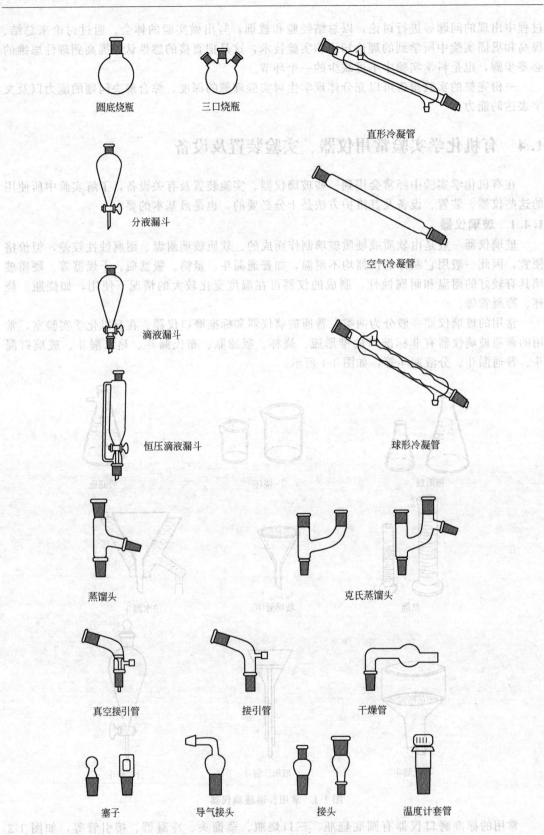

图 1-2 常用标准磨口仪器

标准磨口仪器接口部位的尺寸大小都是统一的，即标准化的。标准磨口仪器根据磨口口径分为 10、14、19、24、29、34、40、50 等号。例如，14 号、19 号、24 号指的就是磨口的最大端直径分别为 14mm、19mm、24mm。由于口径尺寸的标准化、系列化，磨口密合，因此凡属于同类型规格的接口，均可任意互换，各部件能组装成各种配套仪器。对不同类型规格的磨口仪器，还可以通过相应尺寸的大小磨口接头使之相互连接。学生使用的常量仪器一般是 19 号的磨口仪器，半微量实验中采用的是 14 号的磨口仪器。

使用标准磨口仪器应该注意以下几点。

① 在安装仪器时要做到横平竖直，整齐、正确，磨口连接处不应受歪斜的应力，否则易将仪器折断，特别在加热时，仪器受热，应力更大。另外，安装仪器时不可用力过猛，以免仪器破裂。

② 一般情况下，磨口处不必涂润滑剂，以免沾污反应物或产物。但反应中使用强碱时，则要涂润滑剂，以免连接处因碱腐蚀而黏结在一起，无法拆开。减压蒸馏时，应适当涂抹真空脂，以保证装置的密封性。

③ 磨口仪器如果黏结在一起，不可用力拆卸。可用热水煮黏结处或用电吹风对着黏结口处加热，使其膨胀而脱落，还可用木槌轻轻敲打黏结处。

④ 仪器使用完，应及时清洗干净，否则放置时间太久，接口处容易黏结在一起，很难拆开。

⑤ 标准磨口仪器的磨口处要干净，不得黏有固体物质。清洗时，应避免用去污粉擦洗磨口，否则会使磨口连接不紧密，甚至会损坏磨口。

⑥ 带旋塞或具塞仪器（如滴液漏斗）清洗后不用时，应将旋塞与磨口之间用纸片隔开，以免黏结。

常用玻璃仪器的应用范围见表 1-1。

表 1-1 有机化学实验常用玻璃仪器的应用范围

仪 器 名 称	应 用 范 围	备 注
圆底烧瓶	用于反应、回流加热及蒸馏	
三口圆底烧瓶	用于反应，三口分别安装电搅拌、回流冷凝管、温度计等	
冷凝管	用于蒸馏和回流	
蒸馏头	与圆底烧瓶组装后用于蒸馏	
单股接收管	用于常压蒸馏	
双股接收管	用于减压蒸馏	
分馏柱	用于分馏多组分混合物	
恒压滴液漏斗	用于反应体系内有压力时顺利滴加液体	
分液漏斗	用于液体的萃取和分离	也可用于滴加液体
锥形瓶	用于储存液体，混合溶液及加热少量溶液	不能用于减压蒸馏
烧杯	用于加热、浓缩溶液及溶液的混合和转移	
量筒	量取溶液	切勿用直接火加热
吸滤瓶	用于减压过滤	不能用直接火加热
布氏漏斗	用于减压过滤	瓷质
瓷板漏斗	用于减压过滤	瓷质板为活动圆孔板
熔点管	用于测熔点	内装液体石蜡、硅油等
干燥管	装干燥剂，用于无水反应装置	

1.4.2 金属用具

有机化学实验中常用的金属用具有升降架、铁架台、烧瓶夹、冷凝管夹（又称万能夹）、铁圈、S扣、镊子、锉刀、打孔器、水浴锅、热水漏斗、水蒸气发生器、煤气灯、不锈钢刮刀、不锈钢小勺、剪子等。这些仪器应放在实验室规定的地方。要保持这些仪器的清洁，经常在活动部位加上一些润滑剂，以保证灵活不生锈。

1.4.3 常用实验装置

在有机化学实验中，安装好实验装置是做好实验的基本保证。反应装置一般根据实验要求组装。图 1-2 列出了一些常用的标准磨口仪器，利用这些基本"配件"可以搭建出一般常规有机化学实验中所需要的实验装置。常用的实验装置有蒸馏装置、分馏装置、回流反应装置、带有搅拌及回流的反应装置、带有气体吸收的装置、回流分水装置、水蒸气蒸馏装置等。图 1-3～图 1-9 为常见的实验装置图。

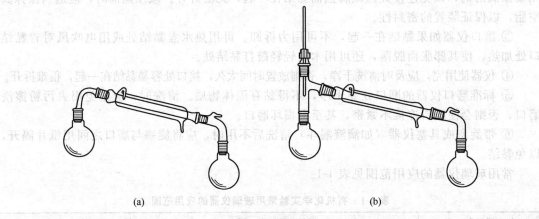

图 1-3 蒸馏装置

图 1-5(a)、(b)、(c) 是一组常见的回流冷凝装置。在室温下，有些反应速率很慢或难于进行，为了使反应尽快地进行，常需要使反应物长时间保持沸腾。在这种情况下，就需要使用回流冷凝装置，使蒸气不断地在冷凝管内冷凝而返回反应器中，以防止反应瓶中的物质逃逸损失。将反应物放在圆底烧瓶中，在适当的热源上或水浴中加热。直立的冷凝管夹套中自下至上通冷凝水，使夹套充满水，水流速度不必很快，能保持蒸气充分冷凝即可。加热程度也需控制，使蒸气上升的高度不超过冷凝管的 1/3。当回流温度不太高时（低于 140℃），通常选用球形冷凝管或直形冷凝管，前者较后者冷凝效果好一些。当回流温度较高时（高于 150℃），就要选用空气冷凝管，因为球形或直形冷凝管在高温下容易炸裂。如果反应物怕受潮，可在冷凝管上端配置干燥管来防止空气中湿气侵入 [见图 1-5(b)]。如果反应中会放出有害气体（如溴化氢），可加接气体吸收装置 [见图 1-5(c)]。图 1-5(c) 的烧杯中可装一些气体吸收液，如酸液或碱液，以吸收反应过程中产生的碱性或酸性气体。

图 1-6(a)、(b) 是一组回流滴加装置。有些反应进行剧烈，放热量大，如将反应物一次加入，会使反应失去控制；有些反应为了控制反应物的选择性，也

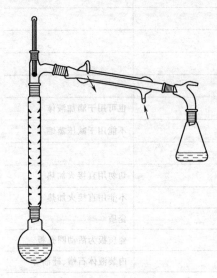

图 1-4 分馏装置

不能将反应物一次加入。在这种情况下，可采用带滴液漏斗的回流冷凝装置（见图 1-6），将一种试剂逐渐滴加进去。可以根据需要，在反应烧瓶外面用冷水浴或冰水浴进行冷却，在某些情况下，也可用水浴加热。

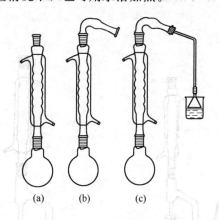

图 1-5　回流冷凝装置

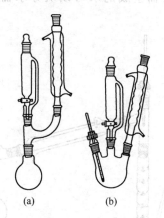

图 1-6　回流滴加装置

图 1-7(a)、(b)、(c) 是一组搅拌回流装置。用固体和液体或互不相溶的液体进行反应时，为了使反应混合物能充分接触，应该进行强烈的搅拌或者振荡。在反应物量小、反应时间短而且不需要加热或温度不太高的操作中，用手摇动容器就可达到充分混合的目的。用回流冷凝装置进行反应时，有时需要间歇的振荡，这时可将固定烧瓶和冷凝管的夹子暂时松开，一只手扶住冷凝管，另一只手拿住瓶颈做圆周运动，每次振荡后应把仪器重新夹好。在搅拌反应中，如果反应混合物量较大，或较黏稠，或含有固体物质，或反应需要较长时间的搅拌，在这些反应的实验中，最好用电动搅拌器。电动搅拌的效率高，节省人力，还可以缩短反应时间。如果反应只需要搅拌、回流和滴加试剂，采用图 1-7(b) 所示装置即可。如果不仅要满足上述要求，而且还要经常测试反应温度，则需采用图 1-7(c) 所示装置，或采用四口烧瓶来装配反应装置。如果反应混合物量较小，或黏稠度较小，或是液体混合物，这时也可使用电磁搅拌器对反应物进行搅拌。在图 1-5 和图 1-6 的反应烧瓶中加入一个长度合适的电磁搅拌子，在烧瓶的下面放电磁搅拌器，调节磁铁转动速度，控制烧瓶中搅拌子的转动速度。

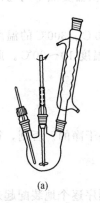
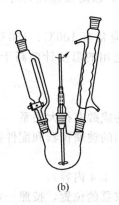
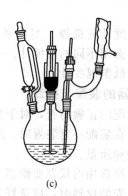

图 1-7　搅拌回流装置

在进行某些可逆平衡反应时，为了使正反应进行到底，可将反应产物之一不断地从反应混合物体系中蒸馏出去，常用与图 1-8、图 1-9 类似的反应装置来进行这种操作。在图 1-8 装置中，反应产物可单独或形成恒沸混合物不断在反应过程中蒸馏出去，并可通过滴液漏斗将一种试剂逐渐地加进去以控制反应速率或使这种试剂消耗完全。在图1-9装置中，有一个分水器，回流下来的蒸气冷凝液进入分水器，分层后，有机层自动被送回烧瓶，而生成的水可以从分水器中放出去，这样可以使某些生成水的可逆反应进行到底。

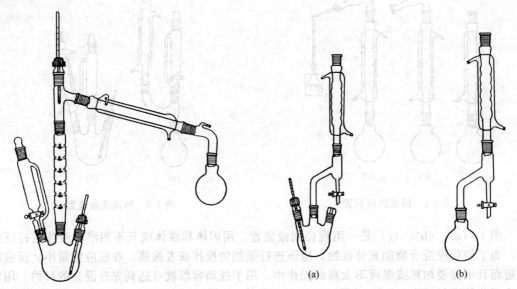

图1-8　滴加蒸出反应装置　　　　　　　　图1-9　回流分水装置

1.4.4　仪器的选择

有机化学实验的各种反应装置都是由一件件玻璃仪器组装而成的，实验中应根据要求选择合适的仪器，选择仪器的原则如下。

(1) 烧瓶的选择　根据液体的体积而定，一般液体的体积应占容器体积的1/3～1/2，最多不超过2/3，也就是说烧瓶容积的大小应是液体体积的1.5倍。进行水蒸气蒸馏和减压蒸馏时，液体体积不应超过烧瓶容积的1/3。

(2) 冷凝管的选择　一般情况下，回流用球形冷凝管，蒸馏用直形冷凝管。但是当蒸馏温度超过140℃时，应该用空气冷凝管，以防温差较大时，由于仪器受热不均匀而造成冷凝管断裂。

(3) 温度计的选择　实验室一般应备有100℃、150℃、200℃、300℃的温度计，根据所测温度可选用不同的温度计。一般选用的温度计要高于被测温度10～20℃。此外，实验室还应备有低温温度计。

1.4.5　仪器的装配与拆卸

仪器装配的正确与否，对于实验的成败有很大关系。

第一，在装配一套装置时，所选用的玻璃仪器和配件都应是干净的。否则，往往会影响产物的产量和质量。

第二，所选用的仪器要恰当（见1.4.4内容）。

第三，安装仪器时，应选好主要仪器的位置，按照一定的顺序逐个地装配起来。要先下后上、从左到右，逐个将仪器边固定边组装。拆卸的顺序与组装相反。拆卸前，应先停止加热，移走加热源，待稍微冷却后，先取下产物，然后再逐个拆掉。拆冷凝管时注意不要将水洒在电热套上。

第四，仪器装配要求做到严密、正确、整齐和稳妥。在常压下进行反应的装置，应与大气相通,不能密闭。

第五,铁夹的双钳应贴有橡皮或绒布,或缠上石棉绳、布条等,切勿让金属与玻璃直接接触,否则容易将仪器夹坏。

总之,在使用玻璃仪器时,最基本的原则是切忌对玻璃仪器的任何部位施加过度的压力或扭歪,实验装置安装马虎不仅看上去使人感觉不舒服,而且也是潜在的危险。因为扭歪的玻璃仪器在加热时会破裂,有时甚至在放置时也会崩裂。

1.4.6 电器设备

实验室有很多电器设备,使用时应注意安全,并保持这些设备的清洁,千万不要将药品洒到设备上。

(1) 烘箱 实验室中使用的恒温鼓风干燥箱,主要用于干燥玻璃仪器或烘干无腐蚀性、热稳定性好的药品。烘干玻璃仪器时,一般将温度控制在100~120℃,鼓风可以加速仪器的干燥。刚洗好的玻璃仪器应尽量倒净仪器中的水,然后把玻璃器皿顺序从上层往下层放入烘箱烘干。器皿口向上,若仪器口朝下,烘干的仪器虽可无水渍,但由于从仪器内流出来的水珠滴到其他已烘干的仪器上,往往易引起后者炸裂。当烘箱已工作时,不能往上层放入湿的器皿,应将烘干的仪器放在上边,湿仪器放在下边,以防止湿仪器上的水滴到热仪器上造成炸裂。带有旋塞或具塞的仪器,如分液漏斗和滴液漏斗,必须拔去盖子,取出旋塞并擦去油脂后才能放入烘箱内干燥。厚壁仪器,如量筒、吸滤瓶、冷凝管等,不宜在烘箱中干燥。把已烘干的仪器取出来时,应放在石棉板上冷却,切不可把烘得很热的玻璃仪器骤然碰到冷水或金属表面,以免破裂。

(2) 气流烘干器 气流烘干器是一种用于快速烘干的仪器设备,如图1-10所示。使用时,将仪器洗干净后,甩掉多余的水分,然后将仪器套在烘干器的多孔金属管上。注意随时调节热空气的温度。气流烘干器不宜长时间加热,以免烧坏电机和电热丝。

(3) 电热套 用玻璃纤维丝与电热丝编织成半圆形的内套,外边加上金属外壳,中间填充保温材料,如图1-11所示。

图1-10 气流烘干器

图1-11 电热套

根据内套直径的大小分为50mL、100mL、150mL、200mL、250mL、1000mL等规格,最大可到3000mL。此设备不用明火加热,使用较安全。由于它的结构是半圆形的,在加热时,烧瓶处于热气流中,因此加热效率较高。恒温电热套可以自动控温,最高加热温度可达250℃。电热套使用方便,控制温度容易,而且不易使有机溶剂着火,是有机化学实验中一种比较理想的加热设备。使用时应注意,不要将药品洒在电热套中,以免加热时药品挥发污染环境,同时避免电热丝被腐蚀而断开。用完后放在干燥处,否则内部吸潮后会降低绝缘性能。

(4) 调压变压器 调压变压器是调节电源电压的一种装置,常用来调节加热电炉、加热板、电热套的温度,调整电动搅拌器的转速等,使用时应注意以下几点。

① 先将调压器调至零点，再接通电源。

② 注意输入端与输出端切勿接错。电源应接到注明为输入端的接线柱上，输出端的接线柱与电炉或搅拌器等的导线连接，切勿接错。使用旧式调压器时，应注意安全，要接地良好，以防外壳带电。

③ 使用时，先接通电源，再调节旋钮到所需要的位置（根据加热温度或搅拌速度来调节）。调节变换时，应缓慢进行。无论使用哪种调压变压器都不能超负荷运行，最大使用量为满负荷的 2/3。

④ 用完后应将旋钮调至零点，并切断电源，放在干燥通风处。应保持调压变压器的清洁，以防腐蚀。

(5) 搅拌器　一般用于反应时搅拌液体反应物，搅拌器分为电动搅拌器和电磁搅拌器。

① 使用电动搅拌器时，应先将搅拌棒与电动搅拌器接好，再将搅拌棒用套管或塞子与反应瓶连接固定好，搅拌棒与套管的固定一般用乳胶管，乳胶管的长度不要太长也不要太短，以免由于摩擦而使搅拌棒转动不灵活或密封不严。在开动搅拌器前，应用手先空试搅拌器转动是否灵活，如不灵活应找出摩擦点，进行调整，直至转动灵活。如是电机问题，应向电机的加油口中加一些机油以保证电机转动灵活，或更换新电机。

② 电磁搅拌器能在完全密封的装置中进行搅拌。它由电机带动磁体旋转，磁体又带动反应器的磁子旋转，从而达到搅拌的目的。电磁搅拌器一般都带有温度和速度控制旋钮，使用后应将旋钮回零，使用时应注意防潮防腐。

(6) 旋转蒸发器　可用来回收、蒸发有机溶剂。在有机化学实验室中，进行合成实验及萃取、柱色谱、高效液相色谱等分离操作时，往往需要使用大量有机溶剂。为了快速蒸发较大体积的溶剂，常使用旋转蒸发器。由于它使用方便，近年来在有机化学实验室中被广泛使用。它利用一台电机带动可旋转的蒸发器（一般用圆底烧瓶）、高效冷凝管、接收瓶，如图 1-12 所示。此装置在常压或减压下使用，可一次进料，也可分批进料。由于蒸发器在不断旋转，可免加沸石而不会暴沸。同时，液体附于瓶壁上形成了一层液膜，加大了蒸发面积，使蒸发速度加快。使用时应注意以下两点。

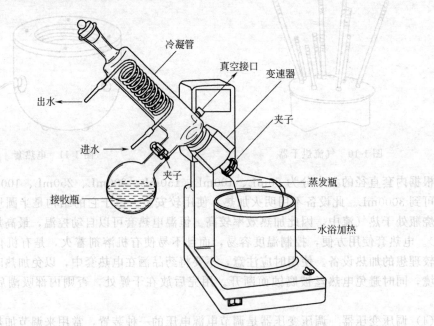

图 1-12　旋转蒸发器

① 减压蒸馏时，当温度高、真空度低时，瓶内液体可能会暴沸。此时，及时转动插管开关，通入冷空气降低真空度即可。对于不同的物料，应找出合适的温度与真空度，以平稳地进行蒸馏。

② 停止蒸发时，先停止加热，再切断电源，最后停止抽真空。若烧瓶取不下来，可趁热用木槌轻轻敲打，以便取下。

1.4.7 其他设备

除上述的电器设备外，有机化学实验室里还有一些其他的辅助设备，如称量设备、减压设备、清洗器、钢瓶、减压表等。使用时应注意正确操作，以保证设备的灵敏度及准确性。

1.4.7.1 电子天平

电子天平是实验室常用的称量设备，尤其在微量、半微量实验中经常使用（见图1-13）。电子天平是一种比较精密的仪器，因此使用时应注意维护和保养。

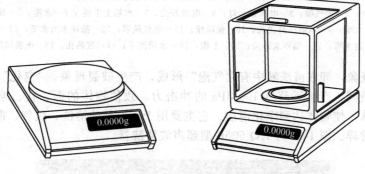

图1-13 电子天平

① 电子天平应放在清洁、稳定的环境中，以保证测量的准确性。勿将其放在通风、有磁场或产生磁场的设备附近，勿在温度变化大、有振动或存在腐蚀性气体的环境中使用。

② 要保持机壳和称量台的清洁，以保证天平的准确性。可用蘸有中性清洗剂（肥皂）的湿布擦洗，再用一块干燥的软毛巾擦干。

③ 将校准砝码存放在安全干燥的场所，不使用时关闭开关，拔掉变压器。

④ 称量时不要超过天平的最大量程。

1.4.7.2 循环水多用真空泵

循环水多用真空泵是以循环水作为流体，利用流体射流产生负压的原理而设计的一种新型多用真空泵，广泛应用于蒸发、蒸馏、结晶干燥、过滤、减压、升华等操作中。由于该泵还可以向反应装置中提供循环冷却水，因此避免了直排水的现象，节水效果明显，特别是在水压不足或缺水的实验室里更显示其优越性。因此，是实验室理想的、常用的减压设备，一般用于对真空度要求不高的减压体系中。图1-14为SHB-Ⅲ型循环水多用真空泵的外观。使用时应注意以下几点。

① 真空泵抽气口最好接一个缓冲瓶，以免停泵时，水被倒吸入反应瓶中，使反应失败。

② 开泵前，应检查是否与体系接好，然后打开缓冲瓶上的旋塞。开泵后，用旋塞调至所需要的真空度。关泵时，先打开缓冲瓶上的旋塞，拆掉与体系的接口，再关泵。切忌相反操作，以免倒吸。

③ 应经常补充和更换水泵中的水，以保持水泵的清洁和真空度。

1.4.7.3 超声波清洗器

超声波清洗器是利用超声波发生器所发出的交频信号，通过换能器转换成交频机械振荡而传播到介质——清洗液中，强力的超声波在清洗液中以疏密相间的形式向被洗物件辐射。

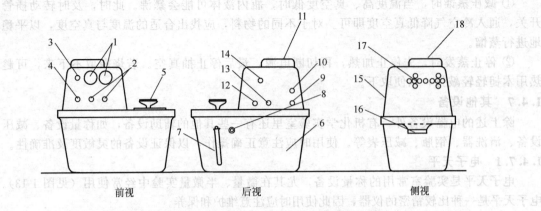

图 1-14 SHB-Ⅲ型循环水多用真空泵外观

1—真空表；2—抽气嘴；3—电源指示灯；4—电源开关；5—水箱上手柄；6—水箱；7—放水软管；
8—溢水嘴；9—电源线进线孔；10—保险座；11—电机风罩；12—循环水出水嘴；13—循环
水进水嘴；14—循环水开关；15—上帽；16—水箱把手；17—散热孔；18—电机风罩

产生"空化"现象，即在清洗液中有"气泡"形成，产生破裂现象。"空化"在达到被洗物体表面破裂的瞬间，产生远超过100MPa的冲击力，致使物体的面、孔、隙中的污垢被分散、破裂及剥落，使物体达到净化清洁。它主要用于小批量的清洗、脱气、混匀、提取、有机合成、细胞粉碎。图 1-15 为 KQ-500B 型超声波清洗器。

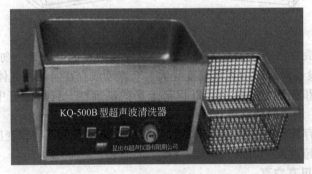

图 1-15 KQ-500B 型超声波清洗器

1.4.7.4 钢瓶

钢瓶又称高压气瓶，是一种在加压下储存或运送气体的容器，应用较广。但若使用不当，将会引发重大事故。若要使用钢瓶，事先应征得指导教师许可，按要求使用。为了防止各种钢瓶在充装气体时混用，全国统一规定了瓶身、横条以及标字的颜色。常用钢瓶的标色见表 1-2。

表 1-2 常用钢瓶的标色

气体类型	瓶身颜色	横条颜色	标字颜色	气体类型	瓶身颜色	横条颜色	标字颜色
氮气	黑	棕	黄	氯气	草绿	白	白
空气	黑	—	白	氨气	黄	—	黑
氧气	天蓝	—	黑	其他一切可燃气体	红	—	白
氢气	深绿	红	红	其他一切不可燃气体	黑	—	黄
二氧化碳	黑	—	黄				

1.4.7.5 减压表

使用钢瓶要用减压表。减压表是由指示钢瓶压力的总压力表、控制压力的减压阀和减压后的分压力表三部分组成。先将减压阀旋到最松位置（即关闭状态），然后打开钢瓶的气阀门，瓶内的气压即在总压力表上显示。慢慢旋紧减压阀，使分压力表达到所需压力。用毕，应先关紧钢瓶的气阀门，待总压力表和分压力表的指针复原到零时，再旋松减压阀。

1.5 常用半微量有机化学实验制备仪器

在半微量有机化学实验中常用半微量实验制备仪器来进行实验。成套的半微量实验制备仪器国内已有许多厂家生产，半微量实验中采用的是14号磨口仪器。常见的半微量实验制备仪器见表1-3，共30种规格，45件。

表1-3 常见的半微量实验制备仪器

编号	中(英)文名称	规格型号	件数
1	三口烧瓶(three-neck flask)	50mL/14×3	1
2	圆底烧瓶(round-bottom flask)	50mL/14	1
		25mL/14	1
		10mL/14	3
		5mL/14	3
3	梨形瓶(pear-shaped flask)	50mL/14	1
		25mL/14	1
		10mL/14	1
4	锥形瓶(erlenmeyer flask)	25mL/14	2
		10mL/14	2
5	直形冷凝管(west condenser)	120mm/14×2	2
6	球形冷凝管(spherical condenser)	120mm/14×2	1
7	空气冷凝管(air-cooled condenser)	120mm/14×2	1
8	分馏柱(fractionating column)	120mm/14×2	1
9	蒸馏头(distillation head)	14×3	2
10	克氏蒸馏头(Claisen head)	14×4	1
11	真空接引管(vacuum adapter)	14×2	2
12	燕尾管(swallowtail-shaped vacuum adapter)	14×4	1
13	分水器(trap for water)	14×2	1
14	U形干燥管(drying tube)	14	1
15	直筒形分液漏斗(cylindrical separatory funnel)	25mL/14×2	1
16	恒压滴液漏斗(pressure-equalized addition funnel)	25mL/14×2	1
17	梨形分液漏斗(pear-shaped funnel)	25mL/14	1
18	大小口接头(reducing or enlarging adapter)	19×14	2
19	空心塞(stopper)	14	4
20	离心管(centrifuge tube)	5mL/14	2
21	具弯管塞(stopper with bent tube)	14	1
22	温度计套管(thermometer adapter)	14	2
23	吸滤瓶(suction filtering flask)	14	1
24	吸滤漏斗(Hirsch funnel)	14	1

1.6 常用玻璃器皿的洗涤和保养

1.6.1 玻璃器皿的洗涤

在进行实验时，为了避免杂质混入反应物中，必须使用清洁的玻璃仪器。实验用过的玻璃器皿必须立即洗涤，应该养成这个习惯。由于污垢的性质在当时是清楚的，用适当的方法进行洗涤是容易办到的。若时间过长，会增加洗涤的难度。

洗刷仪器的一般方法是用水、洗衣粉、去污粉刷洗，刷子是特制的，如瓶刷、烧杯刷、冷凝管刷等。但用腐蚀性洗液时不用刷子。洗刷后，要用清水把仪器冲洗干净。应该注意，洗刷时不能用秃顶的毛刷，也不能用力过猛，否则会戳破仪器。洗涤玻璃器皿时不可用沙子，它能擦伤玻璃乃至龟裂。若难以洗净时，则可根据污垢的性质使用适当的洗液进行洗涤。如果是酸性（或碱性）污垢用碱性（或酸性）洗液洗涤；有机污垢用碱液或有机溶剂洗涤。下面介绍几种常见洗液。

（1）铬酸洗液　焦油状物质和炭化残渣，用去污粉、肥皂、强酸或强碱液常常洗刷不掉，这时常需用铬酸洗液洗涤。这种洗液氧化性很强，对有机污垢破坏力很强。

铬酸洗液的配制方法如下：在一个250mL烧杯内，把5g重铬酸钠溶于5mL水中，然后在搅拌下慢慢加入100mL浓硫酸。加硫酸过程中，混合液的温度将升高到70～80℃。待混合液冷却到40℃左右，把它倒入干燥的、磨口严密的细口试剂瓶中保存起来。

铬酸洗液呈红棕色，经长期使用变成绿色时即失效。当洗液颜色变绿失效时，应该弃去不能倒回洗液瓶中。铬酸洗液是强酸、强氧化剂，具有腐蚀性，使用时应注意安全。

在使用铬酸洗液前，应把仪器上的污物，特别是还原性物质，尽量洗净。尽量把仪器内的水倒净，然后慢慢倒入洗液，转动器皿，使洗液充分浸润不干净的器壁，数分钟后把洗液倒回洗液瓶中。再用少量自来水摇荡后，把洗液倒入废液缸内。最后用清水把仪器冲洗干净。若壁上沾有少量炭化残渣，可加入少量洗液，浸泡一段时间后，再用游动火焰均匀地加热该处，直至冒出气泡，炭化残渣可被除去。

带旋塞和磨口的玻璃仪器，洗净干燥后，在旋塞和磨口之间垫上纸片。

（2）盐酸　用盐酸可以洗去附着在器壁上的二氧化锰、碳酸盐等污垢。

（3）碱液和合成洗涤剂　配成浓溶液即可。用于洗涤油脂和一些有机物，如有机酸。

（4）有机溶剂洗涤液　当胶状或焦油状的有机污垢用上述方法不能洗去时，可选用丙酮、乙醚、苯浸泡，要加盖以免溶剂挥发。或用NaOH的乙醇溶液亦可。有机溶剂价值较高，只有在特殊情况下才使用。用于精制或有机分析的器皿，除用上述方法处理外，还需用蒸馏水刷洗。

器皿是否清洁的标志是：加水倒置，水顺着器壁流下，内壁被水均匀润湿，有一层既薄又均匀的水膜，不挂水珠。

1.6.2 玻璃仪器的干燥

在有机化学实验中，经常需要使用干燥的玻璃仪器，因此在仪器洗净后，还应进行干燥。仪器的干燥与否，有时甚至是实验成败的关键。一般将洗净的仪器倒置一段时间后，若没有水迹，即可使用。但有些实验须严格要求无水，否则阻碍反应正常进行，这时可将已晾干的仪器进一步干燥。干燥玻璃仪器的方法有下列几种。

（1）自然风干　自然风干是指把已洗净的仪器在干燥架上自然风干，这是常用和简单的方法。但必须注意，如玻璃仪器洗得不够干净，水珠便不易流下，干燥就会较为缓慢了。

在有机化学实验中，应尽量采用晾干法于实验前使仪器干燥，故要养成每次实验后马上把玻璃仪器洗净和倒置使之干燥的习惯，以便下次实验时使用，这样就可以节省很多时间。

(2) 在烘箱中烘干　一般用带鼓风机的电烘箱,保持烘箱内的温度在 100～120℃,约 0.5h。具体的方法及注意事项见 1.4.6。

(3) 用气流烘干器吹干　有时仪器洗涤后要尽快使用,可利用气流烘干器吹干。具体的方法及注意事项见 1.4.6。

(4) 用有机溶剂干燥　体积小的仪器,洗涤后急需干燥使用时,可采用此法。首先将洗净的仪器中的水尽量甩干,加入少量乙醇洗涤一次,再用少量丙酮洗涤,倾出溶剂后用压缩空气或电吹风把仪器吹干。先冷风吹 1～2min,待大部分溶剂挥发后,即吹入热风至完全干燥为止,再吹入冷风使仪器逐渐冷却。用过的溶剂应倒入回收瓶中。

1.6.3　常用仪器的保养和清洗

有机化学实验中的各种玻璃仪器的性能是不同的,必须掌握它们的性质、保养和洗涤方法,才能正确使用,提高实验效果,避免不必要的损失。下面介绍几种常用的玻璃仪器的保养和清洗方法。

(1) 温度计　温度计水银球部位的玻璃很薄,容易打破,使用时要特别小心。第一,不能用温度计当搅拌棒使用;第二,不能测定超过温度计的最高刻度的温度;第三,不能把温度计长时间放在高温的溶剂中,否则会使水银球变形,读数不准。

温度计使用后要让它慢慢冷却,特别在测量高温之后,切不可立即用水冲洗,否则会破裂,或水银柱断裂。应悬挂在铁架台上,待冷却后再洗净抹干,放回温度计盒内,盒底要垫一小块棉花。如果是纸盒,放回温度计时要检查盒底是否完好。

(2) 冷凝管　冷凝管通水后很重,所以安装冷凝管时应将夹子夹在冷凝管的中心处,以免翻倒。如内外管都是玻璃质的,则不适用于高温蒸馏。

洗刷冷凝管时要用特制的长毛刷,如用洗涤液或有机溶液洗涤,则用塞子塞住一端。不用时,应直立放置,使之易干。

(3) 分液漏斗　分液漏斗的活塞和盖子都是磨砂口的,若非原配的就可能不严密,所以使用时要注意保护它。各个分液漏斗之间也不要相互调换,用后一定要在活塞和盖子的磨砂口间垫上纸片,以免日久后难以打开。

(4) 砂芯漏斗　砂芯漏斗使用后应立即用水冲洗,否则难以洗净。滤板不太稠密的漏斗,可用强烈的水流冲洗,如果是较稠密的,则用抽滤的方法冲洗。

第 2 章　有机化学实验基本操作

2.1　加热与冷却

2.1.1　加热与热源

一般情况下，升高温度可以使有机化学反应速率加快。通常，反应温度每升高 10℃，反应速率增加一倍。有机化学实验室常用的热源有电热套、煤气灯、电炉、电热板等。加热的方式有直接加热和间接加热，但在有机化学实验室一般不采用直接加热，因为玻璃器皿会因剧烈的温度变化和受热不均匀导致仪器的损坏。同时，由于局部过热还可能引起某些有机化合物的部分分解，甚至燃烧或爆炸。为了避免直接加热可能带来的问题，保证加热均匀，实验室中常根据具体情况应用下列热浴进行间接加热（热浴的液面高度皆应略高于容器中的液面高度）。

（1）空气浴　空气浴就是利用热空气进行间接加热，对于沸点在 80℃ 以上的液体均可采用。把容器放在石棉网上，利用煤气灯隔着石棉网对容器进行加热，石棉网与容器间隔约 1cm，这就是最简单的空气浴。但是，这种加热方式较猛烈，受热仍不均匀，故不能用于回流沸点较低、易燃的液体或者减压蒸馏。

除煤气灯外，电热套也是比较好的、常用的、简便的空气浴加热回流装置（见图 2-1）。因为电热套中的电热丝是用玻璃纤维包裹着的，较安全，一般能从室温加热到 200℃ 左右。电热套主要用于回流加热，蒸馏或减压蒸馏不用为宜，因为在蒸馏过程中随着容器内物质逐渐减少，会使容器壁过热。安装电热套时，要使反应瓶外壁与电热套内壁保持约 2cm 左右的距离，以便利用热空气传热，并防止局部过热等。电热套有各种规格，取用时要与容器的大小相适应。

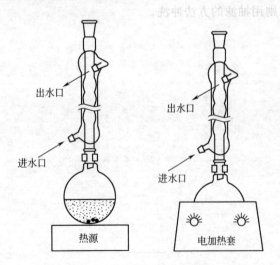

图 2-1　空气浴加热回流装置

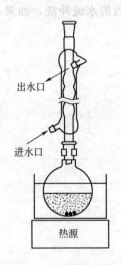

图 2-2　水浴加热回流装置

（2）水浴　水浴是较常用的一种热浴，与空气浴相比加热均匀，温度易控制，比较适合低沸点物质的回流加热。当所需加热温度不超过 100℃ 时，最好使用水浴加热。可将反应容器置入水浴锅中（注意勿使容器接触水浴器壁或底部），水浴液面稍高于容器内的液面，通过煤气灯或电热器对水浴锅进行加热，调节火焰的大小，把水温控制在所需的温度范围内。如果需加热到100℃，可用沸水浴或蒸汽浴。需要注意的是，由于较长时间加热，水浴中的水会不断蒸发，在操作时要适当添加热水，使水浴中水面经常保持稍高于容器内的液面（见图2-2）。

（3）油浴 当加热温度在100~250℃时，可用油浴。油浴的优点是使反应物受热均匀，温度容易控制在一定的范围内。反应物的温度一般低于油浴20℃左右。油浴所能达到的最高温度取决于所用油的种类，常用的油浴液有以下几种。

① 甘油，可以加热到140~150℃，温度过高时则会分解。甘油吸水性强，放置过久的甘油使用前应首先加热蒸去所吸收的水分，之后再用于油浴。

② 植物油，如菜油、蓖麻油、花生油等，可加热到220℃。常在这些植物油中加入1%的对苯二酚等抗氧剂，以增加油在受热时的稳定性。温度过高时会分解，达到闪点时可能燃烧起来，因此使用时要小心。

③ 石蜡，能加热到200℃左右，优点是冷却到室温时凝成固体，便于储藏。但是加热完毕后，在石蜡冷凝成固体前，应先取出浸于石蜡浴中的容器。

④ 液体石蜡，可以加热到220℃左右，温度稍高并不分解，但较易燃烧。

⑤ 硅油，在250℃时仍较稳定，透明度好，只是价格昂贵，一般实验室中较少使用。

在使用油浴加热时，要特别小心，防止着火，当油受热冒烟时，应立即停止加热。万一着火，也不要惊慌，应首先关闭煤气灯，再移去周围易燃物，然后用石棉板盖住油浴口，火即可熄灭。

油量不能过多，否则受热后有溢出而引起火灾的危险。油浴中应挂一支温度计，以便随时观察油浴的温度和有无过热现象，及时调节火焰并控制温度。油浴所用的油中不能溅入水，否则加热时会产生泡沫或引起爆溅。例如，可在回流冷凝管下端套上一个滤纸圈以吸收流下来的水滴。加热完毕取出反应容器时，仍用铁夹夹住反应容器，并令其离开液面悬置在油浴上面，待容器壁上的油滴完后，用纸或干布擦净。

（4）砂浴 当加热温度在220~350℃时，可采用砂浴。砂浴一般是将清洁而又干燥的细砂装在铁盘上，把盛有被加热物料的反应容器半埋在砂中，加热铁盘。由于砂对热的传导能力较差而散热却较快，因此容器底部与砂浴接触的砂层要薄一些，以便于传热。砂浴中应插入温度计，温度计水银球要靠近反应器。使用砂浴时，桌面要铺石棉板，以防辐射热烤焦桌面。由于砂浴温度上升较慢，而散热较快，且不易控制，因而使用不够广泛。

2.1.2 冷却

放热反应进行时，常产生大量的热，能使反应温度迅速升高，如果控制不当，往往会引起反应物的蒸发，逸出反应器，也可能引起副反应，有时甚至会引起爆炸。为了把温度控制在一定的范围内，就需要适当进行冷却。因此，在有机化学实验中，有时需采用一定的冷却剂进行冷却操作，在一定的低温条件下进行反应、分离、提纯等。例如：

① 某些反应要在特定的低温条件下进行，如重氮化反应（一般在0~5℃进行）；
② 沸点很低的有机物，冷却时会减少损失；
③ 要加速结晶的析出；
④ 高度真空蒸馏装置（减压蒸馏用冷阱）。

冷却剂的选择是根据冷却温度和带走的热量来决定的。

（1）水 水廉价，且热容量高，故为常用的冷却剂。但随着季节的不同，其冷却效率变化较大。

(2) 冰-水混合物　也是容易得到的冷却剂，可冷却至0～5℃。要将冰弄得很碎，效果才好。

(3) 冰-盐混合物　通常用冰-食盐混合物，即往碎冰中加入食盐（质量比为3∶1），可冷却至-18～-5℃。实际操作中是按上述质量比把食盐均匀地洒在碎冰上。其他盐类，如$CaCl_2 \cdot 6H_2O$ 5份、碎冰3.5～4份，可冷却至-50～-40℃。

(4) 干冰（固体二氧化碳）　可冷却至-60℃以下。如将干冰加到甲醇或丙酮等适当的溶剂中，可冷却至-78℃，但加入时会猛烈起泡。应将这种冷却剂放在杜瓦瓶（广口保温瓶）中或其他绝热效果好的容器中，以保持其冷却效果。

(5) 液氮　可冷却至-196℃。也应将这种冷却剂放在杜瓦瓶（广口保温瓶）中或其他绝热效果好的容器中，以保持其冷却效果。

应当注意，如果温度低于-38℃时，就不能使用水银温度计，因为水银会凝固（水银的凝固点为-38.9℃），测温应采用内部添加少许颜料的有机液体（如酒精、甲苯、正戊烷）的低温温度计。

2.2　干燥与干燥剂

干燥是指除去附在固体、混杂在液体或气体中的少量水分，或少量有机溶剂的方法。通常有机化合物在进行波谱分析、定性或定量化学分析之前，以及固体有机物在测定熔点前，都必须使其完全干燥，否则将会影响结果的准确性。液体有机物在蒸馏前也要先进行干燥，除去水分，否则沸点前馏分较多，产品损失，甚至沸点也不准。此外，很多有机化学反应需要在"绝对"无水条件下进行，有时不但原料、溶剂和容器需要干燥，而且还要防止空气中的潮气侵入反应容器中。所以，在有机化学实验中，试剂和产品的干燥具有十分重要的意义。

2.2.1　基本原理

有机化合物的干燥方法，大致可分为物理方法和化学方法两种。

(1) 物理方法　物理方法就是不加干燥剂，如烘干、晾干、吸附、分馏、共沸蒸馏、冷冻等。近年来，还应用离子交换树脂、分子筛等方法进行干燥。离子交换树脂是一种不溶于水、酸、碱和有机溶剂的高分子聚合物。分子筛是多水硅铝酸盐的晶体，晶体内部有许多孔径大小均一的孔道和占本身体积一半左右的许多孔穴，它允许小的分子"躲"进去，从而达到将不同大小的分子"筛分"的目的。

(2) 化学方法　化学方法就是向液态有机化合物中加入干燥剂进行除水。根据除水作用原理，干燥剂可分为以下两种。

① 与水可逆地结合生成水合物，例如

$$CaCl_2 + 6H_2O \rightleftharpoons CaCl_2 \cdot 6H_2O$$

② 与水发生不可逆的化学反应而生成一个新的化合物，例如

$$2Na + 2H_2O \longrightarrow 2NaOH + H_2 \uparrow$$

实验室中常用化学方法干燥，应用最广泛的是第一类干燥剂。

2.2.2　液态有机化合物的干燥

(1) 干燥剂的选择　常用干燥剂的种类很多，选择时必须注意以下几点。

① 通常是将干燥剂加入到液态有机化合物中，故所用干燥剂必须不与有机化合物发生化学或催化作用。

② 干燥剂应不溶于液态有机化合物中。

③ 当选用与水结合生成水合物的干燥剂时，必须考虑干燥剂吸水容量和干燥效能。

④ 干燥剂的干燥速度快，且价格低廉。

吸水容量是指单位质量干燥剂吸水量的多少。干燥效能是指达到平衡时液体被干燥的程度。如无水硫酸钠可形成 $Na_2SO_4 \cdot 10H_2O$，即 1g Na_2SO_4 最多能吸收 1.27g 水，其吸水容量为 1.27。但其水合物的蒸气压也较大（25℃时为 255.98Pa），故干燥效能较差。无水氯化钙能形成 $CaCl_2 \cdot 6H_2O$，吸水容量为 0.97。其水合物在 25℃时的蒸气压为 39.99Pa，故无水氯化钙的吸水容量虽然较小，但其干燥效能强。所以在进行干燥操作时，应根据除去水分的具体要求而选择合适的干燥剂。在干燥含水量较大而又不易干燥的（含有亲水性基团）化合物时，常先用吸水量较大的干燥剂除去大部分水分，然后再用干燥效能强的干燥剂进行干燥。

(2) 干燥剂的用量　掌握好干燥剂的用量是很重要的，若用量不足，则可能达不到干燥的目的；若用量太多，则由于干燥剂的吸附而造成液体损失。

通常是在溶解度手册中查出水在液体有机化合物中的溶解度，并根据液体有机化合物的结构、干燥剂的吸水量和干燥效能来估算出干燥剂的用量。但是，干燥剂的实际用量大大超过计算量。以乙醚为例，水在乙醚中的溶解度在室温时为 1%～1.5%，若用无水氯化钙来干燥 100mL 含水的乙醚时，全部转变成 $CaCl_2 \cdot 6H_2O$，其吸水量为 0.97，也就是说 1g 无水氯化钙大约可吸收 0.97g 水，这样无水氯化钙的理论用量为 1g，而实际上远远超过 1g。这是因为醚层的水分不可能完全去净，而且还有悬浮的微细水滴。其次，形成高水合物的时间很长，往往不可能达到应有的吸水容量，因而实际投入的无水氯化钙是大大过量的，常需用 7～10g 无水氯化钙。

实际操作时，一般投入少量干燥剂到液体有机化合物中，进行振摇，如出现干燥剂附着器壁、相互黏结、摇动不易旋转时，则说明干燥剂用量不足，应再添加干燥剂。如投入干燥剂后出现水相，必须用吸管把水吸出，然后再添加新的干燥剂，直到新加的干燥剂不结块、不粘壁，干燥剂棱角分明，摇动时旋转并悬浮，则表示所加干燥剂用量合适。

干燥前液体呈浑浊状，干燥后变澄清，这可简单地作为水分基本除去的标志。

一般干燥剂用量为每 10mL 液体约加入 0.5～1g 干燥剂。由于液体含水量不等、干燥剂质量差异、干燥剂的颗粒大小和干燥时的温度不同等因素，所以较难规定干燥剂的具体用量，上述数量仅供参考。

(3) 常用的干燥剂　常用干燥剂的性能与应用范围见表 2-1。各类有机化合物的常用干燥剂见表 2-2。

表 2-1　常用干燥剂的性能与应用范围

干燥剂	吸水作用	吸水容量	干燥效能	干燥速度	适用范围	备注
氯化钙	$CaCl_2 \cdot nH_2O$ $n=1,2,4,6$	0.97（按 $CaCl_2 \cdot 6H_2O$ 计）	中等	较快，但吸水后易在其表面覆盖液体，应放置较长时间	烃、烯烃、丙酮、醚和中性气体	①价廉；②工业品中含 $Ca(OH)_2$ 或 CaO，故不能干燥酚类；③$CaCl_2 \cdot 6H_2O$ 在 30℃以上易失水；④$CaCl_2 \cdot 4H_2O$ 在 45℃以上失水
硫酸镁	$MgSO_4 \cdot nH_2O$ $n=1,2,4,5,6,7$	1.05（按 $MgSO_4 \cdot 7H_2O$ 计）	较弱	较快	中性,应用范围广,可代替 $CaCl_2$，并可用于干燥酯、醛、酮、腈、酰胺等，并用于不能用 $CaCl_2$ 干燥的化合物	①$MgSO_4 \cdot 7H_2O$ 在 49℃以上失水；②$MgSO_4 \cdot 6H_2O$ 在 38℃以上失水
硫酸钠	$Na_2SO_4 \cdot 10H_2O$	1.25	弱	缓慢	中性，一般用于有机液体的初步干燥	$Na_2SO_4 \cdot 10H_2O$ 在 32.4℃以上失水

续表

干燥剂	吸水作用	吸水容量	干燥效能	干燥速度	适用范围	备注
硫酸钙	$CaSO_4 \cdot 2H_2O$	0.06	强	快	中性硫酸钙经常与硫酸钠配合,作最后干燥之用	①$CaSO_4 \cdot 2H_2O$ 在38℃以上失水;②$CaSO_4 \cdot H_2O$ 在80℃以上失水
氢氧化钠(钾)	溶于水		中等	快	强碱性,用于干燥胺、杂环等碱性化合物(氨、胺、醚、烃)	吸湿性强
碳酸钾	$K_2CO_3 \cdot \frac{1}{2} H_2O$	0.2	较弱	慢	弱碱性,用于干燥醇、酮、酯、胺及杂环等碱性化合物,可代替KOH干燥胺类,可用于酸、酚	有吸湿性
金属钠	$Na+H_2O \longrightarrow \frac{1}{2}H_2+NaOH$		强	快	限于干燥醚、烃、叔胺中痕量水分	忌水
氧化钙(碱石灰,BaO类)	$CaO+H_2O \longrightarrow Ca(OH)_2$		强	较快	中性及碱性气体、胺、醇、乙醚	对热很稳定,不挥发,干燥后可直接蒸馏
五氧化二磷	$P_2O_5+3H_2O \longrightarrow 2H_3PO_4$		强	快,但吸水后表面被黏浆液覆盖,操作不便	适于干燥烃、卤代烃、腈等中的痕量水分;适于干燥中性或酸性气体,如乙炔、二硫化碳、烃、卤代烃	吸湿性很强,用于干燥气体时需与载体相混
硫酸					中性及酸性气体(用于干燥器和洗气瓶中)	不适用于高温下的真空干燥
高氯酸镁			强		包括氯在内的气体(用于干燥器中)	适合于分析用
硅胶					用于干燥器中	吸收残余溶剂
分子筛(硅酸钠铝和硅酸钙铝)	物理吸附	约0.25	强	快	流动气体(温度可高于100℃)、有机溶剂(用于干燥器中)等各类有机化合物	

表 2-2 各类有机化合物的常用干燥剂

液 态 有 机 化 合 物	适 用 的 干 燥 剂
醚类、烷烃、芳烃	$CaCl_2$,Na,P_2O_5
醇类	K_2CO_3,$MgSO_4$,Na_2SO_4,CaO
醛类	$MgSO_4$,Na_2SO_4
酮类	$MgSO_4$,Na_2SO_4,K_2CO_3
酸类	$MgSO_4$,Na_2SO_4
酯类	$MgSO_4$,Na_2SO_4,K_2CO_3
卤代烃	$CaCl_2$,$MgSO_4$,Na_2SO_4,P_2O_5
有机碱类(胺类)	NaOH,KOH

(4) 干燥的温度、时间 已吸水的干燥剂受热后会脱水,其蒸气压随着温度的升高而增加,所以干燥通常在室温下进行效果好。对已干燥的液体在蒸馏前必须把干燥剂滤去。

干燥剂形成水合物需要一定的平衡时间,所以加入干燥剂后必须放置一段时间才能达到脱水效果。

(5) 液态有机化合物干燥的操作步骤　液态有机化合物的干燥操作一般在干燥的锥形瓶内进行，具体步骤如下。

① 干燥前应尽可能除净被干燥液体中的水分，不应有任何可见的水层或悬浮水珠。

② 把待干燥的液体放入锥形瓶中，按照条件选定适量的干燥剂投入液体中（干燥剂的颗粒大小要适宜，太大时吸水很慢，且干燥剂内部不起作用；太小时表面积太大，吸附有机物太多），塞紧塞子（用金属钠作干燥剂时例外，此时塞子中应插入一个无水氯化钙干燥管，使氢气放空，而水汽不会进入），振荡片刻，静置（至少半小时，最好过夜），使所有的水分全部被吸去。若干燥剂用量太少，致使部分干燥剂溶解于水时，可将干燥剂滤出，用吸管吸出水层，再向液体中加入新的干燥剂，放置一段时间，至澄清为止。

③ 将干燥好的液体滤入蒸馏瓶中，然后进行蒸馏。

2.2.3　固体有机化合物的干燥

干燥固体有机化合物，主要是为除去残留在固体中的少量低沸点溶剂，如水、乙醚、乙醇、丙酮、苯等。由于通过重结晶得到的固体常带有水分或有机溶剂，所以应根据化合物的性质选择适当的方法进行干燥。常用的干燥方法如下。

(1) 晾干　固体在空气中自然晾干是最简便、最经济的干燥方法。这种方法适用于在空气中稳定、不易分解、不易吸潮的固体。其方法是将待干燥的固体化合物摊放在表面皿或滤纸上，再用另一张滤纸覆盖起来，让它在空气中自然慢慢晾干。

(2) 加热干燥　对于热稳定性好，且熔点较高的固体化合物，可用恒温烘箱烘干或用恒温真空干燥箱烘干，也可用红外灯烘干。在烘干的过程中，要注意防止过热，加热的温度切忌超过该固体的熔点，以免固体变色或分解。

(3) 吸干　若溶剂难抽干时，应把固体从布氏漏斗中转移到滤纸上，上下均放 2～3 层滤纸，挤压，使溶剂被滤纸吸干。

(4) 干燥器干燥　对于易吸潮或在较高温度干燥时会分解或变色的固体化合物，可放置在干燥器中进行干燥。干燥器分为普通干燥器、真空干燥器和减压恒温干燥器（干燥枪）。

真空干燥器如图 2-3 所示，其底部放置干燥剂，中间隔一个多孔瓷板，把待干燥的物质放在瓷板上，顶部装有带活塞的玻璃导气管，由此处连接抽气泵，使干燥器压力降低，从而提高了干燥效率。使用时注意真空度不易过高，一般以水泵抽至干燥器的盖子推不动即可。使用真空干燥器前必须试压。试压时用网罩或防爆布盖住干燥器，然后抽真空，关上活塞放置过夜。使用时必须十分注意，防止干燥器炸碎，玻璃碎片飞溅伤人。解除器内真空时，开动活塞放入空气的速度宜慢不宜快，以免吹散被干燥的物质。

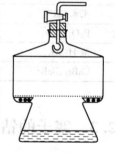

图 2-3　真空干燥器

减压恒温干燥器也称减压恒温干燥枪。当在烘箱或真空干燥器内干燥效果欠佳时，则要使用减压恒温干燥枪（简称干燥枪），如图 2-4(a) 所示。其干燥效率高，特别适用于除去结晶水或结晶醇，但此法仅适用于少量样品的干燥。使用时，将盛有样品的小船放在夹层内，连接上盛有 P_2O_5 的曲颈瓶，然后减压至最高真空度时，停止抽气，关闭活塞，加热溶剂（溶剂的沸点切勿超过样品的熔点），回流。冷溶剂的蒸气充满夹层的外层，这时夹层内的样品就在减压恒温情况下被干燥。在干燥过程中，每隔一定时间应抽气保持应有的真空度。

真空恒温干燥器如图 2-4(b) 所示，与减压恒温干燥枪有相同的功能，差别仅在于加热方式不同。

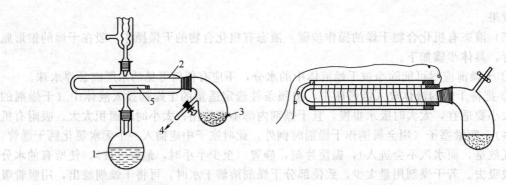

(a) 减压恒温干燥枪　　　　　(b) 真空恒温干燥器

1—盛溶剂的烧瓶；2—夹层；3—曲颈瓶中 P_2O_5；
4—接水真空泵；5—放样品的玻璃或瓷的小船

图 2-4　减压恒温干燥枪

2.2.4　气体的干燥

在有机化学实验室中，常用气体有 N_2、O_2、H_2、Cl_2、NH_3、CO_2，有时实验要求气体中含很少或几乎不含 CO_2、H_2O 等，因此就需要对上述气体进行干燥。气体的干燥主要用吸附法。干燥气体常用仪器有干燥管、干燥塔、U 形管、各种洗气瓶（用来盛液体干燥剂）等。可根据待干燥气体的性质、潮湿程度、反应条件及干燥剂的用量选择不同仪器。气体干燥时常用的干燥剂列于表 2-3 中。

表 2-3　用于气体干燥的常用干燥剂

干　燥　剂	可　干　燥　的　气　体
CaO、碱石灰、NaOH、KOH	NH_3 类
无水 $CaCl_2$	H_2、HCl、CO_2、CO、SO_2、N_2、O_2、低级烷烃、醚、烯烃、卤代烃
P_2O_5	H_2、O_2、CO_2、SO_2、N_2、烷烃、乙烯
浓 H_2SO_4	H_2、N_2、CO_2、Cl_2、HCl、烷烃
$CaBr_2$、$ZnBr_2$	HBr

2.3　塞子的钻孔和简单玻璃工操作

在有机化学实验室中，经常需要用到各种形状的玻璃管、滴管和不同直径的毛细管，要求对玻璃管进行加工，以满足实验的需要。另外，当使用普通玻璃仪器时，还常常需要用到不同规格和形状的玻璃管以及塞子等配件，才能将各种玻璃仪器正确地装配起来。因此，熟练地掌握玻璃工基本操作、塞子的选用及钻孔的方法，是进行有机化学实验必不可少的基本操作。

2.3.1　塞子的钻孔

有机化学实验室常用的塞子有软木塞和橡皮塞两种。软木塞的优点是不易与有机化合物作用，但易漏气、易被酸碱腐蚀。橡皮塞虽然不漏气、不易被酸碱腐蚀，但易被有机化合物侵蚀或溶胀。两种各有优缺点，究竟选用哪一种塞子才合用要由具体情况而定。

（1）塞子的选择　选择塞子的大小应与仪器的口径相适合，塞子进入瓶颈或管颈的部分不能少于塞子本身高度的 1/3，也不能多于 2/3，一般以 1/2 为宜，如图 2-5 所示。否则，就不合用。使用新的软木塞时，只要能塞入 1/3～1/2 即可，因为经过压塞机压紧打孔后就有可能塞入 2/3 左右。

(2) 钻孔器的选择　有机化学实验往往需要在塞子内插入导气管、温度计、滴液漏斗等，这就需要在塞子上钻孔，钻孔用的工具叫钻孔器（也叫打孔器），这种钻孔器是靠手力钻孔的。也有把钻孔器固定在简单的机械上，借机械力来钻孔的，这种工具叫打孔机。每套钻孔器约有 5、6 支直径不同的钻嘴，以供选择。

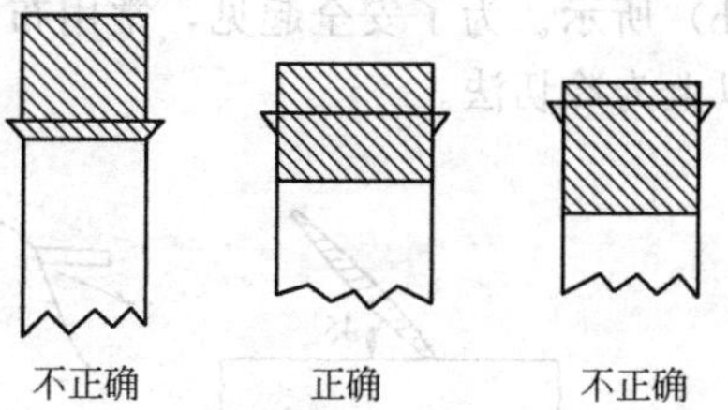

图 2-5　塞子的配置

若在软木塞上钻孔，应选用比欲插入的玻璃管等的外径稍小或接近的钻嘴。若在橡皮塞上钻孔，则要选用比欲插入的玻璃管等的外径稍大的钻嘴，因为橡皮塞有弹性，钻成后会收缩使孔径变小。

总之，塞子孔径的大小，要与所插入孔内的玻璃管、温度计等的直径适宜，要紧密配合，以免漏气。

(3) 钻孔的方法　软木塞在钻孔之前，需在压塞机上压紧，防止在钻孔时塞子破裂。如图 2-6 所示，把塞子小的一端朝上，平放在桌面的一块木板上（其作用是避免当塞子被钻通后钻坏桌面）。钻孔时，左手握住塞子平稳放在木板上，右手持钻孔器的柄，在选定的位置，使劲地将钻孔器以顺时针的方向向下钻动，钻孔器垂直于塞子的平面，不能左右摆动，更不能倾斜，不然钻得的孔道是偏斜的。等钻到塞子的一半时，按逆时针旋转取出钻

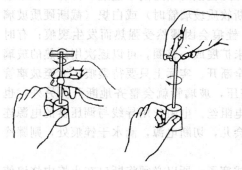

图 2-6　塞子的钻孔

嘴，用钻杆通出钻嘴中的塞芯。然后在塞子大的一面钻孔，要对准小头的孔位，以上述同样的操作钻孔至钻通。拔出钻嘴，通出钻嘴内的塞芯。

为了减少钻孔时的摩擦，特别是对橡皮塞钻孔时，可在钻嘴的刀口搽一些甘油或水。钻孔后，要检查孔道是否合用，如果不费力就能把玻璃管插入，说明孔道过大，玻璃管和塞子之间不能紧密贴合会漏气，不能用。若孔道略小或不光滑，可用圆锉修整。

2.3.2　简单玻璃工操作

(1) 玻璃管（棒）的清洗和干燥　玻璃管（棒）在加工前需要洗净。玻璃管内的灰尘可用水冲洗。对于较粗的玻璃管，可以用两端系有线绳的布条通过玻璃管，来回拉动，擦去管内的脏物。如果玻璃管保存得好，比较干净，仅用布把玻璃管外面擦净，就可以使用。如果管内附着油腻的东西，用水不能洗净，用布条也不能擦净时，可把长玻璃管适当地割短，浸在铬酸洗液里，然后取出用水冲洗。制备熔点管的毛细管和薄板层析点样用的毛细管，在拉制前均应用铬酸洗液浸泡，再用水洗净。

洗净的玻璃管必须干燥后才能加工，可在空气中晾干、用热空气吹干或在烘箱中烘干，但不能用灯火直接烤干，以免炸裂。

(2) 玻璃管的截断　玻璃管的截断操作，一是锉痕，二是折断。锉痕用的工具是小三角锉刀或小砂轮片。锉痕的操作是：把玻璃管平放在桌子的边缘上，左手的拇指按住玻璃管要截断的地方，右手持小三角锉刀把小三角锉刀的棱边放在要截断的地方，如图 2-7(a) 所示。然后用力把锉刀向前或向后拉，同时把玻璃管略微朝相反的方向转动，在玻璃管上刻划出一条清晰、细直的深凹痕，使凹痕约占管周的 1/6。注意锉痕时不要来回拉锉刀，锉痕应在一条直线上，否则不仅锉痕多，容易损坏锉刀，而且会导致玻璃管断茬不整齐。

锉出凹痕后，下一步就是把玻璃管折断。两手分别握住锉痕的两边，凹痕向外，两手的大拇指分别抵住锉痕背面的两侧，用力急速轻轻一压带拉，就可使玻璃管断开，如图 2-7

(b) 所示。为了安全起见，常用布包住玻璃管，同时尽可能远离眼睛，以免玻璃碎粒伤人。以上为冷切法。

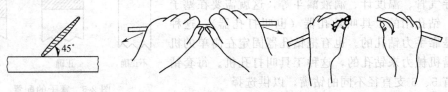

(a) 锉刀锋棱压在玻璃管上　　　(b) 玻璃管的折断

图 2-7　玻璃管的截断

需要折断较粗的玻璃管，或者在玻璃管的近管端处进行截断时，可利用玻璃管（棒）骤然受热或骤然遇冷易裂的性质，使其断裂。可先用锉刀在预切断处割一锉痕，再将一根末端拉细的玻璃棒在煤气灯的氧化焰上加热到红热（截断软质玻璃管时）或白炽（截断硬质玻璃管时），使成珠状，然后把它压触到锉痕的端点处，锉痕会因骤然受强热而发生裂痕；有时裂痕迅速扩展成整圈，玻璃管即自行断开。若裂痕未扩展成一整圈，可以逐次用烧热的玻璃棒的末端压触在裂痕的稍前处引导，直至玻璃管完全断开。实际上只要待裂痕扩大至玻璃管周长的 90% 时，即可用两手稍用力将玻璃管向里挤压，玻璃管就会整齐地断开。另外，也可在粗玻璃管或玻璃管近管端的锉痕处，紧围一根电阻丝。电阻丝用导线与调压器的电源连接。通电后，升高电压使电阻丝呈亮红色。稍呆一会儿，切断电源，滴水于锉痕处，则骤冷自行断裂开。

玻璃管的断口很锋利，容易割破皮肤、橡皮管或塞子，所以必须将断口在火焰中烧熔使其变光滑。方法是将断口放在氧化焰的边缘，不断转动玻璃管，烧到管口微红即可。不可烧得太久，否则管口会缩小。

(3) 玻璃管的弯曲　有机化学实验常常要用到弯成一定角度的曲玻璃管，这就需要实验者自己制作。

弯曲玻璃管的操作如图 2-8 所示，双手持玻璃管，手心向外把需要弯曲的地方放在火焰上预热，然后放进鱼尾形的火焰中加热，受热部分宽约 5cm，在火焰中使玻璃管缓慢、均匀而不停地向同一个方向转动。如果两手用力不均匀，玻璃管就会在火焰中扭歪，造成浪费。当玻璃受热至足够软化时（玻璃管变黄！），立即从火焰中取出，轻轻弯成所需要的角度。为了维持管径的大小，两手持玻璃管在火焰中加热尽量不要往外拉，其次可在弯成角度之后，在管口轻轻吹气（不能过猛！）。弯好的玻璃管整体来看应尽量在同一个平面上，将其放在石棉板上自然冷却，切勿立即与冷的物件接触。例如，不能放在实验台的瓷片上，因为骤冷会使弯好的曲玻璃管破裂，造成浪费。检查弯好的玻璃管的外形，如图 2-9(a) 所示的为合用，即玻璃管弯曲部分的厚度和粗细须保持均匀。如图 2-9(b) 所示的为不合用。

图 2-8　弯曲玻璃管的操作　　　图 2-9　弯好的玻璃管的形状

如要弯成较小角度的玻璃管，可分几次弯成，以免一次弯得过多使弯曲部分发生瘪陷或纠结。分次弯管时，每次的加热部位要稍有偏移，并要等弯过的玻璃管稍冷后再重新加热。

（4）熔点管和沸点管的拉制　这两种管子的拉制实质上就是把玻璃管拉细成一定规格的毛细管。拉制步骤：把一根干净、壁厚 1mm、直径约 8～10mm 的玻璃管，拉成内径约 1mm 和 3～4mm 的两种毛细管，再将内径约 1mm 的毛细管截成 15～20cm 长，把此毛细管的两端在小火上封闭，使用时，再把这根毛细管的中央切断，就成为两根熔点管。

拉毛细管的操作：用两手持住玻璃管的两端，掌心相对，加热方法和曲玻璃管的弯制相同，只不过加热程度要强一些。在玻璃管稍稍变软时，两手轻轻往里挤，以加厚烧软处的管壁。待玻璃管被烧成红黄、很软时，离开火焰，两手平稳地沿水平相反方向趁势拉长，同时两手拇指与食指同向同速捻动，以防拉成扁管。开始拉时要慢一些，逐步加快，拉成内径约 1mm 的毛细管。拉长之后，立刻松开一只手，另一只手提着一端，使管靠垂直重力拉直并冷却定型。待中间部分冷却后，放在石棉板上，以防烫坏实验台。

沸点管的拉制：方法同上，拉成内径 3～4mm 的毛细管，截成 7～8cm 长，在小火上封闭其一端，作为沸点管的外管，另将内径约 1mm 的毛细管截成 8～9cm 长，封闭其一端，作为沸点管的内管，这两根毛细管就可组成沸点管，留作沸点测定的实验使用。

（5）玻璃管插入塞子的方法　先用水或甘油润湿选好的玻璃管的一端（如插入温度计时即水银球的部分），然后左手拿住塞子，右手捏住玻璃管的一端（距管口约 4cm），如图 2-10 所示，稍稍用力转动逐渐插入。必须注意，右手指捏住玻璃管的位置与塞子的距离应保持 4cm 左右，不能太远；其次用力不能过大，以免折断玻璃管刺破手掌，最好用布包住玻璃管则较为安全。插入或者拔出弯曲管时，手指不能捏在弯曲的地方。

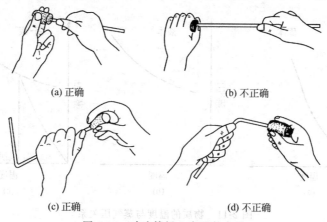

(a) 正确　　　　　　　　　　(b) 不正确

(c) 正确　　　　　　　　　　(d) 不正确

图 2-10　玻璃管插入塞子的方法

思　考　题

1. 选择塞子和钻孔时应注意什么？
2. 截断玻璃管时要注意哪些问题？怎样弯曲和拉细玻璃管？在火焰上加热玻璃管时怎样才能防止玻璃管被拉歪？
3. 弯曲和拉细玻璃管时软化玻璃管的温度有什么不同？为什么不同？弯玻璃管时用力过大、弯得过快，有什么不好？
4. 把玻璃管插入塞子孔道中时要注意什么？

2.4　熔点的测定

通常当结晶物质在大气压下加热到一定温度时，即从固态转变为液态，此时的温度即为该物质的熔点（Melting Point，简记 m.p.）。熔点是鉴定固体有机化合物的重要物理常数，通过测定熔点不仅可以鉴别不同的有机化合物，而且还可以判断其纯度。

2.4.1 基本原理

严格地说，熔点是指在一个大气压下（1atm＝101.325kPa）固体化合物固相与液相平衡时的温度。此时固液两相蒸气压相等。每种纯固体有机化合物都具有固定的熔点。一个纯化合物从开始熔化（始熔）至完全熔化（全熔）的温度范围叫做熔点距，也叫熔点范围或熔程，一般不超过0.5℃。当含有杂质时，熔程较宽，其熔点下降。由于大多数有机化合物的熔点都在300℃以下，较易测定，故利用测定熔点，可以估计出有机化合物的纯度。怎样来理解这种性质呢？可以从分析物质的蒸气压和温度的关系曲线图入手。图2-11(a)是固体的蒸气压随温度升高而增大的情况，图(b)是液体蒸气压随温度变化的曲线，若将图(a)和图(b)两曲线加合，即得到图(c)。可以看出，固相蒸气压随温度的变化速率比相应的液相大，最后两曲线相交于 M 点。在交叉点 M 处，固液两相蒸气压一致，固液两相平衡共存，这时的温度 T_m 即为该物质的熔点。不同的化合物有不同的 T_m 值。当温度高于 T_m 时，固相全部转变为液相；低于 T_m 值时，液相全部转变为固相。只有固液两相并存时，固液两相的蒸气压一致，这就是纯晶体物质具有固定和敏锐熔点的原因。一旦温度超过 T_m，甚至只有几分之一度时，若有足够的时间，固体就可以全部转变为液体。因此，要想精确测定熔点，就要使固体熔化过程尽可能接近于两相平衡状态。在测定熔点过程中，当接近熔点时加热速度一定要慢，并且必须密切注意加热情况，一般以每分钟温度上升约1℃为宜。只有这样，才能使熔化过程接近相平衡条件。图2-12为相随时间和温度而变化的示意图。

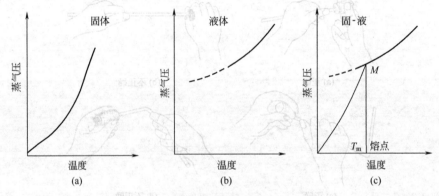

图 2-11 物质的温度与蒸气压关系

需要指出的是，含有杂质的有机化合物熔点比纯有机物的熔点低是普遍情况。但有时（如形成新的化合物或固溶体）两种熔点相同的不同物质混合后熔点并不降低反而升高。少数易分解的有机化合物虽然很纯，但没有固定的熔点。在未到达熔点之前，化合物发生分解，此时有颜色变化或气体产生，这类化合物的熔点实际上就是它们的分解点。

2.4.2 测定熔点的方法

测定熔点的方法较多，目前广泛使用的有毛细管法、熔点测定仪法。

(1) 毛细管法测熔点

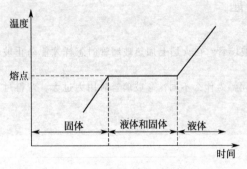

图 2-12 相随时间和温度而变化

① 熔点管 通常用内径约1mm、长约60～70mm、一端封闭的毛细管作为熔点管，这种毛细管的拉制方法见2.3.2简单玻璃工操作。

② 样品的填装　将干燥过的待测试样 0.1~0.2g 放置在干燥洁净的表面皿上，用不锈钢刮刀研细成粉末后聚成小堆，将熔点管的开口端插入样品堆中，使样品挤入管内，然后把熔点管开口端朝上竖立起来，在桌面上墩几下，使样品落入管底（熔点管的下落方向必须与桌面垂直，否则熔点管极易折断），这样重复取样几次，然后使毛细管封口端朝下，在一根长约 40~50cm 直立于表面皿上的玻璃管中自由下落，反复操作几次，使样品粉末紧密堆积在毛细管底部，直至样品高度约 4mm 为止，如图 2-13(a) 所示。为使测定结果准确，操作要迅速，防止样品吸潮，样品一定要研细，填充要均匀且紧密，这样受热时才均匀，如果有空隙，不易传热，将影响结果。

③ 测定熔点的装置　毛细管法测定熔点的装置很多，本实验介绍如下两种常用的装置。
第一种装置，如图 2-13(b) 所示，首先取一个 100mL 的烧杯，置于放有石棉网的铁环上，在烧杯中放入一支玻璃搅拌棒（最好在玻璃棒底端烧一个环，便于上下搅拌），注入载热体（又称浴液或导热液，可根据所测物质的熔点选择）约 60mL；然后将装有样品的毛细管用细橡皮圈固定在温度计上，并使毛细管装样部位于水银球处［见图 2-13(c)］；最后，在温度计上端套一橡皮塞，并用铁夹夹住，将其垂直固定在离烧杯底约 1cm 的中心处。

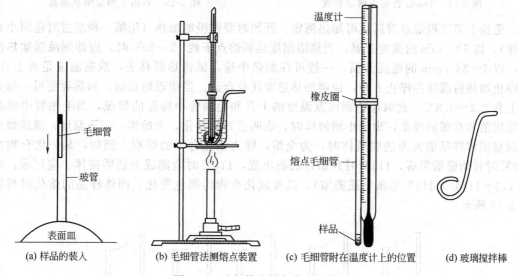

图 2-13　毛细管法测定熔点的装置

第二种装置，如图 2-14 所示，是利用 Thiele 管（又叫 b 形管，也叫熔点测定管）。将 Thiele 管固定在铁架台上，装入导热液于 Thiele 管中至高出上侧管时即可，Thiele 管管口配一缺口单孔软木塞，温度计插入孔中，刻度应向软木塞缺口。把毛细管同前法附着在温度计旁（注意，橡皮圈应在导热液面之上）。温度计插入熔点测定管中深度以水银球恰在熔点测定管的两侧管中部为宜，因为此处对流循环好，温度均匀。加热时，火焰须与熔点测定管的倾斜部分接触。这种装置测定熔点的好处是，管内液体因温差而发生对流作用，省去了人工搅拌的麻烦。但常因温度计的位置和加热部位的变化而影响测定的准确性。

除上述两种装置外，还有图 2-15 所示的双浴式测定熔点装置。

④ 测定熔点的方法　把装置放在光线充足的地点操作。测定熔点的关键之一就是加热速度，使热能透过毛细管，样品受热熔化，令熔化温度与温度计所示温度一致。一般方法是先快速加热（以 5℃/min 为宜），粗测化合物的大概熔点，测定时要注意观察现象并记录样品熔化时的温度，即得试样的粗测熔点。移去火焰，待热浴温度下降至粗测熔点以下大约 30℃时，即可更换一根样品管，并参考粗测熔点进行精测。精测时，先将温度计从热浴中取

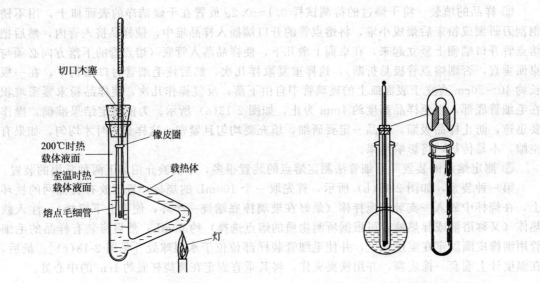

图 2-14　Thiele 管熔点测定装置　　　　图 2-15　双浴式测定熔点装置

出，更换上第二根熔点管后便可加热测定。开始时要慢慢地加热（用第一种装置时还须小心搅拌），以 5℃/min 的速度升温，当热浴温度达到熔点下约 15～20℃ 时，应即刻减缓加热速度，以 1～2℃/min 的速度升温，一般可在加热中途，试将热源移去，观察温度是否上升，如停止加热后温度亦停止上升，说明加热速度比较合适。当接近熔点时，加热要更慢，每分钟上升 0.2～0.3℃，此时应特别注意温度的上升和毛细管中样品的情况，当毛细管中样品开始塌落和有湿润现象，出现小滴液体时，表明已开始熔化，为始熔，记下温度；继续微热至微量固体样品消失为透明液体时，为全熔，即为该化合物的熔程。例如，某一化合物在 112℃ 时开始萎缩塌落，113℃ 时开始有液滴出现，114℃ 时全部成为透明液体。应记录：熔点 113～114℃，112℃ 塌落（或萎缩），以及该化合物的颜色变化。固体样品的熔化过程如图 2-16 所示。

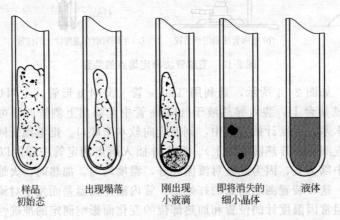

样品　　出现塌落　　刚出现　　即将消失的　　液体
初始态　　　　　　小液滴　　细小晶体

图 2-16　固体样品的熔化过程

对每种试样，熔点测定至少要有两次重复的数据，每一次测定都必须用新的熔点管新装样品，不能使用已测定过熔点的样品管。测定已知物熔点时，要测定两次，两次测定的误差不能大于±1℃。测定未知物熔点时，要测定三次，一次粗测，两次精测，两次精测的误差也不能大于±1℃。

实验完毕，将温度计放好，让其自然冷却至接近室温后用废纸擦去导热液，此时方可用水冲洗，否则容易使水银柱断裂。待导热液冷却后，方可倒回瓶中。

(2) 升华物质熔点的测定　对于升华物质熔点的测定，要用两端封闭的毛细管浸入热浴中测定。

(3) 显微熔点测定仪测定熔点（微量熔点测定法）　毛细管法测定熔点的优点是仪器简单，操作方便，但费时，且不能观察到晶体在加热过程中晶形的变化情况。为了克服这些缺点，可用显微熔点测定仪测定熔点。

这类仪器型号较多，但共同点是样品用量少（2～3粒小结晶），能精确观察样品在受热过程中的变化情况，如晶形的转变、结晶的萎缩、失水等现象，不但可以测微量样品的熔点，还可测高熔点的样品。图 2-17 是 X-5 型显微熔点测定仪（数显控温型）的外观图。图 2-18 是该熔点测定仪的系统图。测量熔点温度范围：室温至 320℃；读数精确度：0.1℃；测量误差：室温至 200℃时，误差为±1℃，200～300℃时，误差为±1.5℃；测试量：熔点样品测试量不大于 0.1mg。

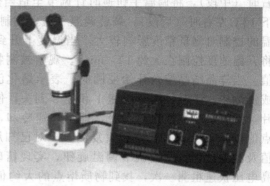

图 2-17　X-5 型显微熔点测定仪

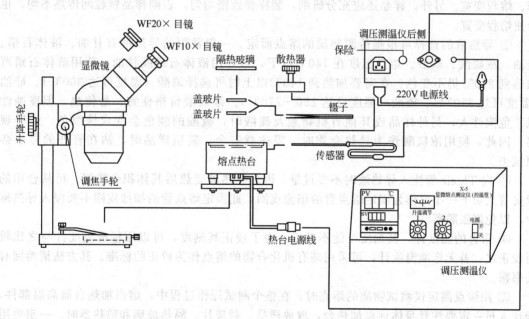

图 2-18　X-5 型显微熔点测定仪的系统

具体操作方法：取两片载玻片，用蘸有乙醚（或乙醚与酒精混合液）的脱脂棉擦拭干净。晾干后，取微量待测物晶粒放在一片载玻片上，并使样品分布薄而均匀，盖上另一片载玻片，轻轻压实，然后放在加热台上，使其位于加热台的中心，盖上保温圆玻璃盖。然后调节调焦手轮，直到能清晰地看到待测物品的像为止。打开熔点测定仪电源开关，接通电源后，仪器显示窗上排"PV"显示 HELO，下排"SV"显示 PASS 字样，表示仪器自检通过。如果显示—HH—，则表示：a. 未接实或未接传感器；b. 传感器热阻开路；c. 超温度量值。自检通过后，系统自动进入工作状态，此时"PV"显示测量值，

"SV"显示上限温度值。根据被测熔点物品的温度值，控制调温手钮1或钮2（1表示升温电压宽量调整，2表示升温电压窄量调整，其电压变化可参考电压表的显示），以期达到在测物质熔点过程中，前段升温迅速，中段升温减慢，后段升温平稳。可先将两调温手钮顺时针调到较大位置，使热台迅速升温，当温度接近待测物体熔点温度以下40℃左右时（中段），将调温手钮逆时针调节至适当位置，使升温速度减慢。在被测物熔点值以下10℃左右时（后段），调整调温手钮，控制升温速度约1℃/min（注意：尤其是后段升温的控制对测量精度影响较大，当温度上升到距待测物熔点值以下10℃左右时，一定要将升温速度控制在大约1℃/min）。观察被测物品的熔化过程，记录初熔和全熔时的温度值。当样品结晶棱角开始变圆时，表示熔化已经开始，结晶形状完全消失表示熔化已完成。测毕，逆时针调节手钮1和钮2到头，使电压调为零，用镊子取下隔热玻璃和载玻片，并将散热器放在热台上，即可快速冷却，以便再次测试或收存仪器。当温度降至熔点值以下40℃时即可进行重复测试。对已知熔点大约值的物质，可根据所测物质的熔点值及测温过程，适当调节调温旋钮，实现精确测量。对未知熔点的物质，可先用中、较高电压快速粗测一次，找到物质熔点的大约值，再根据该值适当调整和精细控制测量过程，最后实现精确测量。在使用这种仪器之前，必须仔细阅读使用指南，严格按操作规程进行。

2.4.3 注意事项

① 待测样品必须经充分干燥后再进行熔点测定。否则，含有水的样品会导致其熔点下降、熔程变宽。另外，样品还应充分研细，装样要致密均匀，否则样品颗粒间传热不均，也会使熔程变宽。

② 导热液的选择可根据待测物质的熔点而定。一般常用的导热液有甘油、液体石蜡、硅油、浓硫酸、磷酸。熔点温度在140℃以下，最好用液体石蜡或甘油，药用液体石蜡可加热到220℃仍不变色。在需要加热到140℃以上时可选择磷酸（温度可达300℃）、硅油（温度可达350℃）、硫酸（温度可达250～270℃）。浓硫酸价格便宜，易传热，但腐蚀性强、危险性大，另外样品或其他有机物触及硫酸时，硫酸的颜色会变成棕黑色，妨碍观察。因此，使用浓硫酸作为导热介质时，需注意安全，装填样品时，沾在管外的样品必须拭去。

③ 向Thiele管注入导热液时不要过量，因为导热液受热后其体积会膨胀。可从合用的橡皮管上切下一小段作为固定熔点管的细橡皮圈，此固定熔点管的细橡皮圈不要浸入导热液中，以免溶胀脱落。

④ 市售的温度计，其刻度可能不准确，为了校正其刻度，可以用标准温度计与之比较而校正之；若无标准温度计，可采用纯有机化合物的熔点作为校正的标准。其方法请参阅有关书籍。

⑤ 用熔点测定仪测试物质的熔点时，在整个测试操作过程中，熔点加热台属高温部件，操作人员一定要注意身体远离加热台，取放样品、载玻片、隔热玻璃和散热器时，一定要用镊子夹持，严禁用手触摸，以免烫伤。用过的载玻片可用乙醚和乙醇混合液擦拭干净，以备下次使用。

⑥ 仪器应放置于阴凉、干燥、无尘的环境中使用与存放。透镜表面有污秽时，可用脱脂棉蘸少许乙醚和乙醇混合液轻轻擦拭，遇有灰尘，可用洗耳球（吹球）吹去。

思 考 题

1. 测定熔点有何意义？
2. 影响熔点测定的因素有哪些？

2.5 沸点的测定

当纯净液体物质受热至其蒸气压与外界压力相等时就会沸腾,此时的温度即为该物质的沸点（Boiling Point,简记 b.p.）。沸点是液体有机化合物的重要物理常数之一,通过测定沸点可以鉴别有机化合物,并判断其纯度。

2.5.1 基本原理

由于分子运动,液体分子有从表面逸出的倾向,这种倾向随着温度的升高而增大。即液体在一定温度下具有一定的蒸气压,液体的蒸气压随温度的升高而增大,而与体系中存在的液体及蒸气的绝对量无关。当液体物质受热时,其蒸气压随温度升高而不断增大,当液体的蒸气压与外界大气压或所给压力相等时,就有大量气泡从液体内部逸出,即液体沸腾,这时的温度称为液体的沸点。显然液体的沸点与所受外界压力有关,外界压力不同,同一液体的沸点会发生变化。当外界压力增大,液体沸腾时的蒸气压增大,沸点升高；相反,当外界压力降低,则液体沸腾时的蒸气压也降低,沸点降低。因此,在说明一个化合物的沸点时,一定要注明测定沸点时的压力条件。蒸气压的量度一般是以汞柱高度来表示。通常所说的沸点是指在 760mmHg❶ 压力下液体的沸腾温度。

在一定压力下,纯的液体有机化合物具有固定的沸点,但当液体不纯时,则沸点有一个温度稳定范围,常称为沸程。通过测沸点、观察沸点范围可以判断液体的纯度,因为纯化合物的沸程一般较窄,约为 0.5~1℃。

需要指出的是,具有恒沸点的液体并不一定都是纯化合物,因为共沸混合物也具有恒定的沸点。因此,测定沸点只能定性地鉴别一个化合物。

2.5.2 测定沸点的方法

一般用于测定沸点的方法有两种,即常量法和微量法。常量法即采用蒸馏法测沸点,此法是当液体的量在 10mL 以上时采用,具体操作方法见 2.6 蒸馏。微量法即采用沸点测定管来测定液体的沸点,当样品量不多时可采用微量法。

微量法测定沸点的方法:取一根内径 3~4mm、长 7~8cm、一端封口的玻璃管作为沸点管的外管,放入欲测定沸点的样品 4~5 滴,在此管中放入一根长 8~9cm、内径约 1mm 的上端封闭的毛细管作为沸点管的内管,即其开口处浸入样品中。把微量沸点管贴于温度计水银球旁,如图 2-19 所示,像熔点测定那样把沸点测定管附在温度计旁,置于浴液中,缓慢加热,慢慢升温。随着温度的升高,管内的气体分子动能增大,气体膨胀,内管中有断断续续的小气泡冒出来,到达样品的沸点时出现一连串的小气泡,此时应停止加热,使液浴的温度下降,气体逸出的速度即渐渐地减慢。仔细观察,最后一个气泡出现而刚欲缩回到内管的瞬间即表示毛细管内液体的蒸气压与大气压平衡时的温度,此时的温度即为该液体化合物在常压下的沸点。

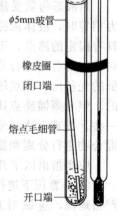

图 2-19 微量法测定沸点装置

微量法测沸点操作练习：测定纯苯的沸点。

2.5.3 注意事项

① 测定沸点时,加热不应过猛,尤其是在接近样品沸点时,升温更要慢一些,否则沸

---

❶ 1mmHg=133.322Pa。

点管内的液体会迅速挥发而来不及测定。外管中的液体要足够多。

② 要重复操作几次，误差应小于±1℃。待浴液温度下降15～20℃后可重新加热再测定。

③ 如果在加热测定沸点过程中，没能观察到一连串小气泡逸出，可能是沸点内管口处没封闭好的缘故。此时，应停止加热，换一根内管，重新测定。

思 考 题

1. 测定沸点有何意义？
2. 用微量法测定沸点，把最后一个气泡刚欲缩回至内管的瞬间的温度作为该化合物的沸点，为什么？

2.6 蒸馏

将液态物质加热沸腾变为蒸气，再将蒸气冷凝为液体，这两个过程的联合操作称为蒸馏(Distillation)。蒸馏是分离和提纯液态有机化合物常用的方法之一，是重要的、必须熟练掌握的基本操作。利用蒸馏可以将沸点相差较大（如30℃以上）的液态混合物分开。纯液态有机化合物在蒸馏过程中沸点范围（沸程）很小（0.5～1℃），而混合物的沸程较大。所以，利用蒸馏可以测定沸点，判定化合物的纯度，定性地鉴定化合物。用蒸馏法测沸点叫常量法，此法样品用量较大，要10mL以上，若样品不多时，可采用微量法测沸点（参见2.5沸点的测定）。

蒸馏操作是有机化学实验中常用的实验技术，一般用于下列几个方面：

① 分离液体混合物，仅在混合物中各成分沸点有较大差别时才能达到有效分离；
② 测定化合物的沸点；
③ 提纯，除去不挥发的杂质；
④ 回收溶剂，或蒸出部分溶剂以浓缩溶液。

2.6.1 基本原理

当液态物质受热时，其蒸气压随温度升高而增大，待蒸气压大到和大气压或所给压力相等时，液体沸腾，这时的温度称为该液体的沸点。纯液态有机化合物在一定压力下具有固定的沸点，不同的物质具有不同的沸点。蒸馏操作就是利用不同物质的沸点差异对液态混合物进行分离和纯化。如蒸馏沸点差别较大（在30℃以上）的液体时，沸点较低的先蒸出，沸点较高的后蒸出，不挥发的留在蒸馏瓶内，这样可达到分离和提纯的目的。但在蒸馏沸点比较接近的混合物时，各物质的蒸气将同时蒸出，只不过低沸点的多一些，故难于达到分离和提纯的目的，这就需要采用分馏操作（参见2.7分馏）对液态混合物进行分离和提纯。

应当指出以下几点。

① 在常压下进行蒸馏时，由于大气压往往不恰好等于101.325kPa（760mmHg），因此严格地说，应该对观察到的沸点加以校正。但由于偏差一般较小，即使大气压相差20mmHg，校正值也不过±1℃左右，因而可忽略不计。

② 当液体中溶入其他物质时，无论这种溶质是固体、液体还是气体，无论挥发性大还是小，液体的蒸气压总是降低的，因而所形成溶液的沸点会有变化。

③ 纯净的液体有机化合物在一定压力下具有恒定的沸点，但具有恒定沸点的液体并非都是纯化合物，因为有些化合物相互之间可以形成二元或三元共沸混合物，形成共沸物时，其液相和气相组成相同，因此在同一沸点下，组成一样。共沸混合物不能通过蒸馏操作进行分离，具体方法见共沸蒸馏（参见2.8共沸蒸馏）。

2.6.2 蒸馏装置

实验室蒸馏装置主要由蒸馏液汽化装置、冷凝装置和接收装置三部分组成。常用的蒸馏装置由蒸馏烧瓶、蒸馏头、温度计套管、温度计、直形冷凝管、接引管、接收瓶等组装而成,如图2-20所示。

(1) 蒸馏烧瓶 为蒸馏容器,液体在瓶内受热汽化,蒸气经蒸馏头支管进入冷凝管。蒸馏前应根据待蒸馏液体的体积,选择合适的蒸馏瓶。通常被蒸馏的液体约占蒸馏瓶容积的1/3~2/3为宜,否则沸腾时液体容易冲出或损失较大。

(2) 冷凝管 蒸气在冷凝管中冷凝成液体。当待蒸馏液体的沸点低于130℃时,应选用直形冷凝管。沸点高于130℃时,应选用空气冷凝管,因为温度高时,如用水作为冷却介质,冷凝管内外温差增大,而使冷凝管接口处局部骤然遇冷容易断裂。液体的沸点很低时,可用蛇形冷凝管,蛇形冷凝管要竖直

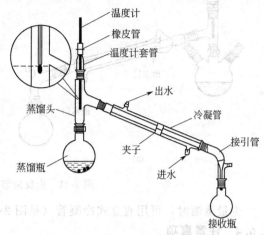

图2-20 常量及半微量蒸馏装置图

装置,切不可斜装,以免使冷凝液停留在其中,阻塞了通道而发生事故。冷凝管下端侧管为进水口,用橡皮管接自来水龙头,上端的出水口套上橡皮管导入水槽中。上端的出水口应向上,才能保证套管内充满水。冷凝管的种类很多,常用的为直形冷凝管,用冷水冷却。

(3) 接收瓶 常用接引管与三角烧瓶或圆底烧瓶连接。如所用的接引管无侧管,则接引管与接收瓶之间应留有空隙,以确保蒸馏装置与外界大气相通。任何蒸馏或回流装置均不能密封,否则当体系受热后液体蒸气压增大时,轻者蒸气冲开连接口,使液体冲出蒸馏瓶,重者会发生装置爆炸引起火灾。

2.6.3 蒸馏操作

(1) 加料 按照图2-20所示,安装好仪器,取下温度计和温度计套管,然后将待蒸馏液体通过放在蒸馏头上的长颈玻璃漏斗慢慢加入到瓶中(长颈玻璃漏斗下口处的斜面应超过蒸馏头支管),投入1~2粒沸石,再配置温度计。温度计水银球上缘恰好与蒸馏头支管下缘在同一水平线上(见图2-20),以保证在蒸馏时整个水银球能完全被液体蒸气浸润。水银球高了,所测沸点会偏低;水银球低了,则所测沸点会偏高。

(2) 加热 加热前,先向冷凝管缓缓通入冷凝水,把上口流出的水引入水槽中。检查仪器装配是否正确,原料、沸石是否加好,一切无误后再开始加热。最初加热宜用小火,以免蒸馏瓶因局部受热而破裂。慢慢增大火力使之沸腾,进行蒸馏。调节火焰或加热套电压,使蒸馏速度以1~2滴/s自接引管滴出馏出液为宜。在蒸馏过程中,应使温度计水银球始终保持有液滴存在,此时温度计的读数就是液体的沸点。要注意温度计读数,记下第一滴馏出液流出时的温度,当温度计读数稳定后,另换一个接收瓶收集所需温度范围的馏出液。如果维持原来加热程度,不再有馏出液蒸出,而温度计会突然下降,这表明该馏分已近蒸完,应停止加热,记下该馏分的沸程和体积(或质量)。注意,即使杂质很少,也不能蒸干,应残留0.5~1mL液体,否则容易发生事故。馏分的温度范围愈小,其纯度就愈高。

(3) 停止蒸馏 馏分蒸完后,如不需要接收第二组分,可停止蒸馏。应先停止加热,待稍冷却后馏出物不再继续流出时,停止通水,拆卸仪器,其程序与装配时相反,即按次序取下接收器、接引管、冷凝管和蒸馏烧瓶。

有时,在有机反应结束后,需要对反应混合物直接蒸馏,此时可以将三口烧瓶作为蒸馏

瓶组装成蒸馏装置直接进行蒸馏,如图 2-21 所示。

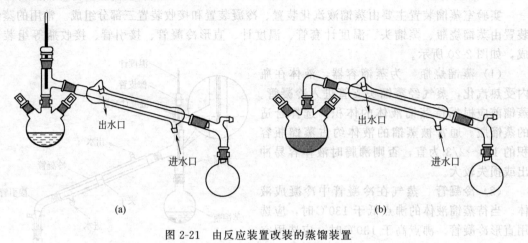

图 2-21 由反应装置改装的蒸馏装置

少量蒸馏时,可用直立式冷凝管(见图 2-22),以减少样品损失。

2.6.4 注意事项

① 加热前要检查装置是否严密、正确,以防事故的发生。整套装置要做到准确端正,不论从侧面还是正面看上去,各个仪器的中心线要在一条直线上。各个铁夹不要夹得太紧或太松,以免弄坏仪器。

② 待蒸馏液的体积约占蒸馏烧瓶体积的 1/3～2/3。

③ 在加热蒸馏前要加入沸石。为了消除在蒸馏过程中的过热现象和保证沸腾的平稳状态,常加入沸石,因沸石能防止加热时的暴沸现象,故把它称为止暴剂。沸石为多孔性物质,刚加入液体中小孔内有许多气泡,当液体受热沸腾时,沸石内的小气泡成为液体的汽化中心,使液体保持平稳沸腾。当加热后,发觉未加沸石或原有沸石失效时,千万不能匆忙地投入沸石,应停止加热,待稍冷却后再加入沸石。另外,也不能在沸腾或接近沸腾的溶液中加入沸石,否则将会引起猛烈的暴沸,液体易冲出瓶口,若是易燃的液体,将会引起火灾。切记!如加热中断,再加热时应重新加入新沸石,以免出现暴沸,因原来沸石上的小孔已被液体充满,不能再起汽化中心的作用。同理,分馏和回流时也要加沸石。

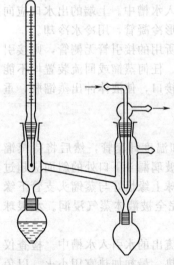

图 2-22 少量蒸馏装置

④ 蒸馏沸点较低易燃液体(如乙醚)时,应用水浴加热,千万不可用明火加热,而且蒸馏速度不能太快,以保证蒸气全部冷凝。如果室温较高,接收瓶应放在冷水中冷却,在接引管支口处连接一根橡皮管,通入水槽的下水管内或引至室外。

⑤ 根据待蒸馏液体的沸点,选择相应的冷凝管,并控制冷凝水的速度。蒸馏物沸点在 70℃ 以下时,水流要快;100～120℃ 时,水流应缓,太冷和太快的水流可能导致冷凝管炸裂;130℃ 以上则应使用空气冷凝管。

⑥ 根据待蒸馏液体的沸点,选择相应的热浴,并控制热浴的温度。热浴的温度一般比待蒸馏物沸点高出 30℃ 为宜,即使蒸馏物沸点很高,也绝不要将浴温超出 40℃。浴温过高会使蒸馏速度过快,蒸馏瓶和冷凝器上部蒸气过大,使大量蒸气逸出,导致突然发生着火或蒸馏物过热发生分解。

⑦ 注意安装、拆卸仪器的顺序。安装的顺序一般是先从热源处开始,然后由下而上,

由左到右（或由右到左），依次安装。拆卸仪器的顺序和安装仪器的顺序相反。

2.6.5 蒸馏操作练习

用蒸馏的方法将工业乙醇提纯为95％乙醇。在25mL蒸馏瓶中加入15mL工业乙醇，进行简单蒸馏。

> **思 考 题**
>
> 1. 从安全和效果两方面考虑，在进行蒸馏操作时应注意哪些问题？
> 2. 蒸馏时，放入沸石为什么能防止暴沸？如果加热后才发觉未加沸石，应该怎样处理才安全？
> 3. 当加热后有馏液出来时，才发现冷凝管未通水，请问能马上通水吗？如果不行，应怎么办？
> 4. 向冷凝管通水是由下而上，反过来效果怎样？把橡皮管套进冷凝管侧管时，怎样才能防止折断其侧管？如果加热过猛，测定出来的沸点是否正确？为什么？
> 5. 为什么蒸馏系统不能封闭？
> 6. 蒸馏时温度计水银球上有无液滴意味着什么？
> 7. 为什么蒸馏时不能将液体蒸干？

2.7 分馏

蒸馏和分馏都是分离提纯液体有机化合物的重要方法，普通蒸馏要求其组分的沸点至少要相差30℃，只有当组分的沸点差达110℃以上时，才能用蒸馏法充分分离。对沸点相近的混合物，用普通蒸馏法就难以精确分离，应当用分馏的方法分离。应用分馏柱对几种沸点相近的液体混合物进行分离和提纯的方法称为分馏（Fractional Distillation）。分馏在化学工业和实验室中被广泛应用，现在最精密的分馏设备已能将沸点相差1~2℃的混合物分开，利用蒸馏或分馏来分离液体混合物的原理是一样的，实际上分馏就是多次的蒸馏。

2.7.1 分馏原理

分馏实际上就是使沸腾着的混合物蒸气通过分馏柱（工业上用分馏塔）进行一系列的热交换。当液体混合物受热沸腾时，其蒸气首先进入分馏柱，由于柱外空气的冷却，蒸气中高沸点的组分就被冷却为液体，流回至烧瓶中，故上升的蒸气中含低沸点的组分就相对地增加。这一过程可以看作是一次简单的蒸馏。当高沸点的冷凝液回流途中遇到新上升的蒸气时，两者之间又进行热交换，上升的蒸气中高沸点的组分又被冷凝，低沸点的组分仍继续上升，易挥发的组分又增加了，这又可看作是一次简单的蒸馏。蒸气就这样在分馏柱内反复地进行着汽化-冷凝-回流等程序，或者说，重复地进行着多次简单蒸馏。因此，当分馏柱的效率相当高且操作正确时，在分馏柱顶部出来的蒸气就接近于纯低沸点的组分。这样，最终便可将沸点不同的物质分离出来。所以在分馏时，柱内不同高度的各段，其组分是不同的，靠近分馏柱顶部，低沸点的含量高，高沸点含量相对低。应用分馏柱的目的是要增大气液两相的接触面，提高分离效率。

了解分馏原理最好是应用恒压下的沸点-组成曲线图（又称为相图，表示这两组分体系中相的变化情况）。通常它是用实验测定在各温度时汽液平衡状况下的气相和液相的组成，然后以横坐标表示组成，纵坐标表示温度而作出的（如果是理想溶液，则可直接由计算作出）。图2-23是大气压下的苯-甲苯体系的温度-组成曲线。从图中可以看出，由苯20％和甲苯80％组成的液体（L_1）在102℃时沸腾，和此液相平衡的蒸气组成约为苯40％和甲苯60％（V_1）。若将此组成的蒸气冷凝成同组成的液体（L_2），则与此液体成平衡的蒸气组成约为苯70％和甲苯30％（V_2）。显然，如此继续重复，即可获得接近纯苯的气相。

采用分馏的分离效果比蒸馏好得多。例如，将20mL甲醇和20mL水混合物分别进行普通蒸馏和分馏，控制蒸出的速度为1mL/3min，每收集1mL馏出液记录温度，以馏出液体积为横坐标,温度为纵坐标,分别得出蒸馏曲线和分馏曲线,如图 2-24 所示,从分馏曲线可以看出,当甲醇蒸出后,温度很快上升,达到水的沸点,甲醇和水可以很好地分开,显然,分馏比普通蒸馏(一次)要好得多。

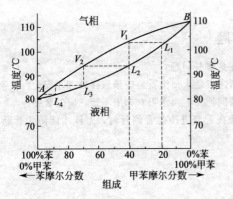

图 2-23 苯-甲苯体系的温度-组成曲线

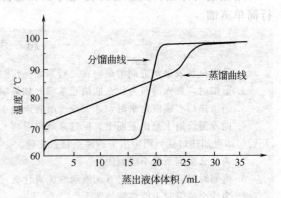

图 2-24 甲醇-水混合物 (1∶1) 的蒸馏和分馏曲线

必须指出,当某两种或三种液体以一定比例混合,可组成具有固定沸点的混合物,将这种混合物加热至沸腾时,在汽液平衡体系中,气相组成和液相组成一样,因此不能用分馏法将其分离出来,只能得到按一定比例组成的混合物,这种混合物称为共沸混合物(或恒沸混合物),它的沸点(高于或低于其中的每一组分)称为共沸点(或恒沸点)。共沸混合物的沸点若低于混合物中任一组分的沸点,则称为低共沸混合物,也有高共沸混合物。

低共沸混合物体系如乙醇-水体系,其相图如图 2-25 所示。常见的共沸混合物见表 2-4。

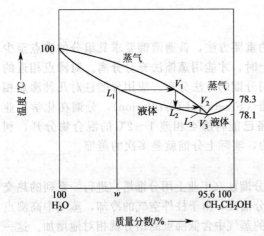

图 2-25 乙醇-水低共沸相图

表 2-4 常见的共沸混合物

	组 成(沸点/℃)	共沸混合物	
		共沸点/℃	各组分质量分数/%
二元共沸混合物	水(100) 乙醇(78.5)	78.2	4.4 95.6
	水(100) 苯(80.1)	69.4	8.9 91.1
	乙醇(78.5) 苯(80.1)	67.8	32.4 67.6
	水(100) 氯化氢(−83.7)	108.6	79.8 20.2
	丙酮(56.2) 氯仿(61.2)	64.7	20.0 80.0
	乙酸乙酯(77.2) 水(100)	70.0	91.0 9.0
	乙醇(78.5) 四氯化碳(76.5)	64.9	16.0 84.0

续表

组　成(沸点/℃)		共沸混合物	
		共沸点/℃	各组分质量分数/%
三元共沸混合物	水(100) 乙醇(78.5) 苯(80.1)	64.6	7.4 18.5 74.1
	水(100) 丁醇(117.7) 乙酸丁酯(126.5)	90.7	29.0 8.0 63.0

由于水能与许多物质形成共沸混合物，所以化合物在进行蒸馏前，必须仔细地用干燥剂除水。

2.7.2 影响分馏效率的因素

(1) 理论塔板　分馏柱效率是用理论塔板来衡量的。分馏柱中的混合物，经过一次汽化和冷凝的热力学平衡过程，相当于一次普通蒸馏所达到的理论浓缩效率，当分馏柱达到这一浓缩效率时，那么分馏柱就具有一块理论塔板。柱的理论塔板数越多，分离效果越好。分离一个理想的二组分混合物所需的理论塔板数与两个组分的沸点差之间的关系见表2-5。

表2-5　二组分的沸点差与分离所需的理论塔板数的关系

沸点差值/℃	108	72	54	43	36	20	10	7	4	2
分离所需的理论塔板数	1	2	3	4	5	10	20	30	50	100

其次，还要考虑理论塔板层高度，在高度相同的分馏柱中，理论塔板层高度越小，则柱的分离效率越高。

(2) 回流比　在单位时间内，由柱顶冷凝返回柱中液体的数量与蒸出物的数量之比称为回流比。若全回流中每10滴收集1滴馏出液，则回流比为9∶1。对于非常精密的分馏，使用高效率的分馏柱，回流比可达到100∶1。

(3) 柱的保温　分馏柱必须进行适当的保温，以便始终维持温度平衡。一般在分馏柱外包扎石棉绳、石棉布等保温材料，以提高分馏效率。

2.7.3 分馏装置

分馏装置与简单蒸馏装置类似，不同之处是在蒸馏瓶与蒸馏头之间加了一根分馏柱，如图2-26所示。分馏柱的种类很多，一般实验室常用的分馏柱有如图2-27所示的几种。

为了提高分馏柱的分馏效率，在分馏柱内装入具有大比表面积的填料，填料的作用是在柱中起到增加蒸气与回流液接触的作用，填充物比表面积越大，越有利于提高分馏效率。填料之间应保留一定的空隙，要遵守适当紧密且均匀的原则，这样就可以增加回流液与上升蒸气的接触机会，如果空隙太小，会导致蒸馏困难。填料有玻璃（玻璃球、短段玻璃管）或金属（金属丝绕成固定形状）两种，玻璃的优点是不会与有机化合物起反应，而金属则可与卤代烷之类的化合物起反应。在分馏柱底部常常放一些玻璃丝以防止填料坠入蒸馏容器内。

实验室中常用的刺型分馏柱［见图2-27(a)，又称韦氏（Vigreux）分馏柱］，是一种柱内呈刺状的简易分馏柱，不需另加填料。半微量实验一般用填料柱，即在一根玻璃管内填上惰性材料，如玻璃、陶瓷或螺旋形、马蹄形等各种形状的金属小片。

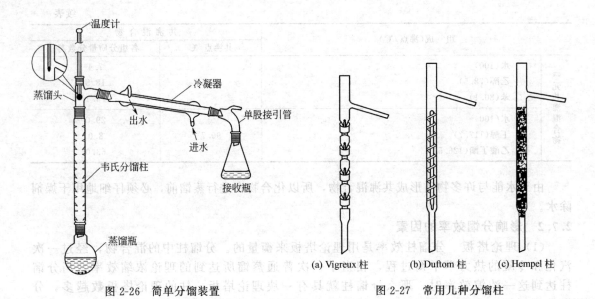

图 2-26 简单分馏装置　　　图 2-27 常用几种分馏柱

2.7.4 操作方法

将待分馏的物质装入圆底烧瓶中,加入 2 粒沸石,依次(分馏柱、温度计、冷凝管、接引管及接收瓶)安装好分馏装置(见图 2-26)。接通冷凝水,选用合适的热浴后开始加热,使液体平稳沸腾。注意控制温度,使蒸气缓缓上升,控制馏出速度维持在 2~3s 一滴。记录第一滴馏出液滴入接收瓶时的温度,然后根据具体要求分段收集馏分,并记录各馏分的沸点范围及体积。

2.7.5 分馏操作练习

(1) 分馏实验　在 50mL 的圆底烧瓶中,加入丙酮和水各 15mL,加 2 粒沸石,安装好分馏装置,开始缓慢加热,控制加热温度,使馏出液速度维持在 2~3s 一滴。分别记录 56~62℃、62~72℃、72~98℃、98~100℃ 时的馏出液体积,直至蒸馏烧瓶中残留液为 1~2mL 时停止加热,待分馏柱内液体流回到烧瓶时测量并记录残留液体积。以柱顶温度(T)为纵坐标,馏出液体积为横坐标(V),作出分馏曲线,讨论分离效率。

(2) 蒸馏实验　取同样体积的水和丙酮,安装好蒸馏装置,加入 2 粒沸石,开始加热。同样记录上述温度段的馏出液体积。在(1)所用的同一张纸上绘出温度-体积的蒸馏曲线图。比较两者的不同及分离效率。

2.7.6 注意事项

① 分馏一定要缓慢进行,控制好恒定的蒸馏速度。在分馏过程中,要注意调节加热温度,使馏出速度适中(2~3s 一滴),如馏出速度太快,就会产生液泛现象,即蒸发速率增至某一程度时,回馏液来不及流回至烧瓶中,上升的蒸气就将下降的液体顶上去,并逐渐在分馏柱中形成液柱,破坏了汽液平衡,这将降低分馏效率。如果出现此现象,应停止加热,待液柱消失后重新加热,使气液达到平衡,再恢复收集馏分。

② 要有相当量的液体自分馏柱流回烧瓶中,即要选择合适的回流比。

③ 必须尽量减少分馏柱的热量散失和波动。在分馏过程中,要防止回流液体在柱内的聚集。为此,要在分馏柱的外面包一定厚度的保温材料,以保证柱内具有一定的温度梯度,防止蒸气在柱内冷凝太快。当使用填充柱时,往往由于填料装得太紧或不均匀,造成柱内液体聚集,这时应重新装柱。

> **思 考 题**
>
> 1. 分馏和蒸馏在原理、装置和操作上有哪些异同？
> 2. 分馏时，温度计应放在什么位置？过高、过低时对分馏有什么影响？
> 3. 为什么分馏时柱身的保温十分重要？
> 4. 为什么分馏时加热要平稳并控制好回流比？
> 5. 含水乙醇为何经过反复分馏也得不到100%乙醇？这是什么原因？

2.8 共沸蒸馏

在 2.6 蒸馏和 2.7 分馏部分已经叙述过，共沸混合物气液相组成一样，因此不能用蒸馏和分馏的方法分离，但可用其他方法破坏共沸组成后，再进行蒸馏或分馏，以达到分离的目的。共沸蒸馏又称恒沸蒸馏（Constant Boiling Distillation），主要用于共沸混合物的分离。共沸混合物是指在一定压力下，具有恒定沸点和组成的混合物。恒沸混合物的沸点比纯物质的沸点更低或更高。

2.8.1 基本原理

在共沸混合物中加入第三组分，该组分与原共沸混合物中的一种或两种组分形成沸点比原来组分和原来共沸物沸点更低的、新的具有最低共沸点的共沸物，使组分间的相对挥发度增大，易于用蒸馏的方法分离。这种蒸馏方法称为共沸蒸馏，加入的第三组分称为恒沸剂或夹带剂。

工业上常用苯作为恒沸剂进行共沸精馏制取无水酒精。常用的夹带剂有苯、甲苯、二甲苯、三氯甲烷、四氯化碳等。

例如，乙醇和水的混合物，由图 2-25 可以看出，95.6%乙醇和4.4%水组成了一个最低共沸混合物，沸点是 78.15℃（乙醇的沸点是 78.3℃，水的沸点是 100℃），不能用分馏的方法分离。如向乙醇和水的共沸混合物中加入第三组分苯，使苯-乙醇-水形成三元共沸混合物（沸点 65℃）蒸出。蒸出后分成上下两层，上层为苯和乙醇，可用分水器使之返回烧瓶中，下层为水和乙醇。这样不断进行，可把水都带出来，然后蒸出苯和乙醇的共沸混合物

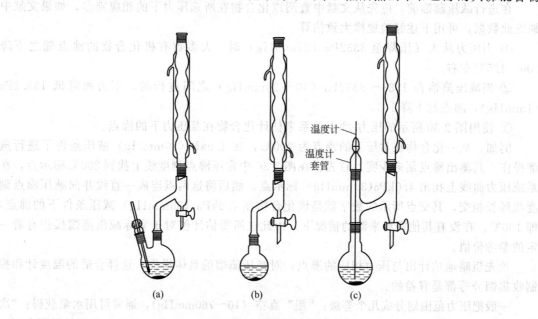

图 2-28 共沸蒸馏装置

（沸点68℃），烧瓶中留下的是相当纯的乙醇。

2.8.2 共沸蒸馏装置

图2-28是实验室常用的共沸蒸馏装置，在蒸馏瓶与回流冷凝管间增加了一个分水器。常用的分水器有几种，如图2-29所示。

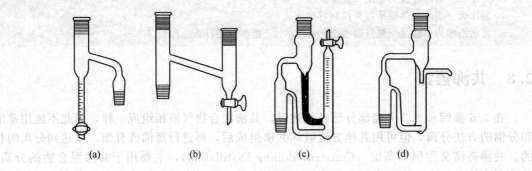

图2-29 常用的几种分水器

2.9 减压蒸馏

减压蒸馏又称真空蒸馏（Vacuum Distillation）是分离和提纯液态或低熔点固体有机化合物的一种重要方法。它特别适用于那些在常压下沸点较高及常压蒸馏时未达沸点即已发生分解、氧化、聚合等反应的热敏性有机化合物的分离提纯。

2.9.1 基本原理

液体化合物的沸点是指它的蒸气压等于外界大气压时的温度。所以，液体化合物的沸点与外界压力有着密切的关系，随着外界施加于液体表面压力的降低，液体沸点就会随之下降。如果用真空泵连接密闭的蒸馏装置，使液体表面的压力降低，即可降低液体的沸点。这种在较低压力下进行蒸馏的操作称为减压蒸馏。

在进行减压蒸馏前，应先从文献中查阅该化合物在所选压力下的相应沸点，如果文献中缺乏此数据，可用下述经验规律大致估算。

① 当压力从大气压降至3332Pa（25mmHg）时，大多数有机化合物的沸点随之下降100~125℃左右。

② 当减压蒸馏在1333~3332Pa（10~25mmHg）之间进行时，压力每降低133.3Pa（1mmHg），沸点约下降1℃。

③ 使用图2-30所示的压力-沸点关系来估计化合物在某压力下的沸点。

例如，某一化合物在常压下的沸点为200℃，在4.0kPa（30mmHg）减压条件下进行蒸馏操作，其蒸出沸点是多少呢？首先，在图2-30中常压沸点刻度线上找到200℃标示点，在系统压力曲线上找出4.0kPa（30mmHg）标示点，然后将这两点连成一直线并向减压沸点刻度线延长相交，其交点所示的数字就是该化合物在4.0kPa（30mmHg）减压条件下的沸点，即100℃。在没有其他资料来源的情况下，由此法所得估计值对于实际减压蒸馏操作有着一定的参考价值。

预先粗略地估计出与压力相应的沸点，对减压蒸馏的具体操作、选择合适的温度计和控制收集馏分等都是有益的。

一般把压力范围划分成几个等级："粗"真空（10~760mmHg），通常可用水泵获得；"次高"真空（0.001~1mmHg），可用油泵获得；"高"真空（<10^{-3}mmHg），可用扩散泵获得。

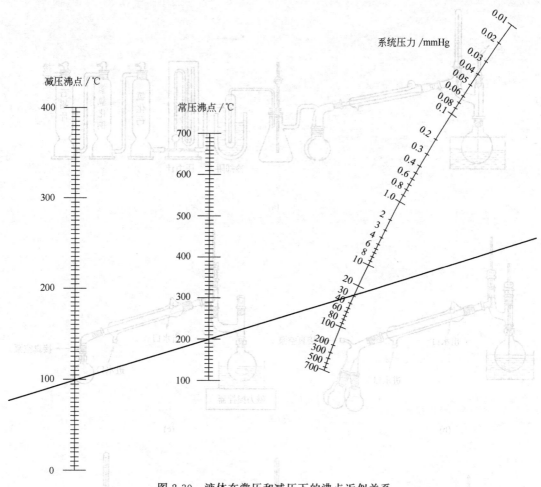

图 2-30　液体在常压和减压下的沸点近似关系
1mmHg＝133.322Pa

2.9.2　减压蒸馏装置

减压蒸馏系统主要是由蒸馏、抽气（减压）以及它们之间的保护和测压装置三部分组成，如图 2-31(a) 所示。

(1) 蒸馏部分　包括蒸馏瓶、克氏蒸馏头、直形冷凝管、真空接引管及接收瓶［见图 2-31(a)］。在圆底烧瓶上插上克氏蒸馏头，在克氏蒸馏头的侧口处插入温度计，直口处插一根毛细管，直至蒸馏瓶底部，距瓶底 1～2mm。毛细管的上端加一节带螺旋夹的橡皮管，螺旋夹用以调节进气量，使抽真空时有极少量的空气进入液体呈微小气泡，起到搅拌和汽化中心的作用，防止液体暴沸，使蒸馏平稳发生。毛细管口要很细，若太粗，进入的空气太多，则会把瓶内液体冲至冷凝管，也会使压力难以降低。蒸馏时若要收集不同的馏分而又不中断蒸馏，则可用两尾或多尾接液管。多尾接液管的几个分支管与作为接收器的圆底烧瓶连接，转动多尾接液管，就可以使不同的馏分进入指定的接收器中［见图 2-31(b)］。

进行半微量或微量减压蒸馏时，如使用能同时加热的电磁搅拌器搅动液体，可以防止液体的暴沸，故可不安装毛细管［见图 2-31(c)］。

蒸馏少量物质时可采用如图 2-31(d)～(f) 所示的减压蒸馏装置。

(2) 抽气部分　实验室里通常用水泵或油泵进行减压。若不需要很低的压力时可用水泵。水泵所能达到的最低压力，理论上相当于当时水温下的水蒸气压力，例如水温在 25℃、

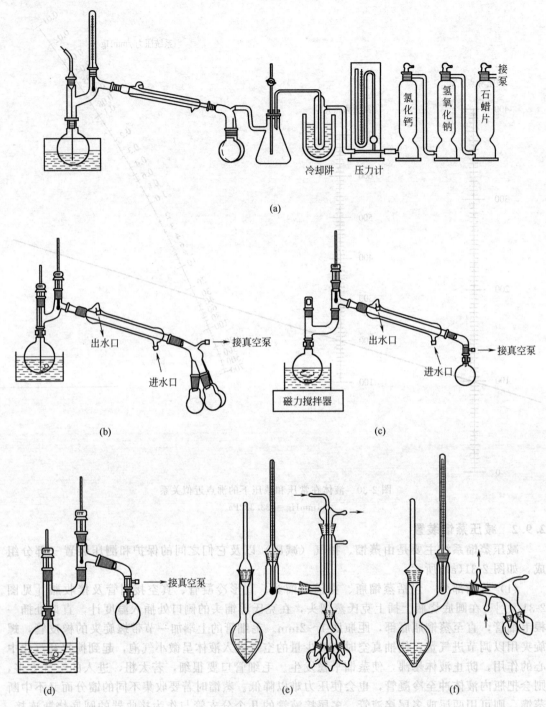

图 2-31 减压蒸馏装置

20℃、10℃时,水蒸气压力分别为 3192Pa、2394Pa、1197Pa(24mmHg、18mmHg、9mmHg)。用水泵进行抽气时,应在水泵前装上安全瓶,以防止水压下降时水流倒吸。停止蒸馏时要先放气,然后关泵。

若需要较低的压力,就要用油泵进行抽气,好的油泵应能抽到 133.3 Pa(1 mmHg)以下。油泵的好坏决定于其机械结构和油的质量,使用油泵时必须把它保护好。如果蒸馏挥发

48

性较大的有机溶剂时，挥发性的有机溶剂蒸气被油吸收后，就会增加油的蒸气压，影响真空效能。而酸蒸气会腐蚀油泵的机件。水蒸气凝结后与油形成浓稠的乳浊液，破坏了油泵的正常工作，因此使用油泵时必须十分注意油泵的保护。

(3) 保护和测压装置部分　当用油泵进行减压时，为了防止易挥发的有机溶剂、酸性物质和水蒸气进入油泵，降低油泵的效率，必须在馏液接收器与油泵之间顺次安装安全瓶、冷却阱、测压计、吸收塔及缓冲瓶（见图2-32）。

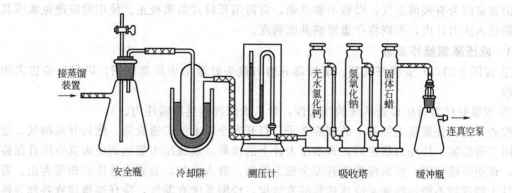

图 2-32　减压蒸馏油泵防护装置

真空接引管上的支口与安全瓶连接，安全瓶的作用不仅是防止压力降低或停泵时油（或水）倒吸流入接收瓶中造成产品污染，而且还可以防止蒸馏时因突然发生暴沸或冲料现象物料进入减压系统。另外，装在安全瓶口上的带旋塞双通管可用来调节系统压力或放气。有时，由于系统内压力突然发生变化，从而导致泵油倒吸，缓冲瓶的设置可以避免泵油冲入气体吸收塔。冷却阱用来冷凝被抽出来的沸点较低的组分，如水蒸气和一些挥发性物质。冷却阱置于盛有冷却剂的广口保温瓶中，冷却剂的选择可视具体情况而定，例如可用冰-水、冰-盐、干冰、液氮等。吸收塔也称干燥塔，一般设2~3个。这些干燥塔中分别装有硅胶（或无水氯化钙）、颗粒状氢氧化钠及片状固体石蜡，用以吸收水分、酸性气体及烃类气体。

测压计的作用是指示减压蒸馏系统内的压力，通常用水银测压计，其结构如图2-33所

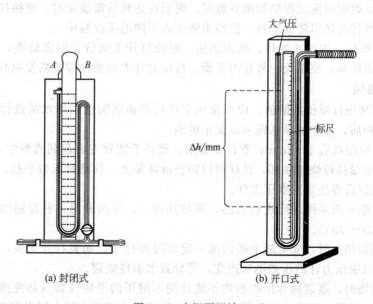

图 2-33　水银测压计

示。图 2-33(a) 为封闭式水银压力计，两臂液面高度之差即为减压蒸馏系统中的真空度。测定压力时可将管后木座上的滑动标尺的零点调整到右臂的汞柱顶端线上，这时左臂的汞柱顶端线所指示的刻度即为系统的真空度。图 2-33(b) 为开口式水银压力计，两臂汞柱高度之差即为大气压力与系统压力之差，因此蒸馏系统内的实际压力为：

实际压力（真空度）＝大气压力(mmHg)－汞柱差

开口式压力计较笨重，读数方式也较麻烦，但准确。封闭式压力计比较轻巧，读数方便，但常常因为有残留空气，以致不够准确，常需用开口式来校正。使用时应避免水或其他污物进入压力计内，否则将严重影响其准确度。

2.9.3 减压蒸馏操作要点

① 按图 2-31(a) 安装好仪器，蒸馏部分磨口接头处涂上少量真空脂，以保证装置密闭和润滑。

② 安装好仪器后应检查系统的气密性，看系统能否达到所需压力。

检查方法：先旋紧毛细管上的螺旋夹子，打开安全瓶上的二通旋塞，然后开泵抽气，逐渐关闭二通旋塞。待压力稳定后，观察压力计上的读数。系统压力能达到所需真空度且保持不变，说明系统密闭。然后慢慢旋开安全瓶上旋塞，放入空气，直到内外压力相等为止。若压力计上的读数不能达到油泵应该达到的真空度，说明系统有漏气，应仔细地检查各部分连接处何处漏气。待排除漏气点后，再重新检查系统的气密性，直至压力稳定并达到所要求的真空度时，方可进行下面的操作。

③ 将待蒸馏的液体放入蒸馏瓶中，液体的体积不超过容器体积的 1/2。

④ 旋紧毛细管上的螺旋夹子，开泵抽气，关闭安全瓶上的二通旋塞，调节毛细管上的螺旋夹，使烧瓶中液体内有连续平稳的小气泡。

⑤ 待系统达到所要求的低压，且压力平稳后，开启冷凝水，选用合适的热浴加热。蒸馏烧瓶至少要有 2/3 浸入浴液中。

⑥ 逐渐升温，控制浴温比待蒸馏液体的沸点约高 20～30℃。密切注意蒸馏的温度和压力，如果不符，则应进行调节。控制馏出速度 1～2 滴/s。当有馏分蒸出时，记录其沸点、相应的压力读数和浴温。纯物质的沸程一般不超过 1～2℃，但有时因压力有所变化，沸程会稍大些。若开始馏出液比收集物沸点低时，则当在达到所需温度时，更换接收器（若用多尾接引管则只需转动接引管的位置，使馏出液流入不同的接收瓶中）。

⑦ 蒸馏完毕后，先停止加热，撤去热浴，慢慢打开毛细管上的螺旋夹，再慢慢地打开安全瓶上的二通旋塞，使系统内外压力平衡，待压力计中的汞柱缓慢恢复原状，再关泵。

2.9.4 注意事项

① 在用油泵进行减压蒸馏前，应在常压下进行普通蒸馏或利用水泵进行减压蒸馏，以除去低沸点的物质，保护油泵系统和油泵中的油。

② 毛细管口距瓶底 1～2mm，管口要很细。此处毛细管主要起到沸腾中心和搅动作用，所以毛细管要尽量接近烧瓶底部，这样可以防止液体暴沸，使沸腾保持平稳。

③ 打开油泵后要注意观察压力计。

④ 减压蒸馏一般采用油浴进行加热，要加热均匀，逐渐升温，并控制油温比待蒸馏液体的沸点高约 20～30℃。

⑤ 减压蒸馏结束时，安全瓶上的活塞一定要缓慢打开，如果打开太快，系统内外压力突然变化，使水银压力计的压差迅速改变，可导致水银柱破裂。

⑥ 减压蒸馏时，蒸馏瓶和接收瓶均不能使用不耐压的平底仪器（如锥形瓶、平底烧瓶等）或有破损的仪器，以防由于装置内处于真空状态，外部压力过大而引起爆炸。

> **思 考 题**
>
> 1. 简述减压蒸馏的原理。
> 2. 在何种情况下采用减压蒸馏？
> 3. 减压蒸馏开始时，为什么要先抽气后加热？结束时为什么要先移开热源，再停止抽气？顺序可否颠倒？为什么？

2.10 水蒸气蒸馏

将水蒸气通入不溶于水的有机化合物中或使有机化合物与水共沸而蒸出，这个操作过程称为水蒸气蒸馏（Steam Distillation）。水蒸气蒸馏是用来分离和提纯液态或固态有机化合物的一种方法，常用在下列几种情况。

① 在常压下蒸馏易发生分解的高沸点有机化合物；
② 混合物中含有大量树脂状杂质或不挥发性杂质，采取蒸馏、萃取等方法都难于分离；
③ 从较多固体反应物中分离出被吸附的液体。

被提纯物质必须具备下列条件：
① 不溶或难溶于水；
② 共沸腾下与水不发生化学反应；
③ 在100℃左右时，必须具有一定的蒸气压，至少666.5～1333Pa（5～10mmHg）。

2.10.1 基本原理

当水与不相混溶的有机物混合共热时，根据道尔顿（Dalton）分压定律，整个体系的蒸气压 p 应为各组分蒸气压之和，即

$$p_{混合物} = p_{水} + p_{有机物}$$

当混合物中各组分蒸气压总和等于外界大气压时，液体沸腾，这时的温度称为该混合物的沸点。显然，混合物的沸点比其中任何单一组分的沸点都低。因此，在常压下应用水蒸气蒸馏，就能在低于100℃的情况下将高沸点组分与水一起蒸出来。

2.10.2 馏出液组成的计算

根据气体方程，蒸出的混合物蒸气中各个气体分压之比（$p_A : p_B$）等于它们的物质的量之比（$n_A : n_B$），即

$$\frac{p_A}{p_B} = \frac{n_A}{n_B}$$

将 $n_A = m_A/M_A$ 和 $n_B = m_B/M_B$ 代入上式得

$$\frac{m_A}{m_B} = \frac{p_A M_A}{p_B M_B}$$

式中 m_A，m_B——各物质蒸气质量；

M_A，M_B——物质 A 和 B 的相对分子质量。

因此，馏出液中有机物与水的质量（m_A 和 m_{H_2O}）之比可按下式计算

$$\frac{m_A}{m_{H_2O}} = \frac{p_A M_A}{p_{H_2O} M_{H_2O}}$$

例 1-辛醇进行水蒸气蒸馏时，1-辛醇与水的混合物在 99.4℃ 沸腾。通过查阅手册得知，1-辛醇的沸点为 195.0℃，1-辛醇的相对分子质量为 130，纯水在 99.4℃ 时的蒸气压为 99.18kPa(744mmHg)。按分压定律，水的蒸气压与 1-辛醇的蒸气压之和等于 101.31kPa(760mmHg)。因此，1-辛醇在 99.4℃ 时的蒸气压为 2.13kPa(16mmHg)，故有

$$\frac{m_A}{m_{H_2O}} = \frac{2.13 \times 10^3 \times 130}{99.18 \times 10^3 \times 18} = 0.155$$

即每蒸出 1g 水便有 0.16g 1-辛醇被蒸出。因此馏出液中 1-辛醇的质量分数为 13%，水的质量分数为 87%。这个数值为理论值，因为实验时有相当一部分水蒸气来不及与被蒸馏物充分接触便离开蒸馏烧瓶，所以实验蒸出水的量往往超过计算值，故计算值仅为近似值。

2.10.3 水蒸气蒸馏装置

图 2-34 是实验室里常用的水蒸气蒸馏装置。包括水蒸气发生器、蒸馏部分、冷凝部分和接收器四个部分。

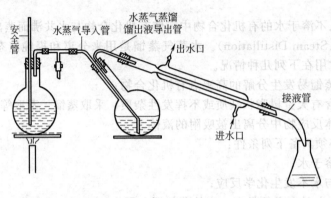

图 2-34 水蒸气蒸馏装置

水蒸气发生器一般是用金属制成的，如图 2-35 所示，也可用短颈圆底烧瓶或三口瓶代替。在由铜或铁金属制成的水蒸气发生器 A 的侧面安装一个水位计 B，以便观察发生器内水位，一般水位不要超过 2/3，最低不要低于 1/3。长玻璃管 C 称为安全管，管的下端接近发生器底部，距底部距离约 1～2cm，可用来调节体系内部的压力并防止系统发生堵塞时出现危险。水蒸气出口与 T 形管 G 连接，T 形管一端与发生器连接，另一端与蒸馏瓶连接。T 形管下口接一段软的橡皮管，用螺旋夹夹住，以便调节蒸气量，除去水蒸气中冷凝下来的水。在蒸馏过程中，如果发生不正常现象，应立即打开夹子，使水蒸气发生器与大气相通。

无论使用哪一种水蒸气发生器，在与蒸馏系统连接时管路越短越好，以减少水蒸气冷凝，否则会降低蒸馏瓶内温度，影响蒸馏效率。

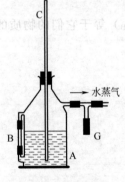

图 2-35 金属制的水蒸气发生器

其他几种水蒸气蒸馏装置如图 2-36 所示。图 2-34 及图 2-36(a)～(d)装置适用于常量物质的水蒸气蒸馏，图 2-36(e)～(f)装置适合于半微量物质的水蒸气蒸馏。

图 2-36(f) 装置的操作方法是：将被提取或分离的混合物放于蒸馏瓶中，加入适量的水及沸石，安好装置，接通冷凝水，开始加热使液体沸腾，蒸气经恒压滴液漏斗的支臂（可事先用保温材料缠绕，以减少热量的散失）到达球形冷凝管，冷却后的液体即储于漏斗筒里。当蒸馏烧瓶中待分离物质全部蒸完后，停止加热。彻底冷却后，即可分出有机相。

这个装置具有许多优点。首先，如果提取出来的物质密度小于水，在恒压滴液漏斗中分层后就浮于上层，而下层的水可以通过活塞放回蒸馏瓶中重复使用，直至所要提取的物质全部富集到漏斗的上层，这样可以节省水和能源。如果提取出来的物质密度大于水，则可预先在漏斗中加入适量的有机溶剂作为萃取剂，这样提取物质就被萃取到上层，下层的水同样

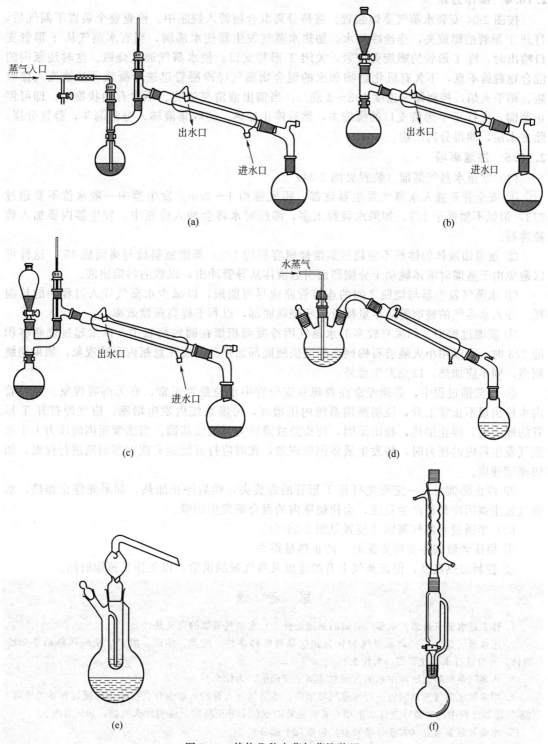

图 2-36 其他几种水蒸气蒸馏装置

可以重复循环,待水蒸气蒸馏完毕,分出上层的有机相,通过蒸馏除去萃取溶剂,即得到所需要的提取物。其次,由于该装置是立式的,具有占实验台面积小、所用仪器均为常规仪器、安装拆卸方便、操作简单的优点。

2.10.4 操作方法

按图 2-34 安装水蒸气蒸馏装置。将待分离混合物转入烧瓶中。检查整个装置不漏气后，打开 T 形管的螺旋夹，连通冷却水，加热水蒸气发生器使水沸腾，当有水蒸气从 T 形管支口喷出时，将 T 形管的螺旋夹拧紧，关闭 T 形管支口，使水蒸气通入烧瓶。这时烧瓶内的混合物翻腾不息，不久有机化合物和水的混合物蒸气经冷凝管迅速冷凝成乳浊液流入接收瓶。调节火焰，控制馏出速度为 2～3 滴/s。当馏出液清亮透明、不含有油状物时，即可停止蒸馏。先打开 T 形管支口的螺旋夹，然后停止加热。将收集液转入分液漏斗，静置分层，除去水层，即得分离产物。

2.10.5 注意事项

（1）常量水蒸气蒸馏（装置见图 2-34）

① 安全管要插入水蒸气发生器底部，距底部约 1～2cm。发生器中一般水位不要超过 2/3，最低不要低于 1/3。如果水装得太多，沸腾时水将会冲入烧瓶中。发生器内要加入数粒沸石。

② 被蒸馏液体的体积不应超过蒸馏烧瓶容积的 1/3。蒸馏瓶斜放与桌面成 45°，这样可以避免由于蒸馏时液体跳动十分剧烈而引起液体从导管冲出，以致沾污馏出液。

③ 水蒸气发生器与烧瓶之间的连接管路应尽可能短，以减少水蒸气导入过程中的热损耗。导入水蒸气的玻璃管应尽量接近圆底烧瓶底部，以利于提高蒸馏效率。

④ 蒸馏过程中，如果有较多的水蒸气因冷凝而积聚在圆底烧瓶中，以致超过烧瓶容积的 2/3 时，可以用小火隔着石棉网在圆底烧瓶底部加热。但要注意瓶内崩跳现象，如果崩跳剧烈，则不应加热，以免发生意外。

⑤ 在蒸馏过程中，必须经常注意观察安全管中水位是否正常，有无倒吸现象。如果管内水柱出现不正常上升，说明蒸馏系统内压增高，可能系统内发生堵塞。应立即打开 T 形管的螺旋夹，停止加热，找出原因，待排除故障后方可继续蒸馏。当蒸馏瓶内的压力大于水蒸气发生器内的压力时，将发生液体倒吸现象，此时应打开螺旋夹或对蒸馏瓶进行保温，加快蒸馏速度。

⑥ 停止蒸馏时，一定要先打开 T 形管的螺旋夹，然后停止加热。如果先停止加热，水蒸气发生器因冷却而产生负压，会使烧瓶内的混合液发生倒吸。

（2）半微量水蒸气蒸馏［装置见图 2-36(f)］

① 恒压滴液漏斗支臂要保温，防止热量损失。

② 控制加热温度，保证蒸气上升的速度及蒸气量的供给，以免拖长蒸馏时间。

思 考 题

1. 什么是水蒸气蒸馏？水蒸气蒸馏的原理是什么？水蒸气蒸馏的意义是什么？
2. 用水蒸气蒸馏来分离或提纯的化合物应具备哪些条件？酯类、酸酐、酰氯、醋酸和邻硝基苯酚（固体）可否进行水蒸气蒸馏，为什么？
3. 水蒸气蒸馏时馏出液中水的含量总是高于理论值，为什么？
4. 用常量水蒸气蒸馏装置进行水蒸气蒸馏时，水蒸气导入管的末端为什么要插入到接近容器的底部？水蒸气蒸馏过程中经常要检查什么事项？若安全管中水位上升很高时，说明什么问题，如何解决？
5. 水蒸气蒸馏装置（常量、半微量）包括几个部分？

2.11 萃取

萃取（Extraction）是有机实验室中常用的分离和提纯有机化合物的操作之一。应用萃取可以从固体或液体混合物中提取出所需要的物质，也可以用来洗去混合物中少量的杂质。通常称前一种为"抽提"或"萃取"，后者为"洗涤"。萃取可分为液-液萃取、液-固萃取、气-液萃取。这里着重介绍液-液萃取。

2.11.1 基本原理

萃取是利用物质在两种互不相溶（或微溶）溶剂中溶解度或分配比的不同来达到分离或提纯目的的一种操作。

萃取的原理是，设溶液由有机化合物 X 溶解于溶剂 A 而成，现如要从溶液中萃取出 X，可以选择一种对 X 溶解度极好，而与溶剂 A 不相混溶且不起化学反应的溶剂 B。把溶液倒入分液漏斗中，加入溶剂 B，充分振荡。静置后，由于 A 与 B 不相混溶，故分成两层。此时 X 在 A、B 两相间的浓度比，在一定温度下，为一常数，称为分配系数，以 K 表示，这种关系叫做分配定律。用公式表示

$$\frac{X 在溶剂 A 中的浓度}{X 在溶剂 B 中的浓度} = K（分配系数）$$

注意：分配定律在假定所选用的溶剂 B 不与 X 起化学反应时才适用。

当用一定量的溶剂萃取有机化合物时，一次萃取好还是多次萃取好呢？依照分配定律，要节省溶剂而提高提取的效率，用一定量的溶剂进行萃取时，分次萃取比一次萃取的效率高。现在用算式来说明。

设在 V mL 被萃取溶液中溶解 W_0 g 溶质 X，每次用 S mL 溶剂 B（与溶剂 A 互不相溶）重复萃取，W_1 为第一次萃取后溶质 X 在溶剂 A 中的剩余量（g），$W_0 - W_1$ 为第一次萃取后溶质 X 在溶剂 B 中的含量（g），则第一次萃取后溶质（X）在溶剂 A 中的浓度（g/mL）和在溶剂 B 中的浓度（g/mL）分别为 W_1/V 和 $(W_0 - W_1)/S$，两者之比等于 K，亦即

$$\frac{W_1/V}{(W_0-W_1)/S} = K \quad 或 \quad W_1 = W_0 \frac{KV}{KV+S}$$

设 W_2 为第二次萃取后溶质 X 在溶剂 A 中的残余量（g），则第二次萃取后

$$\frac{W_2/V}{(W_1-W_2)/S} = K \quad 或 \quad W_2 = W_1 \frac{KV}{KV+S} = W_0 \left(\frac{KV}{KV+S}\right)^2$$

显然，经过 n 次萃取后溶质 X 在溶剂 A 中的残余量 W_n 应为

$$W_n = W_0 \left(\frac{KV}{KV+S}\right)^n$$

当用一定量的溶剂萃取时，人们总是希望溶质在原溶液中的残余量越少越好。由上式可以看出，$KV/(KV+S)$ 恒小于 1，所以 n 越大，W_n 就越小，也就是说用相同量的溶剂分 n 次萃取比一次萃取好，即少量多次萃取效率高。

例 在 100mL 水中含有 4g 正丁酸的溶液，在 15℃时，用 100mL 苯来萃取正丁酸。设已知在 15℃时正丁酸在水和苯中的分配系数 $K=1/3$，若用 100mL 苯一次萃取，则萃取后正丁酸在水溶液中的剩余量为

$$W_1 = 4 \times \frac{\frac{1}{3} \times 100}{\frac{1}{3} \times 100 + 100} = 1.0 \text{g}$$

萃取效率为

$$\frac{4-1}{4} \times 100\% = 75\%$$

若用 100mL 苯分三次来萃取，即每次用 33.3mL 苯来萃取，经过三次萃取后正丁酸在

水溶液中的剩余量为

$$W_3 = 4 \times \left(\frac{\frac{1}{3} \times 100}{\frac{1}{3} \times 100 + 33.3}\right)^3 = 0.5\text{g}$$

萃取效率为

$$\frac{4-0.5}{4} \times 100\% = \frac{3.5}{4} \times 100\% = 87.5\%$$

从上面的计算可知，用同样体积的溶剂，分多次用少量溶剂来萃取，其效率高于一次用全部溶剂来萃取。但是当溶剂的总量保持不变时，萃取次数（n）增加，每次用萃取剂的量（S）就要减小。当 $n > 5$ 时，n 和 S 这两种因素的影响几乎抵消了，再增加萃取次数，W_n/W_{n+1} 的变化很小。所以一般同体积溶剂分为 3~5 次萃取即可。每次所用萃取剂约为被萃取溶液体积的 1/3。

此外，萃取效率还与溶剂的选择密切相关。选择溶剂的基本原则是：溶剂对被提取的物质溶解度较大；与原溶剂不相混溶，并且密度相差较大（便于分离），不起化学反应；沸点低，易于与溶质分开，且便于回收；毒性小；具有良好的化学稳定性，不易分解、聚合等。

一般选择萃取剂时，水溶性较小的物质用石油醚作萃取剂；水溶性较大的物质用苯或乙醚作萃取剂；水溶性极大的物质用乙酸乙酯或类似的物质作萃取剂。常用的萃取剂有乙醚、苯、四氯化碳、石油醚、氯仿、二氯甲烷、乙酸乙酯等。若要从有机物中洗除酸、碱或其他水溶性杂质，可分别用稀碱、稀酸或直接用水洗涤。

2.11.2 操作方法

（1）液-液萃取　液-液萃取是指从溶液中提取所需物质的方法。从溶液中萃取物质最常使用的仪器是分液漏斗。分液漏斗的使用是基本操作之一。常用分液漏斗有球形、筒形、锥形、梨形四种。在有机化学实验中分液漏斗主要应用于：a. 分离两种分层而不起作用的液体；b. 从溶液中萃取某种成分；c. 用水、碱或酸洗涤某种产品；d. 用来滴加某种试剂（即代替滴液漏斗）。

下面以用乙醚从醋酸水溶液中萃取醋酸的实验为例来说明萃取的操作方法，并比较一次萃取和多次萃取的萃取效率。

① 一次萃取法。用移液管准确量取 5mL 冰醋酸与水的混合液（冰醋酸与水以 1:19 的体积比混合），放入 50mL 的分液漏斗中，用 15mL 乙醚萃取。注意近旁不能有火，否则易引起火灾。加入乙醚后，将塞子塞紧，先用右手食指的末节将漏斗上端玻璃塞顶住，再用大拇指及食指和中指握住漏斗，这样漏斗转动时可用左手的食指和中指蜷握在活塞的柄上，使振摇过程中（见图 2-37）玻璃塞和活塞均夹紧。把漏斗放平、前后小心振摇，开始振摇要慢，每隔几秒钟将漏斗的上口向下倾斜，漏斗的活塞部分向上倾斜（朝向无人处），小心打开活塞，及时释放乙醚气体，平衡内外压力，重复操作 2~3 次。然后再用力振摇相当的时间，使乙醚与醋酸水溶液两不相溶的液体充分接触，提高萃取效率，振摇时间太短则影响萃取效率。

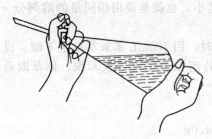

图 2-37　振荡分液漏斗

将分液漏斗置于铁圈上，当溶液分为两层后，小心地旋开活塞，放出下层水溶液于 30mL 三角烧瓶内，加入 2~3 滴酚酞作指示剂，用 0.2mol/L 标准氢氧化钠溶液滴定，记录用去氢氧化钠的体积。计算：a. 留在水中的醋酸量及质量分数；b. 留在乙醚中的醋酸量及质量分数。

② 多次萃取法。准确量取 5mL 冰醋酸和水的混合液于 50mL 的分液漏斗中，用 5mL 乙醚如上法萃取，分去乙醚溶液。将水溶液再用 5mL 乙醚萃取，分去乙醚溶液后，将第二次剩余的水溶液再用 5mL 乙醚萃取。如此前后共计 3 次。最后将用乙醚第三次萃取后的水溶液放入 30mL 的三角烧瓶中，用 0.2 mol/L 氢氧化钠溶液滴定。计算：a. 留在水中的醋酸量及质量分数；b. 留在乙醚中的醋酸量及质量分数。

根据上述两种不同步骤所得数据，比较萃取醋酸的效率。

在使用分液漏斗进行萃取操作时应注意以下几点。

第一，使用分液漏斗前必须检查。检查的事项包括：

a. 分液漏斗的玻璃塞和活塞有没有用塑料线绑住（以免滑出打碎或调错）；

b. 玻璃塞和活塞是否紧密？如有漏水现象，应及时按下述方法处理：脱下活塞，用纸或干布擦净活塞及活塞孔道的内壁，然后用玻璃棒蘸取少量凡士林，先在活塞近把手一端抹上一层凡士林，注意不要抹在活塞的孔中，再在活塞两边也抹上一圈凡士林，然后插上活塞，逆时针旋转至透明时，即可使用。

使用分液漏斗时应注意：

a. 不能把活塞上附有凡士林的分液漏斗放在烘箱内烘干，如需要放在烘箱中烘干时，塞子、旋塞均需卸下，凡士林必须擦净，否则凡士林炭化后很难洗去；

b. 不能用手拿住分液漏斗的下端；

c. 不能用手拿住分液漏斗进行分离液体，应架在铁圈上操作；

d. 上口玻璃塞打开后才能开启活塞；

e. 上层的液体不要由分液漏斗下口放出，以免污染产品。

分液漏斗用后，应用水冲洗干净，玻璃塞和活塞均应用薄纸包裹后再塞回去，以免发生粘连。

第二，所选用的分液漏斗其容积应比待处理液体体积大 1~2 倍。如果漏斗中液体太多，摇动时就会影响液体的接触，使萃取效率降低。萃取溶剂的体积一般为待处理溶液体积的 1/3。

第三，在振荡过程中，特别是溶液呈碱性时，常出现乳化现象；有时由于存在少量轻质沉淀、两液相的相对密度相差较小、两液相溶剂部分互溶等都能引起分层不明显或不分层。此时可加入强电解质（如食盐），使溶液饱和，以降低乳浊液的稳定性，达到破乳的目的。轻轻地旋转漏斗，也可使其加速分层。在一般情况下，长时间静置分液漏斗，可使乳浊液分层。若因溶液呈碱性而产生乳化现象，可加入少量稀酸或采用过滤的方法除去。

第四，在使用低沸点溶剂（如乙醚）作萃取剂时，或使用碳酸钠溶液洗涤含酸液体时，在振荡过程中要不时地放气，以免液体从上口塞处喷出。

第五，液体分层后，可根据两相的密度来判断哪一层是萃取层。密度大的在下层，密度小的在上层。如一时难以分清，最好将上下两层液体都保存起来，待弄清后再弃掉不要的液体。

当待萃取液体体积很小时（仅有 2~3mL，甚至只有几十微升），用分液漏斗显然不很理想，可采用微量萃取技术进行萃取。取一支离心试管，放入待萃取溶液和萃取剂，盖好盖子，用手摇动

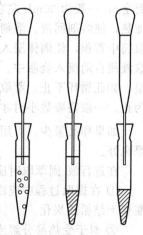

图 2-38 微量萃取法

离心试管或用滴管向液体中鼓气搅动，使液体充分接触，并注意随时开塞放气。静止分层后，用滴管将萃取相吸出，在萃余相中加入新的萃取剂继续萃取，如图 2-38 所示。以后操作如前所述。

（2）液-固萃取　以上介绍的是从溶液中提取物质的方法。下面简单介绍液-固萃取（即从固体中提取物质）的方法。常用的方法有浸取法和连续提取法。

① 浸取法　就是将溶剂加入到被萃取的固体物质中加热，使易溶于萃取剂的物质提取出来，然后再进行分离纯化。当使用有机溶剂作萃取剂时，应使用回流装置。

② 连续提取法　当待提取物的溶解度很小时可采用此法。在实验室中一般使用 Soxhlet 提取器（索氏提取器）来进行，如图 2-39 所示。图 2-40 为简易半微量提取器。Soxhlet 提取器是利用溶剂回流及虹吸原理，使固体物质每一次都能为纯的溶剂所萃取，因而效率较高。

图 2-39　Soxhlet 提取器

图 2-40　简易半微量提取器

使用 Soxhlet 提取器萃取的步骤：在进行提取之前，先将滤纸卷成圆柱状，其直径稍小于提取筒的内径，一端用线扎紧（或用市售的滤纸筒），装入研细的被提取的固体，轻轻压实，上盖以滤纸，放入提取筒中。滤纸筒的高度不要超过虹吸管顶部。从提取筒上口加入溶剂，当发生虹吸时，液体流入蒸馏瓶中，再补加过量溶剂（根据提取时间和溶剂的挥发程度而定），一般为 30mL 左右。向圆底烧瓶中加入几粒沸石，装上冷凝管，通入冷凝水。开始加热，使溶剂回流。溶剂的蒸气从烧瓶进到冷凝管中，冷凝后回流到固体物质里，液体在提取筒中蓄积，使固体浸入液体中。待提取筒中的溶剂液面超过虹吸管上端时，浸泡样品的提取液便自动流入烧瓶中。溶剂受热又会被蒸发，溶剂蒸气经冷凝又回流至提取筒的固体物质里，如此循环不止，萃取物不断地积聚在烧瓶中，直至固体中可溶物质几乎全部被提取出来为止。一般需要数小时才能完成，提取液经浓缩或减压浓缩蒸出溶剂后，即获得提取物。

如果样品量少，可用简易半微量提取器，把被提取固体放于折叠滤纸中，操作方便，效果也好。

在进行液-固萃取时应注意以下几点。

① 在提取过程中应注意调节温度，以免提取出来的溶质较多时，温度过高会使溶质在瓶壁上结垢或炭化。

② 对于受热易分解或易变色的物质不宜采用这一方法。

③ 提取溶剂的沸点较高时不宜采用此方法。

思 考 题

1. 影响萃取效率的因素有哪些？
2. 使用分液漏斗的目的何在？使用分液漏斗时要注意哪些事项？

2.12 重结晶

从有机化学反应中分离出来的固体粗产物往往含有未反应的原料、副产物及杂质，必须加以分离纯化。提纯固体有机化合物最常用的方法之一就是重结晶（Recrystallization）。

2.12.1 基本原理

重结晶的原理是利用混合物中各组分在某种溶剂中的溶解度不同，或在同一溶剂中不同温度时的溶解度不同，而使它们相互分离。

固体有机化合物在任何一种溶剂中的溶解度都是随着温度的变化而变化，一般是温度升高时溶解度增大，温度降低时溶解度减小。若将固体溶解在热的溶剂中，使之达到饱和，冷却时，由于溶解度降低，溶液变成过饱和而析出结晶，这就是重结晶。就同一种溶剂而言，由于被提纯物质和杂质在溶剂中的溶解度不同，可通过热过滤将溶解性较差的杂质滤除，或让溶解性较大的杂质在冷却结晶过程仍保留在母液中，从而达到分离提纯的目的。重结晶一般适用于产品与杂质性质差别较大，产品中杂质含量小于5%的体系。所以将反应粗产物直接重结晶是不适宜的，常用其他方法进行初步提纯，如萃取、水蒸气蒸馏、减压蒸馏等，然后再用重结晶提纯。

重结晶提纯法的一般过程是：选择溶剂→溶解固体→除去杂质→晶体析出→晶体的收集与洗涤→晶体的干燥。

2.12.2 操作方法

(1) 溶剂的选择 选择合适的溶剂是重结晶操作的关键之一。所选择的溶剂应具备下述条件。

① 不与被提纯的有机物发生化学反应。

② 被提纯的有机物应易溶于热溶剂中，而在冷溶剂中几乎不溶。

③ 杂质的溶解度应很大（使杂质留在母液中不随被提纯物的晶体析出，以便分离）或很小（在制成热的饱和溶液后，趁热过滤除去杂质）。

④ 能得到较好的结晶。

⑤ 溶剂的沸点适中。若过低时，溶解度改变不大，难分离，且操作也较难；过高时，附着于晶体表面的溶剂不易除去。

⑥ 价廉易得，毒性低，回收率高，操作安全。

在选择溶剂时，可根据"相似相溶"的原理。溶质往往易溶于结构与其相似的溶剂中。通常极性化合物易溶于极性溶剂中，非极性化合物易溶于非极性溶剂中。借助于查阅有关的手册或辞典，可查出常见有机化合物在不同溶剂中的溶解度。若根据手册或辞典查不到合适的溶剂，应通过实验来确定。其方法是：取0.1g待重结晶的固体样品置于一小试管中，用滴管逐滴加入溶剂，并不断振荡，待加入的溶剂约为1mL后，若固体全部溶解或大部分溶解，则说明此溶剂的溶解度太大，不适宜作此样品的重结晶溶剂；若固体不溶或大部分不溶，但加热至沸腾（溶剂的沸点低于100℃的，则应用水浴加热）时完全溶解，冷却，析出大量结晶，这种溶剂可认为合适；若样品不全溶于1mL沸腾的溶剂中时，则可逐次添加溶剂，每次约加0.5mL，并加热至沸腾，若加入的溶剂总量达3~4mL时，样品在沸腾的溶剂中仍不溶解，则表示这种溶剂不适用，必须寻找其他溶剂。反之，若样品能溶解在3～4mL沸腾的溶剂中，则将试管冷却，观察有没有结晶析出，如没有结晶析出，可用玻璃棒摩擦试管壁或用冰水冷却，以促使结晶析出。若仍未析出结晶，则表示这种溶剂也不适用；若有结晶析出，则以结晶析出的多少来选择溶剂。

按照上述方法逐一试验不同的溶剂，对试验结果加以比较，从中选择最佳的作为重结晶的溶剂。表2-6给出了一些常用的重结晶溶剂。

表 2-6 常用的重结晶溶剂

溶剂名称	沸点/℃	密度/g·cm^{-3}	溶剂名称	沸点/℃	密度/g·cm^{-3}
水	100.0	1.000	二氯甲烷	40.8	1.325
甲醇	64.7	0.792	三氯甲烷	61.2	1.490
95%乙醇	78.1	0.804	四氯化碳	76.5	1.594
冰醋酸	117.9	1.049	乙酸乙酯	77.1	0.901
丙酮	56.2	0.791	二氧六环	101.3	1.030
乙醚	34.5	0.714	四氢呋喃	66.0	0.887
石油醚	30～60	0.68～0.72	N,N-二甲基甲酰胺	153.0	0.950
	60～90		二甲亚砜	189.0	1.101
环己烷	80.8	0.78	乙腈	81.6	0.780
苯	80.1	0.88	硝基甲烷	120.0	1.140
甲苯	110.6	0.867			

当一种物质在一些溶剂中的溶解度很大，而在另一些溶剂中的溶解度又很小，难于找到一种适用的重结晶溶剂时，则可采用混合溶剂。混合溶剂一般由两种能以任何比例互溶的溶剂组成，其中一种溶剂对被提纯物质的溶解度较大（称为良溶剂），而另一种溶剂则对被提纯物质的溶解度较小（称为不良溶剂）。一般常用的混合溶剂有乙醇-水、乙醇-乙醚、乙醇-丙酮、乙醚-石油醚、苯-石油醚等。

(2) 固体的溶解 将待重结晶的粗产物放入锥形瓶或圆底烧瓶中（因为它们的瓶口较窄，溶剂不易挥发，又便于振荡，促进固体物质的溶解），加入比计算量略少的溶剂，加热到沸腾。若仍有固体未溶解，则在保持沸腾下逐渐添加溶剂至固体恰好溶解（要注意判断是否有不溶性杂质，以免误加过多的溶剂）。最后再多加20%左右的溶剂将溶液稀释，以避免在热过滤时由于溶剂的挥发和温度的下降导致溶解度降低而在滤纸上析出结晶造成产品损失。但如果溶剂过量太多，则较难析出结晶，此时需将过量的溶剂蒸出后，再冷却结晶。

如所用的是低沸点易燃或有毒的有机溶剂，必须在瓶口上安装回流冷凝管[见图1-5(a)]，严禁在石棉网上直火加热，应根据溶剂沸点的高低，选用热浴。添加溶剂时，应先把火熄灭，注意安全，防止着火事故的发生。若固体物质在溶剂中溶解速度较慢，需要较长时间时，也要装上回流冷凝管，以免溶剂损失。

在溶解过程中有时会出现油珠状物，这对物质的纯化很不利，因为杂质会被包裹在油珠物中，溶液冷却后，杂质随着晶体同时析出，并夹带少量溶剂，故应尽量避免这种现象的发生。可从下列两方面加以考虑：①所选溶剂的沸点应低于溶质的熔点；②低熔点物质进行重结晶，如选不出沸点较低的溶剂时，则应在比熔点低的温度下溶解固体。

用混合溶剂重结晶时，一般先将被提纯物质溶于沸腾或接近沸腾的适量的良溶剂中。溶液中若有不溶物，可趁热过滤；溶液若有颜色或有某些树脂状物质以及不溶的均匀悬浮体，则要用活性炭煮沸脱色，趁热过滤掉活性炭。将滤液加热至接近沸点，然后向此滤液中慢慢滴加不良溶剂至刚好出现浑浊并不再消失为止，再加热或小心地滴加良溶剂直至滤液恰好变澄清透明，放置冷却，使结晶自溶液中析出。如果冷却后析出油状物，说明两种溶剂比例不对,需重新调整,再进行实验,或另换一对溶剂。若已知两种溶剂的某一定比例适用于重结晶,可事先配好混合溶剂,按单一溶剂重结晶的方法进行。

(3) 杂质的除去

① 趁热过滤。溶液中如有不溶性杂质时,应趁热过滤除去,防止在过滤过程中,由于温度降低而在滤纸上析出结晶。热过滤有两种方法:常压热过滤和减压热过滤(抽滤)。

常压热过滤操作应选用短颈径粗的玻璃漏斗,使用折叠滤纸(菊花形滤纸,折叠方法见图 2-46)和热水漏斗(见图 2-41),以便过滤尽快完成。

将热水漏斗固定在铁架上,向热水漏斗套里注入水,将短颈玻璃漏斗置于热水漏斗套中,并预先加热热水漏斗的侧管。热水漏斗中的水温视所用的溶剂的沸点而定,一般应在所用溶剂的沸点左右。温度过高,导致溶剂沸腾,大量挥发,使结晶在滤纸上析出或堵住漏斗颈,使过滤困难。将折叠滤纸放入漏斗中,用少量热溶剂湿润滤纸,避免干滤纸在过滤时,因吸附溶剂而使结晶析出。漏斗下用三角烧瓶接收(用水作溶剂时方可用烧杯)滤液,漏斗颈紧贴瓶壁,将待过滤的溶液沿玻璃棒小心倒入漏斗中的折叠滤纸内,每次倒入溶液后,都应用表面皿盖上漏斗,以减少溶剂的挥发。若使用的溶剂是水,可边加热边过滤,如果是易燃有机溶剂则务必在过滤时熄灭火焰。过滤完毕,用少量热溶剂冲洗滤纸,把滤纸上析出的少量结晶溶洗下去。若滤纸上析出的结晶较多时,可小心地将结晶刮回原来瓶中,用少量溶剂溶解后再过滤。

减压热过滤时,应事先将布氏漏斗和抽滤瓶用烘箱或气流干燥器烘热,以免漏斗破裂及在漏斗中析出晶体。过滤时动作要快,防止在过滤过程中,由于温度降低而在滤纸上析出结晶。减压热过滤的优点是过滤速度快,缺点是当溶剂的沸点较低时,因减压会使热溶剂蒸发或沸腾,导致溶液浓度变大,使晶体过早析出。减压热过滤的装置如图 2-42 所示。

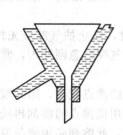

图 2-41 热水漏斗

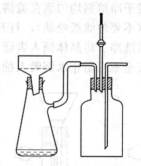

图 2-42 减压热过滤装置

② 活性炭处理。若溶液有颜色或存在某些树脂状物质、悬浮状微粒难以用一般过滤方法过滤时,则要用活性炭处理。活性炭是一种多孔物质,可以吸附色素、树脂状杂质以及均匀的分散物质。活性炭的用量视杂质的多少和颜色的深浅而定,由于它既能吸附杂质又能吸附部分产物,故用量不宜太大,一般用量为固体粗产物的 1%~5%。活性炭对水溶液脱色较好,对非极性溶液脱色效果较差。

具体操作方法:将热的饱和溶液稍冷后加入适量活性炭,在不断搅拌下煮沸 5~10min,然后趁热过滤,如一次脱色不好,可再用少量活性炭处理一次。过滤后如发现滤液中有活性炭时,应重新过滤,必要时使用双层滤纸。

注意:使用活性炭时,决不能向正在沸腾的溶液中加入活性炭,以免溶液暴沸而溅出,造成产品损失。

(4) 晶体的析出 上述饱和滤液冷却后,晶体即析出,但是冷却条件不同时,晶体析出的情况也不同。为了得到形状好、晶体较大而均匀、纯度高的晶体,在晶体析出的过程中应注意将滤液静置，使其缓慢冷却，不要急冷和剧烈搅动，以免晶体过细。否则，晶体颗粒太小，虽然晶体包含的杂质少，但却由于表面积大而吸附杂质多，加大了洗涤的困难。当冷却速度太慢时，晶体颗粒太大，晶体中会夹杂母液，难于干燥。因此，当发现大晶体正在形成时，轻轻摇动使之形成较均匀的小晶体。为使析出结晶更完全，可使用冰水冷却。

如果溶液冷却后仍不结晶，可用玻璃棒摩擦器壁引发晶体形成，或投入"晶种"，以供给晶核，使晶体迅速生成。如果没有该物质的晶体，可用玻璃棒蘸一些溶液稍干后就会析出晶体。晶种加入量不易过多，而且加入后不要搅拌，以免晶体析出太快，使产品的纯度降低。

如果被纯化的物质不析出晶体而析出油状物，其原因之一是热的饱和溶液的温度比被提纯物质的熔点高或接近。虽然油状物长期放置或足够冷却可以固化，但此固体中含杂质较多。可重新加热溶液至澄清后，让其自然冷却，当有油状物出现时，立即剧烈搅拌，使油状物在均匀分散的条件下固化，也可搅拌至油状物完全消失，溶液冷却后析出晶体。

如果结晶不成功，通常必须用其他方法（如色谱法、离子交换树脂法）提纯。

(5) 晶体的收集和洗涤　为了把晶体从母液中分离出来，使留在溶剂中的可溶性杂质与晶体彻底分离，通常采用抽气过滤（或称减压过滤）的方法。使用瓷质的布氏漏斗，布氏漏斗以橡皮塞与抽滤瓶相连，漏斗下端斜口正对抽滤瓶支管，抽滤瓶的支管套上较耐压的橡皮管，与安全瓶连接，再与水泵连接，如图 2-42 所示。在布氏漏斗中铺一张比漏斗底部略小的圆形滤纸，过滤前先用溶剂湿润滤纸，打开水泵，关闭安全瓶活塞，抽气，使滤纸紧紧贴在漏斗上，将待过滤的混合物分批倒入布氏漏斗中，使固体物质均匀分布在整个滤纸面上，用少量滤液将黏附在容器壁上的结晶洗出，合并入布氏漏斗中，继续抽气，并用玻璃钉挤压晶体，尽量除去母液。当布氏漏斗下端不再滴出溶剂时，慢慢旋开安全瓶上的活塞，使其与大气相通，再关闭水泵。滤得的固体，习惯上称滤饼。为了除去结晶表面的母液，应洗涤滤饼。用少量干净溶剂均匀洒在滤饼上，并用玻璃棒或刮刀轻轻翻动晶体，使全部结晶刚好被溶剂浸润（不要使滤纸松动），打开水泵，关闭安全瓶活塞，抽去溶剂，重复操作两次，就可以把滤饼洗净。将晶体倒入表面皿或培养皿中进行干燥。

使用安全瓶可防止水倒吸入抽滤瓶内，在不用安全瓶时，停止抽气前应先拔下橡皮管。禁止用力较猛，否则因空气突然急剧冲入，常会把布氏漏斗中的晶体冲出。

若使用的重结晶溶剂的沸点较高，因其较难挥发，在用原溶剂洗涤一次后，可用低沸点的溶剂再洗一次，使最后的晶体易于干燥（注意：此溶剂必须能与原来的溶剂相混溶，且对晶体是微溶或不溶的）。

过滤少量的结晶（半微量操作）可用玻璃钉漏斗或带滤孔板的小漏斗及抽滤管（见图 2-43）。

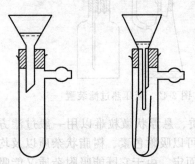

图 2-43　少量物质重结晶过滤装置

(6) 晶体的干燥　用重结晶法纯化后的晶体，其表面还吸附有少量溶剂，为了保证产品的纯度，应根据所用溶剂及结晶的性质选择恰当的方法进行干燥，见 2.2 干燥与干燥剂。

2.12.3　重结晶提纯的操作练习

(1) 常量重结晶　对于待纯化的固体样品量在 1g 以上时，通常采用常量重结晶法进行纯化。

将 2g 粗制的乙酰苯胺及适量的水（约 60mL）加入 100mL 的烧杯中，加热至沸腾，直到乙酰苯胺溶解。若不溶解，可适量添加少量热水，搅拌并加热至接近沸腾使乙酰苯胺溶解。取下烧杯，稍冷后再加入适量的活性炭于溶液中，盖上表面皿，煮沸 5～10min。趁热用放有菊花形滤纸的热水漏斗进行过滤，并用 5mL 沸水洗涤烧杯，用另一干净的烧杯收集滤液。在过滤过程中，热水漏斗和溶液均应用小火加热保温以免冷却。滤液放置至室温，然后将烧杯置于冰水浴中彻底冷却，待晶体全部析出后，抽滤出晶体，并用少量溶剂（冷水）洗涤晶体表面。抽干后，取出产品放在表面皿上晾干或烘干，称量，计算回收率。测定熔点，并与粗乙酰苯胺的熔点比较。乙酰苯胺的 m.p. 为 114℃。

乙酰苯胺在水中的溶解度为：5.5g/100mL(100℃)；0.56g/100mL(25℃)。

（2）半微量重结晶　当待纯化样品较少时（少于 500mg），用常量重结晶方法进行操作样品损失较大，并且用布氏漏斗进行热过滤操作也比较困难，而用图 2-44 所示的 Y 形砂芯漏斗和图 2-45 装置进行操作，则十分方便，且产品损失较小。

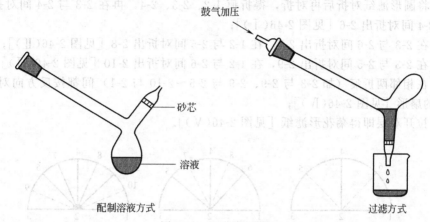

图 2-44　Y 形砂芯漏斗

利用 Y 形砂芯漏斗操作的方法：首先将样品由玻璃管放入球中，加入少许溶剂把落在玻璃管道内的样品冲洗下去，置玻璃球于油浴或热水浴中加热至微沸，再用滴管向球内部加溶剂，直至样品全部溶解。停止热浴，并擦净玻璃球上的油迹或水迹（否则将会污染滤液）。然后，迅速将玻璃球倒置，用橡皮气球通过玻璃管向 Y 形砂芯漏斗内加压，使漏斗内热的饱和溶液经过砂芯漏斗滤入洁净的容器中，静置，结晶。

利用图 2-45 装置进行操作的方法：在 30mL 圆底烧瓶中加入适量的溶剂和沸石，100mL 恒压滴液漏斗（其支臂可事先用保温材料缠绕，以减少热的迅速散失）中装入事先铺有一层活性炭（少量）及一层待纯化的固体样品（0.3~0.5g）的菊花滤纸，然后在恒压滴液漏斗的上方安装一支球形冷凝管并通入冷凝水。加热，回流，溶剂蒸气经恒压滴液漏斗的支臂到达球形冷凝管，冷却后滴下与固体样品接触将其溶解，进入圆底烧瓶中，这样经数次的溶解、回流，直到样品完全溶解为止。待烧瓶冷却后，晶体析出，抽滤出晶体，放置在表面皿上晾干。

Y 形砂芯漏斗需要特殊吹制，并且对于热过滤过程中易析出晶体的化合物，砂芯漏斗易堵塞，给过滤造成困难。图 2-45 装置可以避免这个问题，该装置所用仪器均属常规仪器，无需特殊加工，装置是立式的，占实验台面积小，具有安装拆卸方便，脱色、重结晶、热过滤、浓缩可同时完成等优点。

图 2-45　半微量重结晶装置

2.12.4　注意事项

① 若所用溶剂为有机溶剂，溶解样品时在补加溶剂过程中以及热过滤时，必须先熄灭火焰。

② 用活性炭脱色时，不要把活性炭加入正在沸腾的溶液中。
③ 用布氏漏斗抽滤时，滤纸不应大于布氏漏斗的底面。
④ 热过滤时，操作要迅速，使溶液尽快通过漏斗，以防止由于温度下降使晶体在漏斗上析出。
⑤ 停止抽滤时，应先将抽滤瓶与抽滤泵间连接的橡皮管拆开，或者将安全瓶上的活塞打开与大气相通，再关闭泵，防止水倒流入抽滤瓶内。
⑥ 折叠式滤纸又称菊花形滤纸，因其过滤接触面积大，故过滤速度快，减少了在过滤时析出结晶的机会。其折叠顺序是：

（Ⅰ）将圆形滤纸对折后再对折，得折痕 1-2、2-3、2-4，再在 2-3 与 2-4 间对折出 2-5，在 1-2 与 2-4 间对折出 2-6 [见图 2-46（Ⅰ）]；

（Ⅱ）在 2-3 与 2-6 间对折出 2-7，在 1-2 与 2-5 间对折出 2-8 [见图 2-46（Ⅱ）]；

（Ⅲ）在 2-3 与 2-5 间对折出 2-9，在 1-2 与 2-6 间对折出 2-10 [见图 2-46（Ⅲ）]；

（Ⅳ）在相邻两折痕（如 2-3 与 2-9、2-9 与 2-5～2-10 与 2-1）间都按反方向对折一次，乃呈双层的扇形 [见图 2-46（Ⅳ）]；

（Ⅴ）拉开双层即得菊花形滤纸 [见图 2-46（Ⅴ）]。

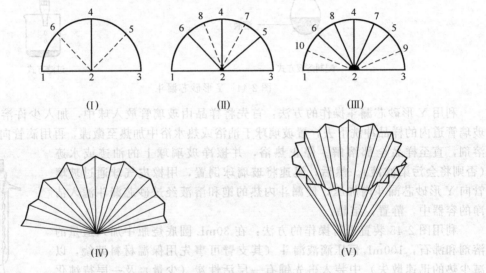

图 2-46　折叠式滤纸的折叠顺序

需要注意的是，每次折叠时，在折纹的集中点切勿重压，以免过滤时破裂。使用时，要将折好的滤纸翻转整理好后放入漏斗，使洁净面接触漏斗壁，避免在折叠过程中被手指弄脏的一面接触滤过溶液。

思 考 题

1. 简述重结晶过程及各步骤的目的。
2. 加活性炭脱色应注意哪些问题？
3. 如何选择重结晶溶剂？
4. 使用有毒或易燃的溶剂进行重结晶时应注意哪些问题？
5. 样品量分别在多少时用常量法或半微量法进行重结晶？
6. 用水重结晶纯化乙酰苯胺时（常量法），在溶解过程中有无油状物出现？油状物是什么？如有油珠出现应如何处理？

2.13 升华

某些物质在固态时具有相当高的蒸气压，当加热时，不经过液态而直接变为蒸气，蒸气受到冷却又直接冷凝成固体，这个过程叫做升华（Sublimation）。升华是提纯固体有机化合物方法之一。利用升华可以除去不挥发的杂质或分离挥发度不同的固体物质。

升华法的优点是得到的产品一般具有较高的纯度，并且操作比重结晶简便。缺点是操作时间较长，产品损失较大，不适合大量产品的提纯。此法只能用于在不太高的温度下有足够大的蒸气压力（在熔点前高于 2mmHg）的固态物质，同时要求固体化合物中杂质的蒸气压较低，因此有一定的局限性。

2.13.1 基本原理

从广义上来说，无论物质的蒸气是由固态直接汽化还是由液态蒸发而得到，只要所产生的蒸气不经过液态而直接转变成固态，这一过程都称为升华。一般说来，具有对称结构的非极性化合物，具有较高的熔点，并且在熔点温度下具有较高的蒸气压，这是因为这类物质的电子云密度分布比较均匀，偶极矩较小，晶体内部静电引力小，所以这类物质易用升华提纯。

图 2-47 是物质的三相平衡图，从此图可以了解如何来控制升华的条件。图中的三条曲线将图分为三个区域，每个区域代表物质的一相。由曲线上的点可读出两相平衡时的蒸气压。图中曲线 GS 表示固相与气相平衡时固相的蒸气压曲线；SY 表示液相与气相平衡时液相的蒸气压曲线；SV 则是固相与液相的平衡曲线。三条曲线相交于 S。S 是物质的三相平衡点，在此状态下物质的气、液、固三相共存。由于不同物质具有不同的液态与固态处于平衡时的温度与压力，因此不同的化合物三相点是不相同的。在三相点温度以下，物质只有固、气两相。升高温度，固态直接转变成蒸气；降低温度，蒸气直接转变成固态，

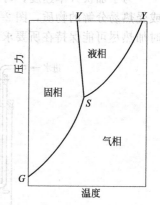

图 2-47 物质的三相平衡图

这就是升华。因此，升华都在三相点温度以下进行，即在固体的熔点以下进行。固体的熔点可以近似地看作是物质的三相点。

与液体化合物的沸点类似，升华点的定义是当固体化合物的蒸气压与外界所施加给化合物表面的压力相等时，该固体化合物开始升华。此时的温度即为该固体化合物的升华点。有些物质在三相点时的平衡蒸气压比较低，在常压下不易升华，这时可利用减压进行升华操作。

2.13.2 实验操作

图 2-48 是常用的常压升华装置。图中 (a) 是实验室中常用的常压升华装置。将粉碎了的待升华的固体样品烘干后置于瓷蒸发皿中，铺匀，上面覆盖一张穿有许多小孔（孔刺向上）的滤纸，以避免升华上来的物质再落到蒸发皿内。用一个直径小于蒸发皿的玻璃漏斗倒置在滤纸上，漏斗颈用棉花塞住，防止蒸气逸出，造成产品损失。将蒸发皿放在砂浴（或其他热浴）上加热，小心调节火焰，控制浴温（低于被升华物质的熔点），让其慢慢升华。蒸气通过滤纸小孔，冷却后凝结在滤纸上或漏斗壁上。如晶体不能及时析出，可在漏斗外壁用湿布冷却。升华完毕，移去热源，稍冷后小心拿下漏斗，轻轻揭开滤纸，用不锈钢刮刀将凝结在滤纸上及漏斗壁上的晶体刮落下来，并收集到干净的器皿内。

当待升华的样品量较大时，可用装置 (b) 分批进行。在空气或惰性气体中进行升华时，可用装置 (c)。

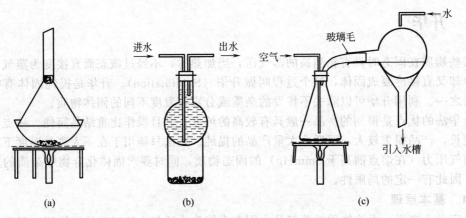

图 2-48 常压升华装置

为了加快升华速度，可在减压下进行升华。减压升华法特别适用于常压下其蒸气压不大或受热易分解的物质，图 2-49 装置用于少量物质的减压升华。通常用油浴加热，以使升华时加热尽可能保持在所要求的温度，并视具体情况采用油泵或水泵抽气。

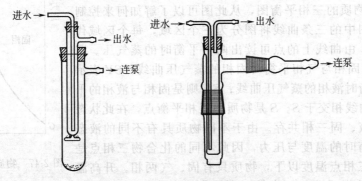

图 2-49 减压升华少量物质的装置

2.14 旋光度的测定

对映异构体的物理性质（如熔点、沸点、相对密度、折射率以及光谱等）和化学性质基本相同，只是对平面偏振光的旋光性能不同。使偏振光振动平面向右旋转的物质称为右旋体，使偏振光振动平面向左旋转的物质称为左旋体。当偏振光通过具有光学活性的物质时，其振动方向会发生偏转，所旋转的角度称为旋光度（Optical Rotation）。

旋光性物质的旋光度和旋光方向可用它的比旋光度来表示。比旋光度是旋光物质特有的物理常数之一，手册、文献上多有记载。测定比旋光度可以鉴定旋光物质的纯度和含量。物质的旋光度除与物质的结构有关外，还与测定时所用溶液的浓度、溶剂、温度、旋光管长度和所用光源的波长等有关，因此常用比旋光度 $[\alpha]_\lambda^t$ 表示各物质的旋光度。

纯液体的比旋光度 $[\alpha]_\lambda^t = \dfrac{\alpha}{Ld}$

溶液的比旋光度 $[\alpha]_\lambda^t = \dfrac{\alpha}{Lc}$

溶液的摩尔旋光度 $[\alpha_M]_\lambda^t = 0.01 M_r \times [\alpha]_\lambda^t$

式中 $[\alpha]_\lambda^t$——旋光性物质在 t℃、光源波长为 λ 时的比旋光度。一般采用钠光（λ 为

589.3nm）；

t——测定时的温度，℃；

d——密度，g/cm³；

λ——光源的光波长，nm；

α——标尺盘转动角度的读数（即旋光度），（°）；

L——旋光管的长度，dm；

c——质量浓度（100mL 溶液中所含样品的质量），g/mL；

M_r——相对分子质量。

在进行不对称合成和拆分具有光学活性的化合物时，得到的常常不是百分之百纯的对映体，而是存在少量镜像异构体的混合物。这时必须用光学纯度或对映体过量值（Enantiomer Excess，缩写为 e.e.）来表示旋光异构体的混合物中一种对映体过量所占的百分率。

光学纯度（P）的定义式为

$$P = \frac{[\alpha]_{D\text{样品}}^{t}}{[\alpha]_{D\text{标准}}^{t}} \times 100\%$$

对映异构体过量值 e.e. 则用下式表示

$$\text{e.e.} = \frac{S-R}{S+R} \times 100\%$$

式中 S——旋光异构体混合物中的主要异构体含量；

R——其对映异构体含量。

根据所得的光学纯度，可以计算试样中两种对映体的相对含量。拆分完全的对映体的光学纯度是100%，若设旋光异构体中（－）对映体光学纯度为 $x\%$，则

$$(-)\text{对映体含量} = \left(x + \frac{100-x}{2}\right) \times 100\%$$

$$(+)\text{对映体含量} = \frac{100-x}{2} \times 100\%$$

例如，已知样品(S)-(－)-2-甲基丁醇的相对密度 $d_4^{23}=0.8$，在20cm 长的盛液管中，其旋光度测定值为$-8.1°$，且其标准 $[\alpha]_D^{23}=-5.8°$（纯）则有

$$\text{比旋光度}[\alpha]_{D\text{样品}}^{23} = \frac{\alpha^{23}}{Ld} = \frac{-8.1°}{2 \times 0.8} = -5.1°$$

$$\text{光学纯度} P = \frac{[\alpha]_{D\text{样品}}^{23}}{[\alpha]_{D\text{标准}}^{23}} \times 100\% = \frac{-5.1}{-5.8} \times 100\% = 88\%$$

$$(-)\text{对映体含量} = \left(88 + \frac{100-88}{2}\right) \times 100\% = 94\%$$

$$(+)\text{对映体含量} = \frac{100-88}{2} \times 100\% = 6\%$$

$$\text{e.e.} = \frac{S-R}{S+R} \times 100 = \frac{94\% - 6\%}{94\% + 6\%} \times 100 = 88\%$$

2.14.1 旋光仪的结构

测定旋光度的仪器叫旋光仪。市售的旋光仪有两种类型：一种是直接目测的；另一种是自动显示数值的。

旋光仪主要由一个钠光源、两个尼科尔棱镜和一个盛有测试样品的盛液管组成。直接目测的旋光仪的基本结构如图 2-50 所示。

光线从光源经过起偏镜（一个固定不动的棱镜），变为在单一方向上振动的平面偏振光，再经过盛有旋光性物质的旋光管时，因物质的旋光性致使偏振光不能通过第二个棱镜，必须

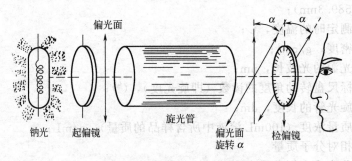

图 2-50　旋光仪示意图

转动检偏镜（一个可转动的棱镜），才能通过。因此，要调节检偏镜进行配光，使最大量的光线通过。由标尺盘上转动的角度，可以指示出检偏镜的转动角度，即为该物质在此浓度时的旋光度。

旋光仪的类型很多，图 2-51 是 WZZ-1S 型数字式旋光仪面板，能直接显示读数，操作方便。

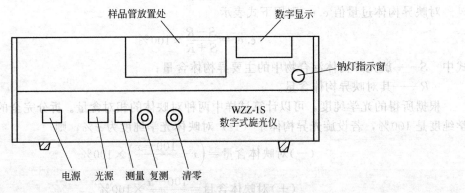

图 2-51　WZZ-1S 型数字式旋光仪面板

2.14.2　旋光度的测定方法

以 WZZ-1S 型数字式旋光仪为例，介绍旋光度测定的操作方法。

① 仪器的安放。旋光仪应在正常照明、室温和湿度条件下使用。防止在高温、高湿条件下使用，避免经常接触腐蚀性气体，否则将影响使用寿命。

② 接通电源和光源。按下"电源"按键，电源指示灯亮。等待 5min 使钠灯发光稳定后再按下光源按键，"光源"指示灯亮。

③ 按下"测量"开关，这时数码管出现数字显示。

④ 清零。在旋光管中放入蒸馏水或配制待测样品的溶剂，放入试样槽中，盖上箱盖，待示数稳定后，按下"清零"按键，使数码管示数为零。一般情况下，本仪器如不放旋光管时读数为零，放入无旋光度溶剂后也应为零。但需防止在测试光束通路上有小气泡，或试管的护片上沾有油污、不洁物，同时也不宜将旋光管护片旋得过紧，这都会影响空白读数。如果读数不是零，必须仔细检查上述因素或用装有溶剂的空白旋光管放入试样槽后再清零。旋光管安放时应注意标记位置和方向。

⑤ 测试。取出旋光管，在旋光管中放入待测样品，按相同的位置和方向放入试样槽中，盖好箱盖。待位于符号管上方的红点亮后再读取读数。

⑥ 复测。按下"复测"按键，读数，取几次测量的平均值为测定结果。注意：须待表示示数稳定后的红点出现后再进行此操作，否则有可能引起下一次测数的误差。

⑦ 仪器使用完毕后，应依次关闭"测量"、"光源"、"电源"开关。

⑧ 根据公式计算比旋光度、对映体过量值等。

2.14.3 注意事项

① 旋光度与光束通路中光学活性物质的分子数成正比。对于比旋光度值较小或溶液浓度小的样品，在配制待测样品溶液时，宜将浓度配高一些，并选用长一点的旋光管，以便观察。

② 温度变化对旋光度具有一定的影响。若在钠光（$\lambda=589.3nm$）下测试，温度每升高1℃，多数光活性物质的旋光度会降低0.3%左右。

③ 测试时，旋光管所放的位置应固定不变，以消除因距离变化所产生的测试误差。

④ 旋光管用后要及时将溶液倒出，洗涤干净，晾干放好。

⑤ 记录所用旋光管的长度、测定温度及所用溶剂（如用水作溶剂则可省略）。

2.15 折射率的测定

折射率（Refractive Index，又称折光率）是有机化合物的重要物理常数之一，固体、液体和气体都有折射率，尤其是液体有机化合物，文献记载更为普遍。通过测定折射率可以判断有机化合物的纯度，也可以用来鉴定未知物。

2.15.1 基本原理

在不同介质中，光的传播速度是不同的。光从一种介质射入另一种介质时，当它的传播方向与两种介质的界面不垂直时，则其在界面处的传播方向会发生改变。这种现象称为光的折射。

根据折射定律，波长一定的单色光在确定的外界条件下（如温度、压力等），从一个介质A射入另一个介质B时，其入射角α与折射角β的正弦之比和两种介质的折射率成反比。

$$\sin\alpha/\sin\beta = n_B/n_A$$

若设定介质A为光疏介质，介质B为光密介质，则$n_A < n_B$，折射角β必小于入射角α，如图2-52所示。

当入射角$\alpha=90°$，即$\sin\alpha=1$时，折射角最大，称为临界角，以β_0表示。折射率的测定都是在空气中进行的，可近似地视为在真空状态中，即$n_A=1$，则

$$n=1/\sin\beta_0$$

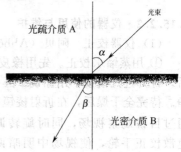

图2-52 光的折射现象

因此，通过测临界角β_0，即可得到介质的折射率（n）。在有机化学实验室里，一般用阿贝（Abbé）折射仪来测定折射率，其工作原理就是基于光的折射现象。

n与物质结构、入射光线的波长、温度、压力等因素有关。通常大气压的变化影响不明显，只是在精密测定时才考虑。使用单色光要比用白光时测得的值更为精确，因此常用钠光（D）（波长589.3nm）作光源。温度可用仪器维持恒定，如可用恒温水浴槽与折光仪间循环恒温水来维持恒定温度。所以，折射率（n）的表示需要注明所用光线波长和测定的温度，常用n_D^{20}来表示，即以钠光为光源，20℃时所测定的n值。

通常温度升高（或降低）1℃时，液态有机化合物的折射率就减少（或增加）$3.5\times10^{-4}\sim5.5\times10^{-4}$，在实际工作中常采用$4\times10^{-4}$为温度变化常数，把某一温度下所测得的折射率换算成另一温度下的折射率。其换算公式为

$$n_D^T = n_D^t + 4\times10^{-4}(t-T)$$

式中 T——规定温度，℃；

t——实验时的温度，℃。

这种粗略计算虽然有一定的误差，但却很有参考价值。

2.15.2 阿贝（Abbé）折射仪

阿贝（Abbé）折射仪是一种操作简便、实验室用来测定折射率的仪器，它是根据临界角折射现象设计的。下面介绍 WYA-$\frac{1}{2}$S 数字阿贝（Abbé）折射仪的结构及使用方法。

2.15.2.1 仪器的工作原理及结构

阿贝（Abbé）折射仪测定物质折射率的原理是基于测定临界角，如图 2-53 所示，由目视望远镜部件和色散校正部件组成的观察部件来瞄准明暗两部分的分界线，也就是瞄准临界角的位置，并由角度-数字转换部件将角度量转换成数字，输入微机系统进行数据处理，而后数字显示出被测样品的折射率。图 2-54 为阿贝（Abbé）折射仪的结构图。

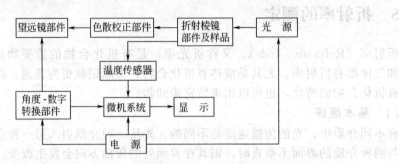

图 2-53 阿贝（Abbé）折射仪工作原理图

2.15.2.2 仪器的使用与维护

(1) 仪器校正 阿贝（Abbé）折射仪必须先用标准试样校正后才能做测定用。

① 用蒸馏水校正。先用橡皮管将折射仪与恒温槽相连接。恒温（一般为 20℃ 或 25℃）后，打开进光棱镜，用擦镜纸蘸少许乙醇或丙酮，顺同一方向把上下两棱镜镜面轻轻擦拭干净。待完全干燥后，在折射棱镜的抛光面上滴 1~2 滴高纯度蒸馏水，盖上上面的进光棱镜，通过目镜观察视场，同时旋转调节手轮和色散校正手轮，使视场中明暗两部分具有良好的反差和明暗分界线具有最小的色散，视场内明暗分界线准确对准交叉线的交点（见图 2-55）。如有偏差则可用钟表螺丝刀通过色散校正手轮中的小孔，小心旋转里面的螺钉，使分划板上交叉线上下移动，使分界线像位移至交叉线的交点，然后再进行测量，直到测数符合要求为止。蒸馏水的折射率为 $n_D^{20} = 1.33299$，$n_D^{25} = 1.33250$。校正完毕后，在以后的测定过程中不允许随意再动此部位。

② 用标准折光玻璃块校正。在折射棱镜的工作表面上滴 1~2 滴溴代萘（$n = 1.66$），再将玻璃块黏附于此镜面上，然后按上述方法进行。测数要符合标准玻璃块上所标定的数据。

(2) 操作步骤及使用方法

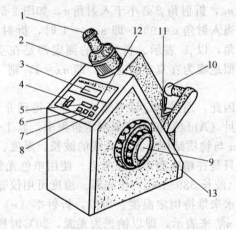

图 2-54 阿贝（Abbé）折射仪的结构图
1—目镜；2—色散校正手轮；3—显示窗；4—"POWER" 电源开关；5—"READ" 读数显示键；6—"BX-TC" 经温度修正锤度显示键；7—"n_D" 折射率显示键；8—"BX" 未经温度修正锤度显示键；9—调节手轮；10—聚光照明部件；11—折射棱镜部件；12—"TEMP" 温度显示键；13—RS232 插口

① 按下"POWER"电源开关（4），聚光照明部件（10）中的照明灯亮，同时显示窗（3）显示 00000。有时显示窗先显示"-"，数秒后显示 00000。

② 打开折射棱镜部件（11），移去擦镜纸。这张擦镜纸是仪器不使用时放在两棱镜之间的，防止在关上棱镜时可能留在棱镜上的细小硬粒弄坏棱镜工作表面。擦镜纸只需用单层。

③ 检查上、下棱镜表面，并用酒精小心清洁其表面。测定每一个样品后也要仔细清洁两块棱镜表面，因为留在棱镜上少量的原来样品将影响下一个样品的测量准确度。

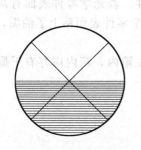

图 2-55　阿贝（Abbé）折射仪在临界角时目镜视野图

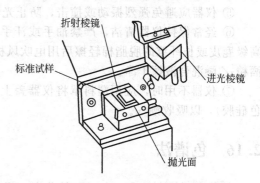

图 2-56　阿贝（Abbé）折射仪折射棱镜部件

④ 将被测样品放在下面的折射棱镜的工作表面上。如样品为液体，可用干净滴管吸 1~2 滴液体样品放在棱镜工作表面上，然后将上面的进光棱镜盖上。如样品为固体，则固体样品必须有一个经过抛光加工的平整表面。测量前需将抛光表面擦净，并在下面的折射棱镜工作表面上滴 1~2 滴折射率比固体样品折射率高的透明的液体（如溴代萘），然后将固体样品抛光面放在折射棱镜工作表面上，使其接触良好。测固体样品时不需将上面的进光棱镜盖上，如图 2-56 所示。

⑤ 旋转聚光照明部件的转臂和聚光镜筒使上面的进光棱镜的进光表面（测液体样品）或固体样品前面的进光表面（测固体样品）得到均匀照明。

⑥ 通过目镜（1）观察视场，同时旋转调节手轮（9），使暗分界线落在交叉线视场中。如从目镜中看到的视场是暗的，可将调节手轮逆时针旋转。看到视场是明亮的，则将调节手轮顺时针旋转。明亮区域在视场的顶部。在明亮视场情况下可旋转目镜，调节视度看清晰交叉线。

⑦ 旋转目镜方缺口里的色散校正手轮（2），同时调节聚光镜位置，使视场中明暗两部分具有良好的反差和明暗分界线具有最小的色散。

⑧ 旋转调节手轮，使明暗分界线准确对准交叉线的交点，如图 2-55 所示。

⑨ 按"READ"读数显示键（5），显示窗中 00000 消失，显示"-"，数秒后"-"消失，显示被测样品的折射率。

⑩ 检测样品温度，可按"TEMP"温度显示键（12），显示窗将显示样品温度。

⑪ 样品测量结束后，必须用酒精或水（样品为糖溶液）小心洗净两镜面，晾干后再关闭保存。

⑫ 仪器折射棱镜部件中有通恒温水结构，如需要测定样品在某一特定温度下的折射率，仪器可外接恒温器，将温度调节到所需温度再进行测量。

(3) 仪器的维护与保养　为确保仪器的精度，防止损坏，应注意维护和保养，并做到以下几点。

① 仪器应放在干燥、空气流通和温度适宜的地方，以免仪器的光学零件受潮发霉。

② 仪器使用前后及更换样品时，必须用丙酮或乙醇清洗干净折射棱镜系统的工作表面并干燥。以防留有其他物质，影响成像清晰度和测量精度。

③ 要保护棱镜，不能在镜面上造成刻痕。在滴加液体样品时，滴管的末端勿触及棱镜。不可测定强酸、强碱等具有腐蚀性的液体。

④ 仪器聚光镜是塑料制成的，为防止带有腐蚀性的样品对其表面产生破坏，必要时用透明塑料罩将聚光镜罩住。

⑤ 仪器应避免强烈振动或撞击，防止光学零件震碎、松动而影响精度。

⑥ 经常保持仪器清洁，严禁油手或汗手触及光学零件。若光学零件表面有灰尘，可用高级鹿皮或长纤维的脱脂棉轻擦后用电吹风机吹去。如光学零件表面粘上了油垢，应及时用酒精-乙醚混合液擦干净。

⑦ 仪器不用时，应用塑料罩将仪器盖上或将仪器放在箱内，箱内应存有干燥剂（如变色硅胶），以吸收潮气。

2.16 色谱法

色谱法（Chromatography）是分离、提纯和鉴定有机化合物的重要方法之一。前面介绍了蒸馏、萃取、重结晶、升华等纯化有机反应粗产物的经典方法，在实验操作中应用这些经典的纯化方法常常会遇到两个问题：第一，要求待分离的混合物具有一定的数量；第二，当混合物中含有物化性质十分相近的两个或两个以上组分时，很难达到预期的分离纯化目的。此时，用色谱法可以达到满意的结果。色谱法的分离效果远比蒸馏、分馏、重结晶等一般方法要好，特别适用于半微量和微量物质的分离提纯。近年来，随着科学技术的飞速发展，色谱分离技术已在化学化工、生物学、医学等领域得到了广泛的应用，并已发展成为分离、纯化和鉴定有机化合物的重要实验技术。

色谱法有许多种类，但基本原理是一致的，即利用待分离混合物中的各个组分在某一物质中（此物质称作固定相）的亲和性差异，如吸附性差异、溶解性（或称分配作用）差异等，让混合物溶液（此相称作流动相）流经固定相，使混合物在固定相和流动相之间进行反复吸附和分配等作用，从而使混合物中的各个组分得以分离。

根据分离过程的原理，色谱法可分为吸附色谱、分配色谱、离子交换色谱等。根据不同的操作条件，色谱法又可分为柱色谱、薄层色谱、纸色谱、气相色谱和高压液相色谱（见表2-7）。现将柱色谱、薄层色谱和纸色谱介绍如下。

表 2-7 常用色谱法的分类

按操作条件分类命名	流动相	固定相	分离原理	应 用 范 围
气相色谱	气体	吸附剂	吸附	快速分离分析微量气体、液体和固体，跟踪反应。不能用于不易挥发固体或对热不稳定化合物的分析
		固定液	分配	
柱色谱	液体	吸附剂	吸附	分离和纯化含有各类官能团的有机化合物
		固定液	分配	
		离子交换树脂	离子交换	适用于离子型物质的分离，例如生物碱、氨基酸、酸、碱和盐类
薄层色谱	液体	吸附剂	吸附	分离和纯化不易挥发的固体和液体，跟踪反应
		固定液	分配	
		离子交换树脂	离子交换	适用于离子型物质的分离，例如生物碱、氨基酸、酸、碱和盐类

续表

按操作条件分类命名	流动相	固定相	分离原理	应用范围
纸色谱	液体	水或固定液	分配	用于氨基酸、有机染料等的分析
高压液相色谱	液体	吸附剂	吸附	适用范围与柱色谱一样广泛,且具有分离速度快、分离效能高、灵敏度高的特性
		固定液	分配	
		凝胶	凝胶渗透	

2.16.1 柱色谱

柱色谱（Colum Chromatography）常用的有吸附柱色谱和分配柱色谱两类，前者常用氧化铝和硅胶作固定相，后者则以附着在惰性固体（如硅藻土、纤维素等）上的活性液体作为固定相（也称固定液）。实验室中最常用的是吸附色谱，因此这里重点介绍吸附色谱。

2.16.1.1 基本原理

柱色谱是分离、提纯复杂有机化合物的重要方法，也可用于分离量较大的有机物。柱色谱是通过色谱柱来实现分离的，图2-57是一般柱色谱的装置。

色谱柱内装有表面积很大、而又经过活化的多孔或粉末状固体吸附剂（固定相），如氧化铝、硅胶等。从柱顶加入样品溶液，当溶液流经吸附柱时，各组分同时被吸附在柱的上端，然后从柱的顶部加入有机溶剂（洗脱剂），当洗脱剂流下时，由于固定相对各组分吸附能力不同，各组分向下移动的速度也不同，形成了不同的层次，即溶质在柱中自上而下按对吸附剂亲和能力的大小分别形成若干色带，如图2-58所示。继续用洗脱剂洗脱时，吸附能力最弱的组分，首先随着溶剂流出，极性强的后流出，分别收集洗脱剂。若各组分是有色物质，则可按色带分开；但若为无色物质，可用紫外光照射后是否出现荧光来检查，或在洗脱时，分段收集一定体积的洗脱液，然后通过薄层色谱逐个鉴定，再将相同组分的收集液合并在一起，蒸除溶剂，即得到单一的纯净物质。

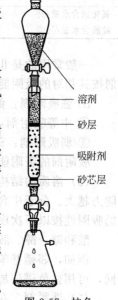

图2-57 柱色谱装置

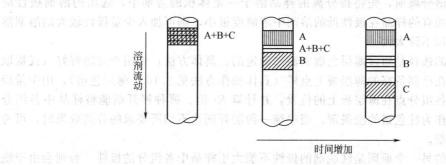

图2-58 色层的展开

2.16.1.2 柱色谱条件的选择

色谱法能否获得满意的分离效果，关键在于色谱条件的选择，下面介绍柱色谱条件的选择。

(1) 吸附剂 常用的吸附剂有氧化铝、硅胶、氧化镁、碳酸钙、活性炭等。选择的吸附剂绝不能与被分离的物质和展开剂发生化学作用，要求吸附剂颗粒大小均匀。颗粒太小，表面积大，吸附能力高，但溶剂流速太慢；颗粒太粗，流速快，分离效果差，因此颗粒大小要适当。通常使用的吸附剂的颗粒大小以 100～150 目为宜。实验室一般使用氧化铝或硅胶，在这两种吸附剂中氧化铝的极性更大一些，它是一种高活性和强吸附的极性物质。市售色谱用的氧化铝可分为酸性、中性和碱性三种。酸性氧化铝是用 1％盐酸浸泡后，用蒸馏水洗至氧化铝的悬浮液 pH 为 4～4.5，适用于分离酸性物质，如有机酸类的分离；中性氧化铝 pH 约为 7.5，适用于分离中性物质，如醛、酮、酯、醌等类化合物；碱性氧化铝 pH 为 9～10，适用于分离碱性有机物质，如生物碱、胺、烃类化合物等。市售的硅胶略带酸性。

吸附剂的活性与其含水量有关，含水量越高，活性越低，吸附剂的吸附能力越弱；反之则吸附能力强。吸附剂的含水量和活性等级关系见表 2-8。

表 2-8 吸附剂的含水量和活性等级关系

活性等级	Ⅰ	Ⅱ	Ⅲ	Ⅳ	Ⅴ
氧化铝含水量/％	0	3	6	10	15
硅胶含水量/％	0	5	15	25	38

一般常用的是Ⅱ级和Ⅲ级吸附剂。Ⅰ级吸附性太强，且易吸水；Ⅴ级吸附性太弱。吸附剂按其相对的吸附能力可粗略分类如下。

① 强吸附剂：低含水量的氧化铝、硅胶、活性炭。
② 中等吸附剂：碳酸钙、磷酸钙、氧化镁。
③ 弱吸附剂：蔗糖、淀粉、滑石粉。

吸附剂的吸附能力不仅取决于吸附剂本身，还取决于被吸附物质的结构。

(2) 溶质的结构和吸附能力　化合物的吸附性与分子的极性有关，分子极性越强，吸附能力越大。分子中含极性较大的基团时，其吸附能力也较强。以氧化铝为例，对各种化合物的吸附性按以下次序递减。

酸和碱＞醇、胺、硫醇＞酯、醛、酮＞芳香族化合物＞卤代物＞醚＞烯＞饱和烃

例如，邻硝基苯胺的偶极矩为 4.45 D，而对位异构体则为 7.1 D，根据它们的极性不同，可用柱色谱法加以分离，邻硝基苯胺先被洗脱下来。

(3) 洗脱剂　在柱色谱分离中，洗脱剂的选择是至关重要的。通常根据被分离物中各组分的极性、溶解度和吸附剂活性来考虑。

在进行柱色谱分离前，先将待分离的样品溶于一定体积的溶剂中，选用的溶剂极性应低，体积要小。如有的样品在极性低的溶剂中溶解度很小，则可加入少量极性较大的溶剂溶解，以使溶液体积不致太大。

一般洗脱剂的选择是通过薄层色谱实验来确定的。具体方法：先用少量溶解好（或提取出来）的样品，在已制备好的薄层板上点样（具体操作方法见 2.16.2 薄层色谱），用少量展开剂展开，观察各组分点在薄层板上的位置，并计算 R_f 值。哪种展开剂能将样品中各组分完全分开，即可作为柱色谱的洗脱剂。当单纯一种展开剂达不到所要求的分离效果时，可考虑选用混合展开剂。

选择洗脱剂的另一个原则是洗脱剂的极性不能大于样品中各组分的极性。否则会由于洗脱剂在固定相被吸附，迫使样品一直保留在流动相中。在这种情况下，组分在柱中移动得非常快，难以建立起分离所要达到的化学平衡，影响分离效果。另外，所选择的洗脱剂必须能够将样品中各组分溶解，但不能同各组分竞争与固定相的吸附。如果被分离的样品不溶于洗脱剂，那么各组分可能会牢固地吸附在固定相上，而不随流动相移动或移动很慢。

色谱层的展开首先使用极性最小的溶剂，使最容易脱附的组分分离，然后逐渐增加洗脱剂的极性，使极性不同的化合物按极性由小到大的顺序自色谱柱中洗脱下来。常用的洗脱剂、洗脱剂的极性及洗脱能力按如下顺序递增：

石油醚＜环己烷＜甲苯＜苯＜二氯甲烷＜氯仿＜环己烷-乙酸乙酯（80∶20）＜二氯甲烷-乙醚（80∶20）＜二氯甲烷-乙醚（60∶40）＜环己烷-乙酸乙酯（20∶80）＜乙醚＜乙醚-甲醇（99∶1）＜乙酸乙酯＜丙酮＜正丙醇＜乙醇＜甲醇＜水＜乙酸

极性溶剂对于洗脱极性化合物是有效的，非极性溶剂对于洗脱非极性化合物是有效的，若分离复杂组分的混合物，通常选用混合溶剂。

所用洗脱剂必须纯粹和干燥，否则会影响吸附剂的活性和分离效果。

吸附柱色谱的分离效果不仅依赖于吸附剂和洗脱剂的选择，而且与制成的色谱柱有关。

(4) 色谱柱的大小和吸附剂的用量　柱色谱的分离效果不仅依赖于吸附剂和洗脱剂的选择，而且还与色谱柱的大小和吸附剂的用量有关。一般要求柱中吸附剂用量为待分离样品量的30～40倍，若需要时可增至100倍，柱高和直径之比一般为8∶1。表2-9列出了它们之间的相互关系，实验者可根据实际情况参照选择。

表 2-9　色谱柱的大小、吸附剂用量和样品量之间的关系

样品质量/g	吸附剂质量/g	色谱柱直径/cm	色谱柱高度/cm
0.01	0.3	3.5	30
0.10	3.0	7.5	60
1.00	30.0	16.0	130
10.00	300.0	35.0	280

2.16.1.3　操作方法

(1) 装柱　装柱是柱色谱中最关键的操作，装柱的好坏直接影响分离效率。装柱之前，先将空柱洗净干燥，然后将柱垂直固定在铁架台上。如果色谱柱下端没有砂芯横隔，就应取一小团脱脂棉或玻璃棉，用玻璃棒将其推至柱底，再在上面铺上一层厚0.5～1cm的石英砂，然后进行装柱。装柱的方法有湿法和干法两种。

① 湿法装柱。将吸附剂（氧化铝或硅胶）用洗脱剂中极性最低的洗脱剂调成糊状，在柱内先加入约3/4柱高的洗脱剂，再将调好的吸附剂边敲打柱身边倒入柱中，同时打开柱子的下旋活塞，在色谱柱下面放一个干净并且干燥的锥形瓶或烧杯，接收洗脱剂。当装入的吸附剂有一定的高度时，洗脱剂流下速度变慢，待所用吸附剂全部装完后，用流下来的洗脱剂转移残留的吸附剂，并将柱内壁残留的吸附剂淋洗下来。在此过程中，应不断敲打色谱柱，以使色谱柱填充均匀并没有气泡。柱子填充完后，在吸附剂上端覆盖一层约0.5cm厚的石英砂或覆盖一片比柱内径略小的圆形滤纸。覆盖石英砂或滤纸的目的是：a. 使样品均匀地流入吸附剂表面；b. 当加入洗脱剂时，可以防止吸附剂表面被破坏。在整个装柱过程中，柱内洗脱剂的高度始终不能低于吸附剂最上端，否则柱内会出现裂痕和气泡。

② 干法装柱。在色谱柱上端放一个干燥的漏斗，将吸附剂倒入漏斗中，使其成为细流连续地装入柱中，并轻轻敲打色谱柱柱身，使其填充均匀，再加入洗脱剂湿润。也可先加入3/4的洗脱剂，然后倒入干的吸附剂。

由于氧化铝和硅胶的溶剂化作用易使柱内形成缝隙，所以这两种吸附剂不易使用干法装柱。

(2) 加样及洗脱　液体样品可以直接加入到色谱柱中，如浓度低可浓缩后再进行分离。固体样品应先用少量的溶剂溶解后再加入到柱中。在加入样品时，应先将柱内洗脱剂排至稍低于石英砂表面后停止排液，用滴管沿柱内壁把样品一次加完。在加入样品时，应注意滴管尽量向下靠近石英砂表面。样品加完后，打开下旋塞，使液体样品进入石英砂层后，再加入少量的洗脱剂将壁上的样品洗脱下来，待这部分液体的液面和吸附剂表面相齐时，即可打开安置在柱上装有洗脱剂的滴液漏斗的旋塞，加入洗脱剂，进行洗脱，直至所有色带被展开。

(3) 分离成分的收集　如果样品中各组分都有颜色时，可根据不同的色带用锥形瓶分别进行收集，然后分别将洗脱剂蒸除得到纯组分。但大多数有机物质是没有颜色的，只能分段收集洗脱液，再用薄层色谱或其他方法鉴定各段洗脱液的成分，成分相同者可以合并。

柱色谱方法是费时的，但由于操作方便，分离量可以大至几克，小至几十毫克，仍显示其价值。

在装柱、加样和洗脱时应注意以下几点。

① 如果装柱时吸附剂的顶面不呈水平，将会造成非水平的谱带。若吸附剂表面不平整或内部有气泡时会造成沟流现象（谱带前沿一部分向前伸出的现象叫沟流），如图2-59和图2-60所示。所以，吸附剂要均匀装入管内，装柱时要轻轻不断地敲击柱子，以除尽气泡，不留裂痕，防止内部造成沟流现象，影响分离效果。装柱完毕后，在向柱中添加溶剂时，应沿柱壁缓缓加入，以免将表层吸附剂和样品冲溅泛起，造成非水平谱带，覆盖在吸附剂表层的石英砂也是起这个作用的。

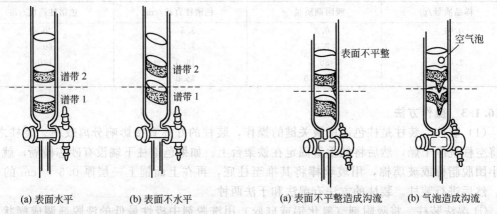

(a) 表面水平　　(b) 表面不水平　　　　　　　(a) 表面不平整造成沟流　(b) 气泡造成沟流

图 2-59　水平的和非水平的谱带前沿的对比　　　　　图 2-60　沟流现象

② 洗脱剂应连续平稳地加入，不能中断，不能使柱顶变干，因为湿润的柱子变干后，吸附剂可能与柱壁脱开形成裂沟，结果显色不匀，也产生不规则的谱带。

③ 洗脱剂的流速对柱色谱分离效果具有显著影响。在洗脱过程中，样品在柱内的下移速度不能太快，如果溶剂流速较慢，则样品在柱中保留的时间长，各组分在固定相和流动相之间能得到充分的吸附或分配作用，从而使混合物，尤其是结构、性质相似的组分得以分离。但样品在柱内的下移速度也不能太慢（甚至过夜），因为吸附剂表面活性较大，时间太长有时可能造成某些成分被破坏，使色谱带扩散，影响分离效果。因此，层析时洗脱速度要适中。通常洗脱剂流出速度为每分钟5~10滴，若洗脱剂下移速度太慢可适当加压或用水泵减压，以加快洗脱速度。

④ 当色谱带出现拖尾时，可适当提高洗脱剂的极性。

2.16.1.4　柱色谱操作练习

甲基橙和亚甲基蓝的分离。甲基橙和亚甲基蓝均为指示剂，它们的结构式是

NaO_3S-⟨phenyl⟩-N=N-⟨phenyl⟩-$N(CH_3)_2$　　　　$(CH_3)_2N$-⟨phenothiazine⟩-$N(CH_3)_2$

甲基橙　　　　　　　　　　　　　　　　　　亚甲基蓝

由于它们的结构不同、极性不同，吸附剂对它们的吸附能力不同，洗脱剂对它们的解吸速度也不同。极性小、吸附能力弱、解吸速度快的亚甲基蓝先被洗脱下来，而极性大、吸附

能力强、解吸速度慢的甲基橙后被洗脱下来，从而使两种物质得以分离。

（1）装柱　选一合适色谱柱，洗净干燥后垂直固定在铁架台上，色谱柱下端置一锥形瓶。在柱底铺一层脱脂棉，轻轻压紧，然后再铺上一层石英砂（约0.5cm厚），用木制试管夹敲打柱身，使砂子上层水平。关闭柱下部旋塞，向柱内倒入95%乙醇至柱高的3/4处，打开活塞，控制乙醇流出速度为每秒钟1滴。然后将用乙醇溶剂调成糊状的一定量的吸附剂（中性氧化铝）通过一只干燥的粗柄短颈漏斗从柱顶加入，同时打开色谱柱下端的旋塞，使溶剂慢慢流入锥形瓶。填充吸附剂的过程中要敲打柱身，使装入的氧化铝紧密均匀，顶层水平。当装柱至3/4时，再在上面加一层0.5cm厚的石英砂。操作时一直保持上述流速，但要注意不能使砂子顶层露出液面，柱顶变干。

（2）加样　把1mg甲基橙和5mg亚甲基蓝溶于2.2mL乙醇中。打开色谱柱的旋塞，将其顶部多余的溶剂放出。当液面降至离石英砂顶层还有1mm高时，关闭活塞，将上述溶液用滴管小心地加入柱内。打开旋塞，待液面降至石英砂层时，用滴管取少量溶剂洗涤色谱柱内壁上沾有的样品溶液。注意不要把柱内吸附剂平面冲得凹凸不平。

（3）洗脱与分离　样品加完并混溶后，开启旋塞，当液面下降至和砂层顶层相近时，便可滴加洗脱剂（95%乙醇）进行洗脱，流速控制在每秒钟1滴，这时亚甲基蓝谱带和甲基橙谱带分离。亚甲基蓝因极性较小，向柱下部移动，极性较大的甲基橙留在柱的上端。继续加入足够量的95%的乙醇，使亚甲基蓝的色带全部从柱子里洗下来。待洗出液呈无色时，更换一只接收器，改用水为洗脱剂，这时甲基橙向柱子下部移动，用容器收集，同样至洗出液呈无色为止。

思　考　题

1. 色谱柱的底部和上部装石英砂的目的何在？
2. 装柱不均匀或者有气泡、裂缝，将会造成什么后果，为什么？
3. 为什么洗脱的速度不能太快，也不宜太慢？
4. 极性大的组分为什么要用极性较大的溶剂洗脱？试举一例加以说明。

2.16.2　薄层色谱

薄层色谱（Thin Layer Chromatography，TLC）。常用的有吸附色谱和分配色谱两类，本节介绍的是薄层吸附色谱。

2.16.2.1　基本原理

薄层色谱的原理及分离过程与柱色谱相似，吸附剂的性质和洗脱剂的相对洗脱能力在柱色谱中适用的同样适用于TLC中。与柱色谱不同的是，TLC中的流动相沿着薄板上的吸附剂向上移动，而柱色谱中的流动相则沿着吸附剂向下移动。薄层色谱的特点是：所需样品少（几微克到几十微克），分离时间短（几分钟到几十分钟），效率高，是一种微量、快速和简便的分离分析方法。可用于精制样品、化合物鉴定、跟踪反应进程和柱色谱的先导（即为柱色谱摸索最佳条件）等方面。薄层色谱不仅可以分离少量样品（几微克），而且也可以分离较大量的样品（可达500mg），特别适用于挥发性较低、或在高温下易发生变化而不能用气相色谱进行分离的化合物。

薄层色谱是将吸附剂均匀地涂在玻璃板（或某些高分子薄膜）上作为固定相，经干燥、活化后，用管口平整的毛细管把少量待分离的样品"点"在薄板一端约1cm处，晾干或吹干后置薄板于盛有展开剂（流动相）的展开槽内，浸入深度为0.5cm。当展开剂在吸附剂上展开时，由于吸附剂对各组分吸附能力不同，展开剂对各组分的解吸能力不同，各组分向前移动的速度不同。其结果是吸附能力强的组分相对移动得慢些，而吸附能力弱的移动得快些。当展开剂上升到一定程度，停止展开时，各组分便停留在薄板的不同部位，从而使混合物的各组分得以分离（见图 2-61）。

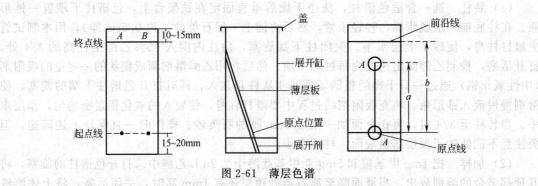

图 2-61　薄层色谱

将色谱板取出，如果各组分本身有颜色，则薄层板干燥后会出现一系列高低不同的斑点，如果本身无色，则可用各种显色方法使之显色，或在紫外灯下显色，以确定斑点位置。记录原点至斑点中心及展开剂前沿的距离，计算 R_f 值。

$$R_f = \frac{溶质的最高浓度中心至原点中心的距离}{溶剂前沿至原点中心的距离} = \frac{d_{斑点}}{d_{溶剂}}$$

在薄板上混合物的每个组分上升的高度与展开剂上升的前沿之比称为该化合物的 R_f 值，又称比移值（见图 2-62）。

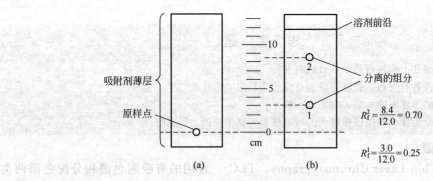

图 2-62　计算 R_f 值示意图

R_f 值随被分离化合物的结构、固定相和流动相的性质、温度以及薄层板本身的因素而变化。当固定相、流动相、温度、薄板厚度等实验条件固定时，各物质的 R_f 值是恒定的，因此可利用 R_f 值对未知物进行定性鉴定。但由于影响 R_f 值的因素很多，在鉴定时采用标准样品对照。通过比较两者的 R_f 值，可对样品作出定性鉴定，还可以通过比较未知物和标准物的色斑大小或颜色深浅来定量判定甚至定量测定其含量。良好的分离 R_f 值应在 0.15～0.75 之间，否则应该调换展开剂重新展开。

在 TLC 中所用的吸附剂颗粒比柱色谱中用的要小很多，一般为 260 目以上。当颗粒太大时，表面积小，吸附量少，样品随展开剂移动速度快，斑点扩散较大，分离效果不好；当颗粒太小时，样品随展开剂移动速度慢，斑点不集中，效果也不好。TLC 最常用的吸附剂是硅胶和氧化铝，它们的分子中都含具有孤偶电子的氧原子和能形成氢键的—OH 基。

薄层吸附色谱和柱吸附色谱一样，化合物的吸附能力与它们的极性成正比，具有较大极性的化合物吸附较强，因而 R_f 值较小。因此利用化合物极性的不同，用硅胶或氧化铝薄层色谱可将一些结构相近的物质或顺、反异构体分开。各类有机化合物与硅胶或氧化铝的亲和力大小次序如下：

羧酸＞醇＞伯胺＞酯、醛、酮＞硝基化合物＞醚＞烯＞卤代烃＞烷

硅胶是无定形多孔物质，略具酸性，适用于酸性和中性物质的分离和分析，薄层色谱用的硅胶分为：

硅胶 H——不含黏合剂；

硅胶 G——含黏合剂（煅石膏，$2CaSO_4 \cdot H_2O$），标记 G 代表石膏（Gypsum）；

硅胶 HF_{254}——含荧光物质，可在波长 254nm 的紫外光下发出荧光；

硅胶 GH_{254}——既含煅石膏，又含荧光剂。

同样，氧化铝也分为氧化铝 G、氧化铝 HF_{254} 及氧化铝 GF_{254}。氧化铝的极性比硅胶大，适用于分离极性小的化合物。

黏合剂除煅石膏外，还有淀粉、聚乙烯醇和羧甲基纤维素钠（CMC）。使用时，一般配成百分之几的水溶液。如羧甲基纤维素钠的质量分数一般为 0.5%～1%，最好是 0.7%；淀粉的质量分数为 5%。通常将薄板按加黏合剂和不加黏合剂分为两种，加黏合剂的薄板称为硬板，不加黏合剂薄板称为软板。

2.16.2.2 操作方法

(1) 薄板的制备　薄板的制备方法有两种：一种是干法制板；另一种是湿法制板。

干法制板　一般用氧化铝作吸附剂，涂层时不加水，将氧化铝倒在玻璃板上，取直径均匀的一根玻璃棒，将两头用胶布缠好，在玻璃板上滚压，把吸附剂均匀地铺在玻璃板上。这种方法简便，展开快，但是样品展开点易扩散，制成的薄板不易保存。

湿法制板　是实验室最常用的。对湿板按铺层的方法不同可分为平铺法、倾注法和浸涂法三种。制板前首先将吸附剂制成糊状物。称取 3g 硅胶 G，边搅拌边慢慢加入到盛有 6～7mL 0.5%～1% CMC 清液的烧杯中，调成糊状（3g 硅胶约可铺 7.5cm×2.5cm 载玻片 5～6 块）。

注意：硅胶 G 糊易凝结，所以必须现用现配，不宜久放。

① 平铺法。用购置或自制的薄层涂布器（见图 2-63）进行制板，涂层既方便又均匀，是科研中常用的方法。当大量铺板或铺较大板时常用此法。

② 倾注法。将调好的糊状物倒在玻璃板上，用手左右摇晃，使其表面均匀平整，然后放在水平的平板上晾干。这种制板的方法厚度不易控制。

③ 浸涂法。将两块干净的载玻片对齐紧贴在一起，浸入盛有糊状物的容器中，使载玻片上涂上一层均匀的吸附剂，取出分开，晾干。

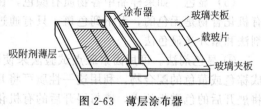

图 2-63　薄层涂布器

薄板制备的好与坏直接影响色谱的分离效果，在制备过程中应注意以下几点。

① 要制备均匀而又不带块状的糊状涂层浆料，应把硅胶加到溶剂中去，边加边搅拌混合物。若把溶剂加到吸附剂中，常会产生团块。

② 涂板前，一定要将玻璃板洗净、擦干，涂布速度要快。

③ 铺板时，尽量均匀，不能有气泡、颗粒等，而且厚度（0.25～1mm）要固定。否则，在展开时溶剂前沿不齐，色谱结果也不易重复。

④ 湿板铺好后，应放在水平的平板上晾干，千万不要快速干燥，否则薄层板会出现

(2) 薄板的活化　把涂好的薄板置于室温自然晾干后，再放在烘箱内加热活化，进一步除去水分。活化时需慢慢升温。硅胶板一般在105～110℃的烘箱中活化0.5h即可。氧化铝板在200℃烘4h可得到活性Ⅱ级的薄层板，150～160℃烘4h可得到活性Ⅲ～Ⅳ级的薄层板。活化后的薄板应保存在干燥器中备用。

(3) 点样　在距薄板的一端10mm处，用铅笔轻轻画一条横线作为点样时的起点线，在距薄层板的另一端5mm处，再画一条横线作为展开剂向上爬行的终点线（画线时不能将薄层板表面破坏），如图2-64所示。

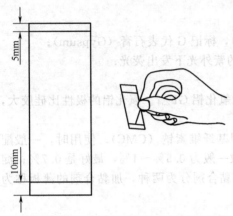

图2-64　薄板及薄板点样方法

将样品溶于低沸点溶剂（如甲醇、乙醇、丙酮、氯仿、苯、乙醚及四氯化碳）中配成1%左右的溶液，用内径1mm管口平齐的毛细管，吸取少量的样品点样，垂直轻轻地点在起点线上。若溶液太稀，一次点样不够，则可待前一次点样的溶剂挥发后再重新点样，但每次点样都应点在同一圆心上，点样的次数依样品溶液的浓度而定，一般为2～5次。点样后斑点直径不超过2mm，点样斑点过大，往往会造成脱尾、扩散等现象，影响分离效果。若在同一板上点几个样品，则几个样品应点在同一直线上，样点间距为1～1.5cm。点样结束待样品干燥后，方可进行展开。点样要轻不可刺破薄层板。

(4) 展开　薄层色谱展开剂的选择和柱色谱一样，主要根据样品的极性、溶解度、吸附剂的活性等因素来考虑。溶剂的极性越大，则对化合物的洗脱力也越大，也就是R_f值也越大。薄层色谱用的展开剂绝大多数是有机溶剂，各种溶剂的极性参见柱色谱部分。

薄层的展开需要在密闭的容器中进行，先将选择的展开剂放在层析缸中（液层高度约0.5cm），使层析缸内溶剂蒸气饱和5～10min，再将点好样品的薄板按图2-61所示放入层析缸中进行展开。注意：展开剂液面的高度应低于样品斑点。在展开过程中，样品斑点随着展开剂向上迁移，当展开剂前沿至薄层板上边的终点线时，立刻取出薄层板。将薄层板上分开的样品点用铅笔圈好，计算比移值。

(5) 显色　如果样品中各物质有颜色，在展开后就可以清楚地看到各个色斑。但大多数有机化合物是无色的，看不到色斑，只有通过显色才能使斑点显现。常用的显色方法有显色剂法和紫外光显色法。

① 显色剂法。可用卤素斑点试验法来使薄层色谱斑点显色。许多有机化合物能与碘生成棕色或黄色的配合物。利用这一性质可将几粒碘置于密闭容器中，待容器充满碘蒸气后，将展开后的色谱板放入，碘与展开后的有机化合物可逆地结合，在几秒钟到数分钟内化合物斑点的位置呈黄棕色。色谱板自容器取出后，呈现的斑点一般在2～3s内消失，因此必须用铅笔标出化合物的位置。碘熏显色法是观察无色物质的一种有效方法，因为碘可以与除烷烃和卤代烃以外的大多数有机物形成有色配合物。此外，还可使用腐蚀性显色剂，如浓盐酸、浓硫酸等。

② 紫外光显色法。用硅胶GF_{254}制成的薄板，由于加入了荧光剂，在三色紫外灯光下观察，展开后的有机化合物在亮的荧光背景上呈暗色斑点，此斑点就是样品点。

用各种显色方法使斑点出现后，应立即用铅笔圈好斑点的位置，并计算R_f值。以上这些显色方法在柱色谱和纸色谱中同样适用。

思 考 题

1. 展开剂的液面高出薄板的斑点，将会产生什么后果？
2. 用薄层色谱分析混合物时，如何确定各组分在薄层上的位置？如果斑点出现拖尾现象，这可能是什么原因所引起的？
3. 点样时，所用毛细管管口要平整，为什么？

2.16.3 纸色谱

纸色谱（Paper Chromatography）和薄层色谱一样，主要用于反应混合物的分离和鉴定。此法一般适用于微量有机物质（5～500μg）的定性分析，分离出来的色点也能用比色方法定量。纸色谱的优点是操作简单，价格便宜，所得到的色谱图可以长期保存。缺点是展开时间较长，因为展开过程中，溶剂的上升速度随着高度的增加而减慢。由于纸色谱对亲水性较强的组分分离效果较好，故特别适用于多官能团或高极性化合物（如糖、氨基酸等）的分离。

2.16.3.1 基本原理

纸色谱属于分配色谱的一种。通常用特殊的滤纸作为固定相——水的支持剂，流动相则是含有一定比例水的有机溶剂，通常称为展开剂。其原理和吸附色谱不同，是根据混合物的各组分在两种互不相溶的液相间分配系数的不同而达到分离目的。分配色谱法在原则上与液-液连续萃取方法相同。流动相亲脂性较强的溶剂在含水的滤纸上移动时，流动相中各组分在滤纸上受到两相溶剂的影响，产生了分配现象。亲脂性稍强的组分在流动相中分配较多，移动速度较快，有较高的 R_f 值。反之亲水性的组分在固定相中分配较多，移动慢些，从而得到分离。

2.16.3.2 纸色谱的装置

纸色谱的装置如图2-65所示，由展开缸、橡皮塞、钩子组成的。钩子被固定在橡皮塞上，展开时将滤纸挂在钩子上。

2.16.3.3 操作方法

先将层析滤纸在展开溶剂蒸气中放置过夜，在滤纸一端2～3cm处用铅笔划好起点线。然后将要分离的样品溶液用毛细管滴在起点线上，待溶剂挥发后，将滤纸的另一端悬挂在展开槽的玻璃钩上，使滤纸下端与展开剂接触。由于毛细管作用展开剂沿滤纸条上升，当展开剂前沿接近滤纸上端时，将滤纸取出，记下溶剂前沿位置，晾干。若被分离物中各组分是有色的，滤纸条上就有各种颜色的斑点显出，计算各化合物的比移值 R_f。比移值的计算及影响比移值的因素可见2.16.2薄层色谱。

图2-65 纸色谱装置

对于无色混合物的分离，通常将展开后的滤纸晾干或吹干，置于紫外灯下观察是否有荧光，或者根据化合物的性质，喷上显色剂，观察斑点位置。

2.17 谱学分析技术

鉴定有机化合物并确定其结构是有机化学的一项重要任务。随着科学技术的发展，有机化学工作者已广泛使用红外光谱（Infrared Spectroscopy，IR）、核磁共振谱（Nuclear Magnetic Resonance，NMR）、紫外光谱（UV）、质谱（MS）等波谱技术来鉴定和测定有机化合物的结构。波谱分析方法具有分析速度快、用样量少等优点。近年来，国内外有机化学实验教学中已将红外光谱仪、核磁共振仪、紫外光谱仪等现代分析仪器引入其中，用来测定有机化合物的结构。本书中着重介绍红外光谱和核磁共振谱的相关知识，因为这两种仪器是有机化学实验中使用较为普遍的仪器。

2.17.1 红外光谱

几乎所有具有共价键的化合物，无论是有机物还是无机物，都会在电磁波谱的红外区吸收不同频率的电磁波。电磁波的红外区位于可见光（400～800nm）和无线电波（1cm）之间。红外光的波长在 $0.7～1000\mu m$（波数 $14000～10cm^{-1}$）范围内，通常又把这个区域划分为近红外区 $[0.7～2.5\mu m(14000～4000cm^{-1})]$、中红外区 $[2.5～25\mu m (4000～400cm^{-1})]$ 和远红外区 $[25～1000\mu m (400～10cm^{-1})]$ 三个区域。用于有机化合物结构分析的是中红外区，因为分子振动的基频在此区域。红外区与其他电磁波之间的相互关系如图 2-66 所示。

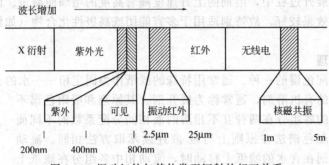

图 2-66 振动红外与其他类型辐射的相互关系

2.17.1.1 基本原理

红外光谱的产生是基于分子中原子的振动。当用一束红外光照射样品分子时，样品分子吸收红外光后，体系能量增加，产生振动能级的跃迁。有机分子中不同的键（如 C—C、C=C、C=O、C—H、O—H 和 N—H 等），由于振动能级不同，所吸收的红外光的频率也不同，因此可以通过红外光谱的特征吸收频率来鉴定这些键是否存在。

分子振动主要有两种形式：伸缩振动（Stretching）和弯曲振动（Bending）。伸缩振动是指沿着键轴方向的振动；弯曲振动是指键角交替地发生变化的振动。以亚甲基为例，分子振动的形式如图 2-67 所示。在这些振动中，只有那些在振动时发生偶极矩变化的才能吸收红外光。这是因为振动引起电荷分布的改变所产生的电场，与红外辐射的电磁场发生共振而引起吸收。当用不同波长的红外光照射样品分子时，只有那些频率与特定的振动形式一致的红外光才能被吸收。在振动能级发生变化时，常伴随着一系列转动能级的改变，因而在 2～$25\mu m$ 波长范围内观察有机化合物的红外光谱时，所看到的吸收谱带是连续的、峰谷相间的，而不是断续的线形红外光谱。因此，红外光谱是分子的振动-转动光谱。

红外光谱图中的横坐标表示波长 λ（单位是 μm，$1\mu m = 10^{-6} m$）或波数 ν（单位是

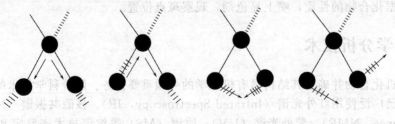

图 2-67 分子振动形式

cm^{-1}），它们之间的换算关系如下。

$$\nu = \frac{1}{\lambda(cm)} = \frac{10^4}{\lambda(\mu m)}$$

红外光谱图中的纵坐标表示透光率 T 或光密度 A，可用下式计算。

$$T = \frac{I}{I_0}$$

式中　I_0——入射光的强度；
　　　I——入射光被样品吸收一部分后，透过光的强度。

$$A = \lg\frac{1}{T} = \lg\frac{I_0}{I} = kcl$$

式中　k——消光系数；
　　　c——样品浓度；
　　　l——样品厚度。

红外光谱中特征吸收带通常以波数（cm^{-1}）表示。习惯上常常把特征吸收带称为特征频率，因此本书中所述特征频率、振动频率皆指波数而言。

大量实验事实表明，同一种化学键或基团，在不同化合物中吸收谱带的位置大致相同（变化范围较窄）。化学键和基团的特征频率是定性分析的依据。表 2-10 列举了一些常见的化学键和基团的特征频率。更详细的数据可查阅相关专著。

表 2-10　红外光谱中一些常见化学键和基团的特征频率

区域	基团	特征频率/cm^{-1}	振动形式	吸收强度	说明
第一频率区	—OH（游离）	3650~3580	伸缩	m,sh	判断有无醇类、酚类和有机酸的重要依据
	—OH（缔合）	3400~3200	伸缩	s,b	判断有无醇类、酚类和有机酸的重要依据
	—NH$_2$，—NH（游离）	3500~3300	伸缩	m	
	—NH$_2$，—NH（缔合）	3400~3100	伸缩	s,b	
	—SH	2600~2500	伸缩		
	C—H 伸缩振动				
	不饱和 C—H				不饱和 C—H 伸缩振动出现在 3000cm^{-1} 以上
	≡C—H（叁键）	3300 附近	伸缩	s	
	=C—H（双键）	3010~3040	伸缩	s	=CH$_2$ 出现在 3085cm^{-1} 附近
	苯环中 C—H	3030 附近	伸缩	s	强度上比饱和 C—H 稍弱，但谱带较尖锐
	饱和 C—H				饱和 C—H 伸缩振动出现在 3000cm^{-1} 以下（3000~2800cm^{-1}），取代基影响小
	—CH$_3$	2960±5	反对称伸缩	s	
	—CH$_3$	2870±10	对称伸缩	s	
	—CH$_2$	2930±5	反对称伸缩	s	三元环中的 CH$_2$ 出现在 3050cm^{-1}
	—CH$_2$	2850±10	对称伸缩	s	—C—H 出现在 2890cm^{-1}，很弱
第二频率区	—C≡N	2260~2220	伸缩	s 针状	干扰少
	—N=N	2310~2135	伸缩	m	
	—C≡C—	2260~2100	伸缩	v	R—C≡C—H 2100~2140cm^{-1}；R—C≡C—R'，2190~2260cm^{-1}；若 R=R'，对称分子无红外谱带
	—C=C=C—	1950 附近	伸缩	v	

续表

区域	基团	特征频率/cm^{-1}	振动形式	吸收强度	说明
第三频率区	C=C	1680~1620	伸缩	m,w	
	芳环中 C=C	1600,1580	伸缩	v	苯环的骨架振动
		1500,1450			
	—C=O	1850~1600	伸缩	s	其他吸收带干扰少,是判断羰基(酮、酸、酯、酸酐等)的特征频率,位置变动大
	—NO$_2$	1600~1500	反对称伸缩	s	
	—NO$_2$	1300~1250	对称伸缩	s	
	S=O	1220~1040	伸缩	s	
第四频率区	C—O	1300~1000	伸缩	s	C—O 键(酯、醚、醇类)的极性很强,故强度大,常成为谱图中最强的吸收
	C—O—C	900~1150	伸缩	s	醚类中 C—O—C 的 $v_{as}=(1100\pm50)$cm^{-1} 是最强的吸收。C—O—C 对称伸缩在 900~1000cm^{-1},较弱
	—CH$_3$,—CH$_2$	1460±10	CH$_3$ 反对称变形	m	大部分有机化合物都含有 CH$_3$、CH$_2$ 基,因此该峰经常出现;很少受取代基的影响,且干扰少,是 CH$_3$ 基的特征吸收
			CH$_2$ 变形		
	—CH$_3$	1370~1380	对称变形	s	
	—NH$_2$	1650~1560	变形	m~s	
	C—F	1400~1000	伸缩	s	
	C—Cl	800~600	伸缩	s	
	C—Br	600~500	伸缩	s	
	C—I	500~200	伸缩	s	
	=CH$_2$	910~890	面外摇摆	s	
	—(CH$_2$)$_n$— (n>4)	720	面内摇摆	v	

注:s—强吸收,b—宽吸收带,m—中等强度吸收,w—弱吸收,sh—尖锐吸收峰,v—吸收强度可变。

一般化学键和基团的特征频率在 4000~1300cm^{-1} 区域中,这个区域叫做特征区,基团的鉴定工作主要在特征区进行。一般,在 3700~3200cm^{-1} 的谱带可能是羟基或氨基;在 3100~2800cm^{-1} 的谱带是 C—H 伸缩振动(3000cm^{-1} 是一个分界线,高于 3000cm^{-1} 时是不饱和 C—H 伸缩振动,低于 3000cm^{-1} 时是饱和 C—H 伸缩振动);在 2200cm^{-1} 附近的谱带可能是叁键;在 1900~1650cm^{-1} 的谱带是羰基;在 1650~1600cm^{-1} 的谱带是 C=C 双键;在 1600cm^{-1} 和 1500cm^{-1} 附近的谱带是芳环;低于 1300cm^{-1} 区域的吸收大多是一些单键(如 C—C、C—N、C—O 等的伸缩振动和各种变形振动)。这些单键的强度相差不大,在分子中又连在一起,互相影响,变动范围大,特征性差,故称之为指纹区。指纹区的特征性虽差,但对分子结构十分敏感,分子结构的细微变化会引起指纹区光谱的明显改变。因此,指纹区的信息对化合物的鉴定也是十分重要的。

2.17.1.2 实验方法

红外光谱测定的一大优点就是对气态、液态和固态样品都能进行分析。由于金属卤化物不吸收红外光,所以常用 KBr、NaCl 制样品池和分光棱镜。

(1) 液体样品的制备 测定液体样品的红外光谱都采用液膜法。液体样品的红外光谱一般常采用可拆卸液体池进行测定,如图 2-68 所示。在池架的底座上先放一块橡皮垫片,然后放上一块抛光的盐片,滴上一些干燥的液体样品到该盐片上,再用另一块抛光的盐片盖上(两盐片间可以不放铝垫片),并轻轻旋转滑动,使样液涂布均匀,以保证形成的液膜无气泡。然后再放上另一垫片,压上池架垫板,用螺母拧紧整个吸收池,并安放在红外光谱仪

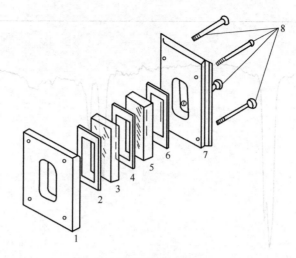

图 2-68 可拆卸液体池
1—池架前板；2，6—橡皮垫片；3，5—KBr 窗片；4—控制光程长度的
铝垫片（有 0.025～1mm 各种规格）；7—池架后板；8—固定螺杆

中，记录红外光谱。

制备时应注意以下几点。

① 在 2～15μm 范围内记录红外光谱采用氯化钠片，若在较长的波长范围（12～25μm）内记录红外光谱则应采用溴化钾片。

② 使用盐片时必须注意防潮，以免抛光的盐片表面因吸收水汽而变得模糊，从而引起入射光的散射。

③ 盐片必须保存在干燥器中，使用后先用软纸吸去样品，然后用二氯甲烷、四氯化碳或氯仿清洗盐片。待干燥后，放回干燥器。

④ 对于易挥发的液体样品应使用密封池。

(2) 固体样品的制备　固体样品的测试一般可采用卤盐压片法、液体石蜡（Nujol，精制的矿物油）研糊法、溶液法和薄膜法。

① 卤盐压片法。这是固体样品测试的最常用方法。将预先干燥好的分析纯 KBr 100～200mg 与干燥固体样品 2～3mg 放在玛瑙研钵中，混合研磨成极细粉末，并将其装入金属模具中。轻轻振动模具，使混合物在模具中分布均匀。然后把模具放在压片机上，慢慢加压到 20MPa，保持 2min 左右，再慢慢减压，使压力降到零。从压片机上取下模具，从中取出压好的含有样品的透明的 KBr 片，放置在盐片支架上，并安放在红外光谱仪上进行测试，记录红外光谱。

② 液体石蜡研糊法。将研细的干燥固体样品 3～5mg 放在玛瑙研钵中，加入 2～3 滴液体石蜡研磨成糊状，然后将糊状物涂抹在盐片上，并另用一块盐片覆盖在上面。再将该盐片放置在盐片支架上，并安放在红外光谱仪中，记录红外光谱。

用研糊法应注意以下三点。

第一，液体石蜡糊薄膜应是半透明的，以防止辐射的散射，若不透明或有细颗粒必须重磨。

第二，液体石蜡为碳氢化合物，因此在 3030～2830cm^{-1} 处有 C—H 伸缩振动，在 1460～1375cm^{-1} 附近有 C—H 弯曲振动（见图 2-69）。

第三，盐片应保存在干燥器中，使用完后应将盐片用软纸抹干净，再用二氯甲烷清洗，干燥后仍放回干燥器中。

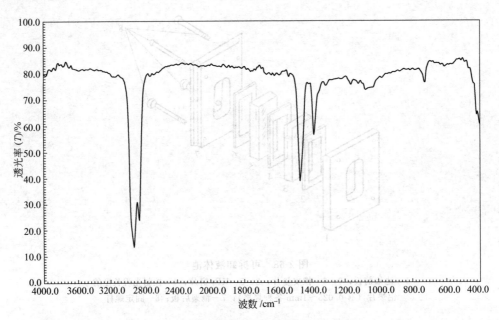

图 2-69 液体石蜡（Nujol）的红外光谱图

③ 溶液法。选择合适的溶剂将固体样品溶解配成溶液，然后注入液体样品池进行测定。选用的溶剂应合适：一般要求溶剂对溶质的溶解度要大，红外光范围内无吸收，不腐蚀窗片，分子简单，极性小，对溶质没有强的溶剂化效应。如 CS_2、CCl_4、$CHCl_3$ 等，它们本身的吸收峰可通过以溶剂为参比来校正。

④ 薄膜法。某些材料难以用前面几种方法测试，也可以试用薄膜法。一些高分子膜常常可以直接来测试，而更多的情况是要将样品制成膜。

a. 熔融法：熔点低、对热稳定的样品可以放在窗片上用红外灯烤，使其受热成流动性液体加压压成膜。

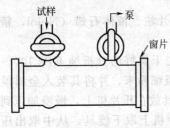

图 2-70 气体池

b. 溶液成膜法：样品溶于低沸点溶剂后，涂在平板上使其成膜。

c. 切片成膜法：不溶、难溶、又难粉碎的固体可以用机械切片法成膜。

(3) 气体样品　气体样品的红外测试可采用气体池进行，如图 2-70 所示。在样品导入前先抽真空，样品池的窗口多用抛光的 NaCl 或 KBr 晶片。常用的样品池长 5cm 或 10cm，容积为 50～150mL。吸收峰强度可通过调整气池内样品的压力来达到。对强吸收气体，只要注入 666.6Pa 的样品；对弱吸收气体，需注入 66.66kPa 的样品。因为水蒸气在中红外区有强吸收峰，所以气体池一定要干燥。样品测完后，用干燥的氮气流冲洗气体池，并将其保存在干燥器中。

2.17.1.3 谱图解析

测定出样品的红外光谱以后，对出现的谱带进行解析，进而推测样品的官能团和化学结构。谱图解析主要依靠对光谱与化学结构关系的理解和经验积累，灵活运用基团特征吸收峰及其变化规律。

红外光谱定性分析，一般采用两种方法：一种是用已知标准物对照；另一种是标准图谱查对法。已知物对照应由标准品和被检物在完全相同的条件下，分别绘出其红外光谱进行对照，谱图相同，则肯定为同一化合物。标准图谱查对法是一个最直接、可靠的方法。根据待测样品的来源、物理常数、分子式以及谱图中的特征谱带，查对标准谱图来确定化合物。常用标准图谱集为萨特勒红外标准图谱集（Sadtler, Catalog of Infrared Standard Spectra）。

在用未知物图谱查对标准谱时，必须注意以下几点。

① 比较所用仪器与绘制的标准图谱在分辨率与精度上的差别，可能导致某些峰的细微结构有差别。

② 未知物的测绘条件一致，否则图谱会出现很大差别。当测定溶液样品时，溶剂的影响大，必须要求一致，以免得出错结论。若只是浓度不同，只会影响峰的强度，而每个峰之间的相对强度是一致的。

③ 必须注意引入杂质吸收带的影响。如 KBr 压片可能吸水而引进了水的吸收带等。应尽可能避免杂质的引入。

一般图谱的解析大致步骤如下。

① 先从特征区入手，找出化合物所含主要官能团。

② 分析指纹区，进一步找出官能团存在的依据。因为一个基团常有多种振动形式，所以确定该基团就不能只依靠一个特征吸收，必须找出所有的吸收带才行。

③ 对指纹区谱带位置、强度和形状仔细分析，确定化合物的可能结构。

④ 对照标准图谱，配合其他鉴定手段，进一步验证。

2.17.1.4 注意事项

① 选择适当的试样浓度和厚度。使最高谱峰的透光率在 1%～5%，基线在 90%～95%，大多数的吸收峰透光率在 20%～60%。

② 由于水在 $3710cm^{-1}$ 和 $1630cm^{-1}$ 处有强吸收峰，因此在作红外光谱分析时，待测样品及盐片均需充分干燥处理。

③ 固体样品经研磨（在红外灯下）后仍应随时注意防止吸水，否则压出的片子易沾在模具上。可拆式液体池的盐片应保持干燥透明，每次测定前后均应反复用无水乙醇及滑石粉抛光（在红外灯下），但切勿用水洗。

④ 多组分试样的红外光谱测绘前应预先分离。

⑤ 液体石蜡为碳氢化合物，在 $3030\sim2830cm^{-1}$ 处有 C—H 伸缩振动，在 $1460\sim1375cm^{-1}$ 处有 C—H 弯曲振动，故在解析红外光谱时应注意将这些峰划去，以免对图谱的正确解析产生干扰（参见图 2-69）。

⑥ 实验完毕后，应将玛瑙研钵、刮刀、模具等接触样品部件用乙醇擦洗，红外灯烘干，冷却后放入干燥器中。样品池或样品仓应卸除，以防止样品污染或仪器腐蚀。最后关闭红外光谱仪，冷却至室温后，将仪器罩上，登记操作时间和仪器状况，经指导教师允许后方可离去。

2.17.2 核磁共振谱

核磁共振谱（Nuclear Magnetic Resonance），简称 NMR。1945 年，以 F. Bloch 和 E. M. Purcell 为首的两个研究小组同时独立发现了核磁共振现象，观察到了核磁共振的色散信号和吸收信号，利用核磁共振可以直接观察磁矩在外磁场中的能级分裂，也使核磁矩的测定比传统的光谱方法提高了 $10^5\sim10^6$ 的精度。1951 年，Proctr 等人发现了化学位移，从而开辟了核磁共振新的前景，高分辨率核磁共振谱仪应运而生，不断完善发展，成为分析测定分子结构强有力的手段。核磁共振分析能够提供 4 种结构信息：化学位移 δ、耦合常数 J、各种核的信号强度比和弛豫时间。通过分析这些信息，可以了解特定原子（如 ^1H、^{13}C 等）的化学环境、原子个数、邻接基团的种类及分子的空间构型。所以 NMR 在化学、生物学、医学和材料科学领域的应用日趋广泛，在有机化合物的结构研究中是一种重要的剖析工具。

核磁共振源于能产生磁场的核自旋，对于自旋量子数 $I>0$ 的核，如^1H、^2H、^{13}C、^{15}N、^{19}F、^{29}Si、^{31}P 这类核有磁矩，都能产生核磁共振；而自旋量子数 $I=0$ 的核没有磁矩，如^{12}C、^{16}O 和^{32}S 没有核自旋，不能用 NMR 谱来研究。在有机化学中最有用的是氢核和碳核。氢同位素中，^1H 质子的天然丰度比较大，磁性也比较强，比较容易测定。组成有机化合物的元素中，氢是不可缺少的元素，本节仅简单介绍^1H NMR 谱。

2.17.2.1 基本原理

很多原子核具有自旋的性质，自旋同时产生磁矩。核自旋量子数 $I\neq 0$ 的原子核具有磁性。当磁性核置于外磁场 H_0 中时，由于外磁场 H_0 与磁性核的相互作用，磁性核在外磁场中要有一定的排列，共有 $2I+1$ 个取向，每个取向可由一个磁量子数（m）表示。氢原子（^1H）

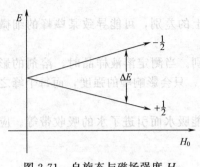

图 2-71 自旋态与磁场强度 H_0 能量差之间的相互关系

原子核的自旋量子数 $I=1/2$，因此^1H 核在磁场中只能有 $2\times\frac{1}{2}+1=2$ 个取向，即两个能级（见图 2-71）。当 $m=+1/2$ 时，自旋取向与外加磁场方向同向，能量较低；当 $m=-1/2$ 时，自旋取向与外加磁场方向相反，能量较高。两个能级之间的能级差 ΔE 与外磁场强度 H_0 成正比。

$$\Delta E=h\gamma H_0/2\pi$$

式中 γ——核的磁旋比；
h——普朗克常数；
H_0——外磁场强度。

如果在与外磁场 H_0 垂直的方向，用一定频率的电磁波作用到氢核上，当电磁波的能量 $h\nu$ 恰好等于质子两种取向的能级差 $h\nu=E_2-E_1$ 时，处于低能态 E_1 的核吸收射频能量跃迁到高能态 E_2，称为核磁共振（NMR）。所以核磁共振须满足下列条件。

$$h\nu=\Delta E=h\gamma H_0/2\pi$$

即
$$\nu=\gamma H_0/2\pi$$

式中 ν——电磁波的频率。

图 2-71 表明了自旋态与磁场强度 H_0 能量差之间的相互关系，可以看出自旋态之间的能量差与 H_0 成正比。

质子的共振频率不仅由外加磁场和核的磁旋比决定，而且还受到质子周围的分子环境的影响。在实际的分子环境中，氢核外面是被电子云所包围的，当氢核处在外加磁场中时，其外部电子在外加磁场相垂直的平面上绕核旋转的同时，将产生一个与外加磁场相对抗的附加磁场，附加磁场使外加磁场对核的作用减弱。这种核外电子削弱外加磁场对核的影响的作用称为屏蔽。若以 σ 表示屏蔽常数，外加磁场为 H_0，这个屏蔽作用的大小为 σH_0，所以核实际感受到的磁场强度不是 H_0，而是 $H=H_0-\sigma H_0$。故核磁共振的条件应表达为

$$\nu=\frac{\gamma H}{2\pi}=\frac{\gamma}{2\pi}H_0(1-\sigma)$$

σ 是核的化学环境的函数。因为各种氢核所处的化学环境不同，所以 σ 值也不同，故各种核在不同磁场强度下共振，产生了化学位移。也就是说，不同类型的质子，在外磁场的作用下其共振频率并不相同，从而导致图谱上信号的位移。由于这种位移是因质子周围的化学环境不同而引起的，故称为化学位移。由它可以了解分子的结构。化学位移用 δ（单位 10^{-6}）表

示，定义式为

$$\delta = \frac{\nu_{样品} - \nu_{标准}}{\nu_0} \times 10^6$$

式中 $\nu_{样品}$ ——样品的共振频率；

$\nu_{标准}$ ——标准物的共振频率；

ν_0 ——所用波谱仪的频率。

^{1}H NMR 测定中通常用四甲基硅烷（Tetramethylsilane，TMS）作为标准物，将它的质子共振位置定为零。由于它的屏蔽比一般的有机分子大，故大多数有机化合物中质子的共振信号出现在它的左侧。具体测定时一般把 TMS 溶于被测溶液中，称为内标法。TMS 不溶于重水，当用重水作溶剂时，将装有 TMS 的毛细管置于被测重水中测定，称为外标法。一些常见有机官能团质子的化学位移见图 2-72。

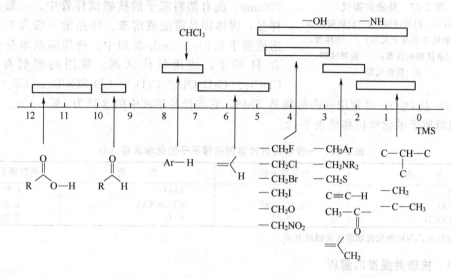

图 2-72 一些常见有机官能团质子的化学位移

影响化学位移的主要因素有相邻基团的电负性、各向异性效应、范德华效应、溶剂效应及氢键作用。表 2-11 列出了连在不同基团上氢原子的化学位移。

表 2-11 与不同基团相连氢原子的化学位移

氢原子的类型	化学位移(δ)	氢原子的类型	化学位移(δ)
TMS	0	I—C—H	2~4
环丙烷	0~1.0	HO—C—H	3.4~4
RCH$_3$	0.9	R—O—C—H	3.3~4
R$_2$CH$_2$	1.3	—(O)$_2$—C—H	5.3
R$_3$CH	1.5	R—COO—C—H	3.7~4.1
—C=C—H	4.6~5.9	RO—CO—C—H	2~2.6
—C=C—CH$_3$	1.7	HO—CO—C—H	2~2.6
—C≡C—H	2~3	R—COO—H	10.5~12
—C≡C—CH$_3$	1.8	R—CO—C—H	2~2.7
Ar—H	6~8.5	R—CO—H	9~10
Ar—C—H	2.2~3	R—CO—N—H	5~8
F—C—H	4~4.45	R—O—H	4.5~9
Cl—C—H	3~4	Ar—O—H	4~12
Cl$_2$—C—H	5.8	R—NH$_2$	1~5
Br—C—H	2.5~4	O$_2$N—C—H	4.2~4.6

2.17.2.2 仪器简介与核磁共振样品的制备

核磁共振仪根据电磁波的来源，可分为连续波和脉冲-傅里叶变换两类；如按磁场产生的方式，可分为永久磁铁、电磁铁和超导磁铁三种；也可按磁场强度不同，分为 60MHz、90MHz、100MHz、200MHz、500MHz 等多种型号，一般磁场强度越高，仪器分辨率越好。

核磁共振仪主要由磁铁、射频振荡器和线圈、扫描发生器和线圈、射频接收器和线圈以及示波器和记录仪等部件组成，如图 2-73 所示。

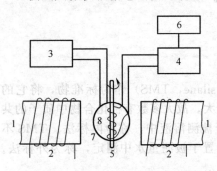

图 2-73 核磁共振仪
1—磁铁；2—扫场线圈；3—射频振荡器；
4—射频接收器及放大器；5—试样管；
6—记录仪和示波器；7—射频线圈；
8—接收线圈

用于测定核磁共振谱的有机化合物必须是纯品。在核磁共振测定时，将样品装在内径为 5mm、长为 200mm、配有塑料塞子的核磁试样管中。一般是液体样品，固体样品需配成溶液。样品量一般为 5~10mg 溶质溶于 0.5~1.0mL 溶剂中。所用溶剂本身必须不含 H 质子，或用氘代试剂。常用的溶剂有 CCl_4、$CDCl_3$、C_2D_5OD、C_6D_6、CD_3SOCD_3、CF_3COOD、CD_3COCD_3 及 D_2O。最常用的内标物是 TMS，它在样品溶液中的含量为 1‰~4‰。一些氘代溶剂残留质子的化学位移见表 2-12。

表 2-12 一些常用氘代溶剂残留质子的化学位移 (δ)

溶 剂	残留质子的 δ	溶 剂	残留质子的 δ
$CDCl_3$	7.27	D_2O	4.7①
CD_3OD	3.35,4.8①	CD_3SOCD_3	2.50
CD_3COCD_3	2.05	C_6D_6	7.20

① 变动较大，与所测化合物浓度及温度有关。

2.17.2.3 核磁共振谱的解析

核磁共振谱的解析，一般说来，首先要根据谱中所出现的信号数目确定分子中含有几种类型的质子。其次，要根据谱图中各类质子的化学位移（δ）值判断质子的类型。通过测量积分曲线的阶梯高度，以确定各类质子之间的比例。最后，观察和分析各组峰的裂分情况，根据耦合常数 J 和峰形确定彼此耦合的质子，了解邻位碳原子上氢的数目，从而可推知化合物的结构。若是已知化合物，还可以与标准的核磁共振图谱对照，以确定化合物结构是否正确、有否杂质等。对于复杂的有机化合物，很多情况下还要结合其他波谱信息及有关的物理、化学性质作综合分析才能获得正确的分子结构。

第3章 半微量有机化合物的合成实验

3.1 烷烃和烯烃的制备

脂肪烃分为三类：烷烃，烯烃和炔烃。

烷烃分子中只含有单键，因此也称之为饱和烃，它们是比较稳定的。烯烃分子中至少含有一个双键，而炔烃分子中至少含有一个叁键，因此烯烃和炔烃也称为不饱和烃。它们是比较活泼的。这三类烃与某些试剂的反应是不同的。在许多情况下可用于区别这三类化合物。下面的实验说明了典型的饱和烃与不饱和烃的一些反应。

烯烃是重要的有机化工原料。工业上主要通过石油裂解的方法制备烯烃，有时也利用醇在氧化铝等催化剂作用下，进行高温催化脱水来制取。实验室则主要用醇的脱水或卤代烃的脱卤化氢来制备烯烃。

实验1 甲烷的制备与性质

【反应式】

主反应

$$CH_3COONa + NaOH \xrightarrow{\triangle} CH_4\uparrow + Na_2CO_3$$

副反应

$$2CH_3COONa \xrightarrow{\triangle} CH_3COCH_3 + Na_2CO_3$$

【主要试剂】

无水醋酸钠 5.00g（61.0mmol），碱石灰 3.00g，氢氧化钠 2.00g（50.0mmol），1％溴的四氯化碳溶液，0.1％高锰酸钾溶液，浓硫酸，10％硫酸，庚烷。

【实验步骤】

1. 甲烷的制备

将 5g NaAc（无水）、3g 碱石灰、2g NaOH（s），放在研钵中快速研细混合均匀后用锡纸包成筒状，置于硬质大试管中，根据图 3-1 组装仪器。检查装置不漏气后，用煤气灯小火

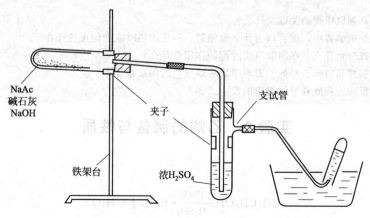

图 3-1 制备甲烷的装置图

徐徐均匀地加热整支试管，再用较大的火焰强热靠近试管口的反应物，使该处的反应物反应后，逐渐将火焰往试管底部移动。估计空气排尽后，用排水集气法收集甲烷，并做下列性质实验。

2. 性质实验

(1) 卤代反应　在装有甲烷的两支试管中各加入1%溴的四氯化碳溶液2~3滴，用胶塞塞紧，其中一支用黑布或黑纸包裹好，振荡后把包裹好的一支试管放在实验台柜内，另一支试管则放在阳光（或日光灯）下光照15~20min。试比较这两支试管中液体的颜色是否相同。有什么变化？为什么？

(2) 高锰酸钾试验　向另一支装有甲烷的试管中，加入0.1%高锰酸溶液2~3滴和10%硫酸1mL，用胶塞塞紧，振荡，观察颜色有何变化？这说明了什么问题？

取正庚烷0.5mL，按照(1)、(2)两项所列步骤进行烷烃的性质实验，并观察有何结果。

(3) 可燃性　将装有点燃实验用的甲烷气体大支管倒置，再将充满水的滴液漏斗迅速安装到大支管管口上，然后正立大支管（见图3-2），用夹子固定在铁架台上后，开启滴液漏斗的活塞和止水夹，并在试管的尖嘴处点燃气体。观察记录火焰的特征。

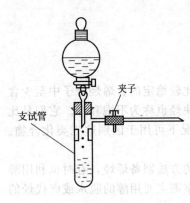

图3-2　点燃甲烷的装置图

【注意事项】

① 药品要干燥无水，操作迅速、防吸潮，装置严密。

② 锡纸要平整，没有损坏。

③ 无水醋酸钠、碱石灰和氢氧化钠的混合物要研细铺平于锡纸上；包药品时药品要放松散，不要包实；将锡纸卷成筒形，留好出气口。

④ 加热要移动，由前至后，防止集中加热，烧坏试管。

⑤ 收集完气体后，或随时中断反应前，要先将导气管移出水面，断开支试管与硬质大试管之间的导气管，才能熄火，以防倒吸。

⑥ 气体点燃前，安装滴液漏斗要迅速，滴液的同时打开螺旋夹，然后立刻点燃尖嘴处的气体。

⑦ 实验完成后，剩余物要小心倒入废液桶中。

思　考　题

1. 实验室顺利制取甲烷的关键是什么？
2. 在制取甲烷的装置中，试管口为什么要倾斜？停止加热时应如何正确操作？
3. 碱石灰的成分是什么？在制取甲烷时起的作用是什么？
4. 用锡纸包裹样品有什么好处？影响甲烷火焰颜色的因素是什么？
5. 用游离基反应历程解释甲烷与溴的反应。

实验2　乙烯的制备与性质

【反应式】

主反应

$$CH_3CH_2OH \xrightarrow[H_2SO_4]{170℃} C_2H_4 \uparrow + H_2O$$

副反应

$$CH_3CH_2OH + H_2SO_4 \begin{cases} CO\uparrow + CO_2\uparrow + SO_2\uparrow + H_2O \\ CH_3COOH + SO_2\uparrow + H_2O \end{cases}$$

【主要试剂】

95%乙醇 4.00mL（68.5mmol），浓硫酸 12.00mL（225mmol），1%溴的四氯化碳溶液，0.1%高锰酸钾溶液，10%硫酸，10%氢氧化钠溶液。

【实验步骤】

1. 乙烯的制备

在125mL干燥的蒸馏烧瓶中加入4mL 95%乙醇、12mL浓硫酸（边加边摇边冷却）和几粒沸石，塞上带有温度计（300℃）的塞子。根据图3-3（a）组装仪器。检查装置不漏气后，强热反应物，使反应物的温度迅速地上升到160～170℃，调节火焰，保持此范围的温度，使乙烯气流均匀地发生，待空气被排尽后，作下列性质实验。

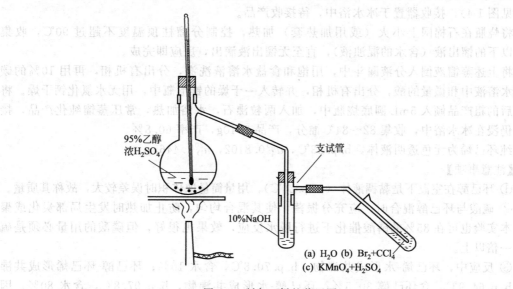

图 3-3　制备乙烯的装置图

2. 性质实验

（1）加成反应　在一个干燥的试管中加入1%溴的四氯化碳溶液0.5mL，将制得的乙烯气体通入试管中［见图3-3(b)］，并观察有何现象发生。

（2）氧化反应　在一小试管中加入0.1%高锰酸钾溶液0.1mL和10%硫酸2mL，振摇使混合均匀后，将制得的乙烯气体通入试管中［见图3-3(c)］，并观察有何现象发生。

（3）可燃性　在制备乙烯装置的尖嘴处点燃气体，观察燃烧情况，同时与甲烷燃烧情况作比较，并说明原因。

【注意事项】

① 浓硫酸加入乙醇中时，要边加边振摇边冷却。
② 反应温度要迅速上升，超过140℃以上，并保持温度在170℃左右。
③ 实验完成后要及时断开烧瓶与洗气瓶之间的导管，然后再熄火，以防倒吸。
④ 实验完成后，剩余物要小心倒入废液桶中。

思　考　题

1. 制备乙烯时温度为什么要迅速上升到160～170℃，如果不迅速升高温度结果如何？

2. 制备乙烯时有哪些杂质生成？它们分别在装置中哪一部分被除去？
3. 乙烯的火焰与甲烷的火焰有何不同？

实验3 环己烯的合成

【反应式】

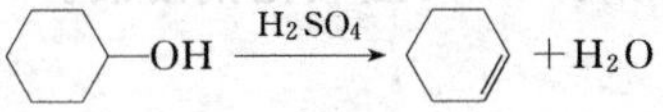

$$\text{C}_6\text{H}_{11}\text{OH} \xrightarrow{\text{H}_2\text{SO}_4} \text{C}_6\text{H}_{10} + \text{H}_2\text{O}$$

【主要试剂】

环己醇 5mL（4.81g，48mmol），浓硫酸 0.5mL。

【实验步骤】

在 10mL 干燥的圆底烧瓶中加入 4.81g 环己醇，慢慢滴加 0.5mL 浓硫酸，边加边摇边冷却烧瓶，使其混合均匀，加入两粒沸石。在烧瓶上口安装好分馏柱，分馏柱接普通蒸馏装置（见图 1-4），接收器置于冰水浴中，待接收产品。

将烧瓶在石棉网上小火（或用加热套）加热，控制分馏柱顶温度不超过 90℃，收集 85℃以下的馏出液（含水的混浊液），直至无馏出液滴出，反应即完成。

将上述蒸馏液倒入分液漏斗中，用饱和食盐水溶液洗涤，分出有机相，再用 10%的碳酸钠水溶液中和微量的酸，分出有机相，并转入一干燥的锥形瓶中，用无水氯化钙干燥。将干燥后的粗产品倾入 5mL 圆底烧瓶中，加入两粒沸石，水浴加热，常压蒸馏纯化产品，接收器仍浸在冰水浴中，收集 82～84℃馏分，产品 2.40g，产率 60.8%。

纯环己烯为无色透明液体，b. p. 83℃，d_4^{20} 0.8102，n_D^{20} 1.4465。

【注意事项】

① 环己醇在室温下是黏稠液体（m. p. 24℃），用量筒量取体积时误差较大，故称其质量。

② 硫酸与环己醇混合时，应充分振荡，使其混合均匀，防止加热时发生局部炭化或聚合。本实验也可在 85%的磷酸催化下进行脱水反应，效果也很好，但磷酸的用量必须是硫酸的一倍以上。

③ 反应中，环己烯-水形成共沸物，b. p. 70.8℃，含水 10%；环己醇-环己烯形成共沸物，b. p. 64.9℃，含环己醇 30.5%；环己醇-水形成共沸物，b. p. 97.8℃，含水 80%。因此，加热时温度不可过高，蒸馏速度不易过快，以 2～3s 一滴为宜，以尽量避免未反应的环己醇被蒸出。

④ 在收集和转移环己烯时，应保持其充分冷却，以避免因挥发而造成的损失。

⑤ 在蒸馏已干燥的产物时，所用仪器均需干燥无水。

思 考 题

1. 本实验采取什么措施提高产率？哪一步骤操作不当会降低收率？本实验的操作关键是什么？
2. 在粗产品环己烯中，加入氯化钠饱和水溶液于馏出液中的目的是什么？
3. 用无水氯化钙作干燥剂有何优点？
4. 反应时柱顶温度控制在何值最佳？
5. 试写出环己醇在酸催化下脱水形成环己烯的反应机理。

3.2 卤代烃的制备

卤代烃是一类重要的有机合成中间体。通过卤代烃的取代反应，能制备多种有用的化合物，如腈、胺、醚等。在无水乙醚中，卤代烃和镁作用生成 Grignard 试剂 RMgX，后者和羰基化合物如醛、酮、二氧化碳等作用，可制取醇和羧酸。由醇与氢卤酸反应制备卤代烃，是卤代烃制备中的一个重要方法。

实验 4　溴乙烷的合成

【反应式】
主反应
$$NaBr + H_2SO_4 \longrightarrow HBr + NaHSO_4$$
$$C_2H_5OH + HBr \rightleftharpoons C_2H_5Br + H_2O$$

副反应
$$2C_2H_5OH \xrightarrow{H_2SO_4} C_2H_5OC_2H_5 + H_2O$$
$$C_2H_5OH \xrightarrow{H_2SO_4} CH_2=CH_2 + H_2O$$
$$2HBr + H_2SO_4(浓) \longrightarrow Br_2 + SO_2 + 2H_2O$$

【主要试剂】

溴化钠 5.00g (48.6mmol)，95％乙醇 3.3mL (2.60g, 56.5mmol)，浓硫酸 6.3mL (11.60g, 116mmol)。

【实验步骤】

在 30mL 圆底烧瓶中加入 3.3mL 95％乙醇及 3mL 水，在不断振摇和冷水冷却下，慢慢加入 6.3mL 浓硫酸。混合物冷至室温后，在冷却下加入 5.00g 研细的溴化钠，稍加振摇混合后，加入几粒沸石，按图 1-3 安装成常压蒸馏装置。接收器内放入少量冰水，并将其置于冰水浴中，以防止产品的挥发损失。接引管的支管用橡皮管导入下水道或室外（为什么?）。通过石棉网用小火（或用加热套）加热烧瓶，使反应平稳进行，直至无油状物馏出为止（约 40min）。

将馏出液倒入分液漏斗中，将有机层（哪一层?）转移至一个干燥的锥形瓶中，并将其浸在冰水浴中冷却，边振摇边滴加浓硫酸，以除去乙醚、乙醇、水等杂质，直至溶液有明显的分层为止。再用干燥的分液漏斗分去硫酸层（哪一层?）。将溴乙烷粗产品转入干燥的蒸馏瓶中（如何转入?），加入沸石，水浴加热蒸馏。将已称量的干燥的接收器外用冰水浴冷却，收集 37~40℃的馏分，产品 3.20g，产率 60.4％。

纯溴乙烷的 b.p. 38.4℃，d_4^{20} 1.4604，n_D^{20} 1.4239。

【注意事项】

① 加少量的水可以防止反应进行时产生大量泡沫，减少副产物乙醚的生成和避免氢溴酸的挥发。

② 反应开始时会产生大量的气泡，故应严格控制反应温度，使其平稳地进行，防止反应物冲出蒸馏瓶。溴乙烷沸点较低，蒸馏时一定要慢慢加热，否则蒸气来不及冷却而逸失，造成产品损失。

③ 为防止产物挥发，在接收粗产品时接收器内外应用冰水冷却。

④ 在产品纯化时，应尽可能将水分除净，否则当用浓硫酸洗涤时，由于放热，会使产品挥发损失。

⑤ 浓硫酸用来除去乙醚、乙醇、水等杂质，当洗涤不够时，馏分中仍可能含极少量水及乙醇，它们与溴乙烷分别形成共沸物（溴乙烷-水，b.p. 37℃，含水约 1％；溴乙烷-乙醇，b.p. 37℃，含醇 3％）。

思　考　题

1. 造成本实验产率不高的主要原因是什么?

2. 本实验根据哪种原料计算产率？
3. 粗产品中有哪些杂质？如何去除？
4. 为减少溴乙烷的挥发损失，以提高产率，本实验中采取了哪些措施？
5. 试写出生成溴乙烷的反应机理。

实验5　正溴丁烷的合成

【反应式】
主反应

$$NaBr + H_2SO_4 \longrightarrow HBr + NaHSO_4$$
$$n\text{-}C_4H_9OH + HBr \rightleftharpoons n\text{-}C_4H_9Br + H_2O$$

副反应

$$2C_4H_9OH \xrightarrow{\text{浓}H_2SO_4} C_4H_9OC_4H_9 + H_2O$$
$$2C_4H_9OH \xrightarrow[\triangle]{\text{浓}H_2SO_4} CH_3CH_2CH=CH_2 + CH_3CH=CHCH_3 + 2H_2O$$

【主要试剂】
溴化钠 3.33g（32.4mmol），正丁醇 2.5mL（2.02g，27.3mmol），浓硫酸 4.0mL（7.36g，73.6mmol）。

【实验步骤】
在30mL圆底烧瓶中加入3.3mL水，慢慢地加入4.0mL浓硫酸，混合均匀并冷却至室温。加入2.5mL正丁醇及3.33g溴化钠，振摇后，加入几粒沸石，装上回流冷凝管，冷凝管上端接溴化氢吸收装置［见图1-5(c)，注意：使漏斗口恰好接触水面，切勿浸入水中，以免倒吸］，用5%氢氧化钠溶液作吸收剂。

将烧瓶在石棉网上小火（或用加热套）加热回流0.5h，回流过程中适当摇动烧瓶，以使反应物充分接触。反应完毕，稍冷却后改为蒸馏装置，蒸出正溴丁烷粗品，至馏出液清亮为止。

用毛细滴管将馏出液移入分液漏斗中，加入3.3mL水洗涤，将有机层（哪一层？）转入到另一干燥的分液漏斗中，用2.0mL浓硫酸洗涤一次，分出硫酸层（哪一层？）。有机层再依次用水、饱和碳酸氢钠溶液及水各5mL洗涤。将正溴丁烷移入干燥的锥形瓶中，用无水氯化钙干燥，间歇摇动锥形瓶，直至液体透明。将干燥后的产物小心地转入到干燥的蒸馏烧瓶中。在石棉网上加热蒸馏，收集99～103℃的馏分。产品1.65g，产率44.1%。测定折射率及红外光谱。

纯正溴丁烷为无色透明液体，b.p. 101.6℃，d_4^{20} 1.2760，n_D^{20} 1.4399。

纯正溴丁烷的红外光谱见图3-4。

【注意事项】
① 正溴丁烷粗品是否蒸完，可从以下三种方法来判断：第一，馏出液是否由浑浊变为澄清；第二，蒸馏瓶中上层的油层是否消失；第三，取一支试管收集几滴馏出液，加入少许水摇动，如无油珠出现，说明正溴丁烷已蒸完。

② 分液时，根据液体的密度来判断产物在上层还是下层，如果一时难以判断，应将两相全部留下来。

③ 用水洗涤产物后，如有红色，说明含有溴，是由浓硫酸氧化生成的，可用少量饱和亚硫酸氢钠溶液洗去。

$$2NaBr + 3H_2SO_4(\text{浓}) \longrightarrow Br_2 + SO_2 + 2NaHSO_4 + 2H_2O$$
$$Br_2 + 3NaHSO_3 \longrightarrow 2NaBr + NaHSO_4 + 2SO_2 + H_2O$$

④ 浓硫酸可溶解粗产物中少量的正丁醇、正丁基醚及丁烯，所以应使用干燥的分液漏

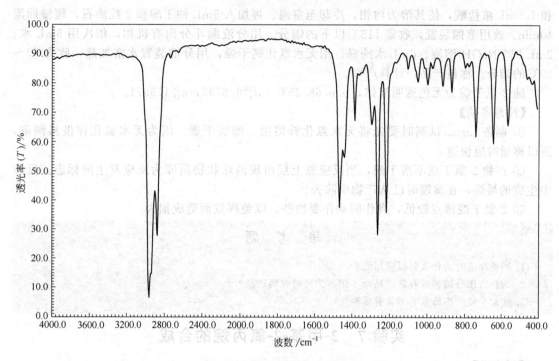

图 3-4 正溴丁烷的红外光谱图

斗,以防漏斗中残余水分冲稀硫酸而降低洗涤效果。正丁醇与正溴丁烷可以形成共沸物(b.p.98.6℃,含质量分数为13%的正丁醇),蒸馏时很难除去。因此在用浓硫酸洗涤时,应充分振荡。分液时硫酸应尽量分干净。

⑤ 各次水洗时应注意观察水面上是否有悬浮的油状产物。如有,则在放下油层后,轻轻旋转分液漏斗促使悬浮物离心下沉,再将其放出,并入有机层,以减少损失。

思 考 题

1. 本实验中,浓硫酸起何作用?其用量及浓度对实验有何影响?
2. 本实验可能有哪些副反应?如何减少副反应的发生?
3. 加料时,如不按实验操作中的加料顺序,如先使溴化钠与浓硫酸混合然后再加正丁醇和水,将会出现什么现象?
4. 从反应混合物中分离出粗产品正溴丁烷时,为什么用蒸馏的方法,而不直接用分液漏斗分离?
5. 后处理时,各步洗涤的目的是什么?为什么用饱和碳酸氢钠水溶液洗涤之前,要先用水洗涤?
6. 为什么蒸馏前一定要把无水氯化钙干燥剂除去?

实验 6 2-氯丁烷的合成

【反应式】

$$\underset{\underset{OH}{|}}{C_2H_5CHCH_3} + HCl \xrightarrow{ZnCl_2} \underset{\underset{Cl}{|}}{C_2H_5CHCH_3} + H_2O$$

【主要试剂】

无水氯化锌 16g (117.4mmol),仲丁醇 5mL (4.03g,54.4mmol),浓盐酸 7.5mL (约 9g,90mmol)。

【实验步骤】

在 50mL 圆底烧瓶上装好回流冷凝器及气体吸收装置,向反应瓶中加入 16g 无水氯化锌和 7.5mL 浓盐酸，使其溶为均相，冷却至室温。再加入 5mL 仲丁醇和 2 粒沸石，缓慢回流 40min。改用蒸馏装置，收集 115℃ 以下的馏分。用分液漏斗分出有机相，依次用 6mL 水、2mL 5% NaOH 溶液、6mL 水洗涤。用无水氯化钙干燥，用分馏装置水浴加热，收集 67~69℃ 的馏分。称量产品，计算产率。

纯 2-氯丁烷为无色透明液体，b.p. 68.25℃，d_4^{20} 0.8732，n_D^{20} 1.3971。

【注意事项】

① 制备 Lucas 试剂时要先将无水氯化锌熔融、彻底干燥。因为无水氯化锌极易潮解，所以称量时应快速。

② 产物 2-氯丁烷不溶于酸，当反应瓶上层出现油珠状物质即为反应发生的标志。反应中生成的烯烃，在蒸馏时已从产物中除去。

③ 2-氯丁烷沸点较低，操作时动作要快些，以免挥发而造成损失。

思 考 题

1. 回流反应时为什么要缓慢回流？
2. 为什么用分馏装置收集产品而不用蒸馏装置收集产品？
3. 实验中使产率降低的因素有哪些？

实验 7　2-甲基-2-氯丙烷的合成

【反应式】

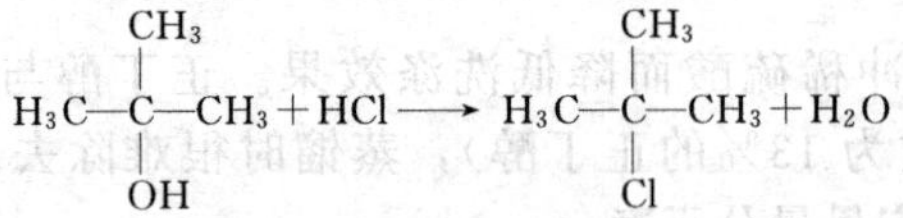

$$H_3C-\underset{\underset{OH}{|}}{\overset{\overset{CH_3}{|}}{C}}-CH_3 + HCl \longrightarrow H_3C-\underset{\underset{Cl}{|}}{\overset{\overset{CH_3}{|}}{C}}-CH_3 + H_2O$$

【主要试剂】

叔丁醇 3.93mL (3.1g, 42mmol)，浓盐酸 10.5mL (12.39g, 126mmol)。

【实验步骤】

在 50mL 分液漏斗中加入 3.93mL 叔丁醇和 10.5mL 浓盐酸，不断振摇 10~15min 后，静止。待明显分层后，分去水层。有机层分别用水、5%碳酸氢钠和水各 3mL 洗涤。产品用无水氯化钙干燥后转入蒸馏烧瓶中，加入沸石，接收瓶置于冰水浴中。在水浴上蒸馏收集 50~51℃ 的馏分。称量产品，计算产率。

纯 2-甲基-2-氯丙烷为无色透明液体，b.p. 52℃，d_4^{20} 0.847，n_D^{20} 1.3877。

【注意事项】

① 叔丁醇凝固点为 25℃，温度较低时呈固态，需在温热水浴中熔化后取用。

② 盐酸倒入含醇的分液漏斗中后，盖好塞子，静止放置分层后再用力振摇，振摇时注意放气。

③ 产物沸点较低，操作时动作要快些，以免挥发而造成损失。

思 考 题

1. 产率较低的原因是什么？
2. 在洗涤粗产物时，碳酸氢钠溶液浓度过高，洗涤时间过长，将对产物有何影响？

3.3　醚的制备

醚的制备方法很多，如醇脱水、硫酸二烷基酯和酚盐作用、威廉姆逊反应（Williamson

Reaction)等。醇脱水制取醚的方法常用于制备单醚（也称对称醚），如甲醚、乙醚等。若用两种不同的醇经脱水制备混合醚（也称不对称醚），则会生成多种醚的混合物，分离较困难。除了制取芳基烷基醚外，一般很少用脱水法制备混合醚。由于硫酸二甲酯、硫酸二乙酯等烷基化试剂毒性很大，因而采用威廉姆逊反应制备混合醚最常见。威廉姆逊合成法利用醇（酚）钠与卤代烃作用，它既可以合成单醚，也可以合成不对称醚，主要用于合成不对称醚，特别是制备芳基烷基醚时产率最高。这种合成方法的反应机理是烷氧（或酚氧）负离子对卤代烷或硫酸酯的亲核取代反应（即 S_N2 反应）。

醚一般为无色、易挥发、易燃、易爆的液体。由于醚的自氧化作用，会生成少量遇振或受热易爆炸的过氧化物，故在使用久存的醚时，必须先检验有无过氧化物。

醚能溶解多数的有机化合物，有些有机反应必须在醚中进行，如 Grignard 反应，因此醚是有机合成中常用的溶剂。

实验 8　乙醚的合成

【反应式】

$$CH_3CH_2OH + H_2SO_4 \xrightleftharpoons{100\sim130℃} CH_3CH_2OSO_2OH + H_2O$$

$$CH_3CH_2OSO_2OH + CH_3CH_2OH \xrightleftharpoons{135\sim145℃} CH_3CH_2OCH_2CH_3 + H_2SO_4$$

总反应

$$2CH_3CH_2OH \xrightleftharpoons[H_2SO_4]{140℃} CH_3CH_2OCH_2CH_3 + H_2O$$

副反应

$$CH_3CH_2OH \xrightarrow{H_2SO_4} \begin{array}{l} \xrightarrow{170℃} H_2C=CH_2 + H_2O \\ \xrightarrow{[O]} CH_3CHO + SO_2 + H_2O \end{array}$$

$$CH_3CHO \xrightarrow{H_2SO_4} CH_3COOH + SO_2 + H_2O$$

$$SO_2 + H_2O \longrightarrow H_2SO_3$$

【主要试剂】

95%乙醇 18.5mL（317mmol），浓硫酸 6mL（108mmol），5%氢氧化钠溶液，饱和氯化钠溶液，饱和氯化钙溶液，无水氯化钙。

【实验步骤】

在干燥的 50mL 三口瓶中，放入 6mL 95%乙醇，在冷水浴冷却下，边摇动烧瓶边缓缓加入 6mL 浓硫酸，混合均匀，并加入几粒沸石。按图 3-5 安装仪器。滴液漏斗内盛有 12.5mL 95%乙醇，漏斗脚下端和温度计的水银球必须浸入液面以下，距离瓶底约 0.5~1cm 处。用作接收器的烧瓶应置于冰水浴中冷却，接引管的支管接上橡皮管通入下水道或室外。

将反应瓶放在石棉网上（或加热套中）加热，使反应液温度较迅速地上升到 140℃。开始由滴液漏斗慢慢滴加乙醇，控制滴入速度与馏出液速度大致相等（1 滴/s），并维持反应温度在 135~145℃，约 30~45min 滴加完毕，再继续加热 10min，当温度上升到 160℃ 时，撤掉热源，停止反应。

将馏出液转入分液漏斗中，依次用 6mL 5%氢氧化钠溶液、6mL 饱和氯化钠溶液洗涤，最后用 6mL 饱和氯化钙溶液洗涤 2 次。

分出醚层，用无水氯化钙干燥（注意：容器外仍需用冰水冷却）。当瓶内乙醚澄清时，将其小心地转入蒸馏烧瓶中，加入沸石，按图 1-3（b）的蒸馏装置，在预热过的热水浴上（50℃左右）蒸馏，收集 33~38℃ 的馏分，产量 3.5~4.5g，产率约 35%。

纯乙醚为无色透明液体，b.p. 34.5℃，d_4^{20} 0.7137，n_D^{20} 1.3526。

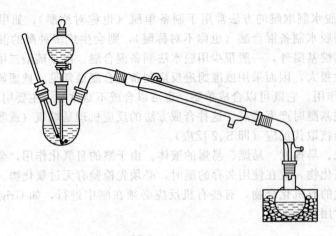

图 3-5 乙醚制备装置

【注意事项】

① 装置要严密。

② 乙醚很易挥发并容易着火。乙醚蒸气与空气的混合物在一定比例范围内遇火易爆炸，故需要特别注意安全。反应结束后要先停止加热，冷却后再拆下接收器。精制乙醚时，实验台附近严禁明火。

③ 精制乙醚时的热水浴必须在别处预先准备好热水（或用恒温水浴锅），使其达到所需温度，切勿一边用明火加热一边蒸馏。

④ 控制好反应温度及滴加乙醇的速度（1 滴/s），若滴加速度明显超过馏出速度，不仅乙醇未作用就被蒸出，而且会使反应液的温度骤降，减少醚的生成。

⑤ 洗涤时注意洗涤顺序，并且室内要无明火。

⑥ 氢氧化钠洗涤后，常会使醚层碱性太强，若接下来直接用氯化钙溶液洗涤，将会有氢氧化钙沉淀析出。为减少乙醚在水中的溶解度，在用氯化钙洗涤以前先用饱和氯化钠洗，这样既可以洗去残留的碱，还可以除去部分乙醇。

⑦ 每次洗涤后分液要彻底。最后用无水氯化钙干燥 30min 左右。氯化钙和乙醇能形成复合物 $CaCl_2 \cdot 4CH_3CH_2OH$，因此未作用的乙醇可以被除去。

⑧ 乙醚与水形成共沸物（b.p. 34.15℃，含水 1.26%），馏分中还含有少量乙醇，故沸程较长。

思 考 题

1. 本实验中，要把混在粗制乙醚里的杂质一一除去，应采用哪些措施？
2. 反应温度过高或过低对反应有何影响？
3. 为什么温度计的水银球及滴液漏斗的末端均应浸于反应液中？
4. 为什么在开始时乙醇和浓硫酸的混合液加热到乙醇沸点以上，乙醇不会蒸出来？
5. 为什么要用无水氯化钙作干燥剂？干燥时间过短有什么影响？
6. 洗涤粗乙醚时为什么要用饱和食盐水，而不用清水？

实验9 正丁醚的合成

【反应式】
主反应

$$2CH_3CH_2CH_2CH_2OH \xrightleftharpoons[134\sim135℃]{H_2SO_4} (CH_3CH_2CH_2CH_2)_2O + H_2O$$

副反应

$$CH_3CH_2CH_2CH_2OH \xrightleftharpoons[>135℃]{H_2SO_4} CH_3CH_2CH=CH_2 + H_2O$$

【主要试剂】

正丁醇 5.2mL（4.21g，56.8mmol），浓硫酸 0.8mL（1.47g，15mmol），50%硫酸，无水氯化钙。

【实验步骤】

在干燥的 50mL 两颈瓶中加入 5.2mL 正丁醇，加入 0.8mL 浓硫酸，边加边摇边冷却，混合均匀，加入两粒沸石。按装置图 3-6，一瓶口装上温度计，温度计的水银球必须浸入液面以下。另一瓶口装上分水器，分水器上端接回流冷凝管。先在分水器内加入 (V−0.7)mL 水（V 为分水器的容积），然后将烧瓶在石棉网（或加热套）上先用小火加热 10min，使瓶内液体微沸，但不到回流温度（约 100～115℃），然后加热保持回流。随着反应的进行，分水器中液面不断增高，反应液的温度也不断上升。分水器中液面增高是由于反应生成的水以及未反应的正丁醇经冷凝管冷凝后聚集于分水器内，因相对密度不同，水在下层，而较水轻的有机相浮于上层，积至分水器支管时即可返回到反应瓶中，继续加热至瓶内温度升高到 134～135℃。待分水器全部被水充满时，表示反应已基本完成（约需 1h），停止加热。

反应物冷却后，把混合物连同分水器里的水一起倒入内盛 8mL 水的分液漏斗中，充分振摇，静止后，分出产物粗制正丁醚。用 50% H_2SO_4 洗涤两次（3mL×2），再用 5mL 水洗涤一次。分出有机层，用无水氯化钙干燥产品。

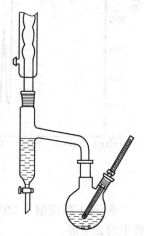

图 3-6　正丁醚制备装置

将干燥后的粗产品倒入圆底烧瓶中蒸馏（注意：不要把氯化钙倒入瓶中），收集 139～142℃的馏分。称量产品，计算产率，测定折射率及红外光谱。

纯正丁醚为无色透明液体，b.p. 142℃，d_4^{20} 0.773，n_D^{20} 1.3992。

纯正丁醚的红外光谱见图 3-7。

【注意事项】

① 加料时，正丁醇和浓硫酸如不充分摇动混匀，硫酸局部过浓，加热后易使反应溶液变黑。

② 按反应式计算，生成水的量约为 0.5g 左右，但是实际分出水的体积要略大于理论计算量，因为有单分子脱水的副产物生成。所以，在实验以前预先在分水器中加入 (V−0.7)mL 水，加上反应生成的水正好充满分水器，而使气化冷凝后的醇正好溢流返回反应瓶中，从而达到自动分离的目的。

③ 本实验通过恒沸混合物蒸馏方法，采用分水器将反应生成的水不断从反应中除去。在反应液中，正丁醚和水形成恒沸物，b.p. 94.1℃，含水 33.4%。正丁醇和水形成恒沸物，b.p. 93℃，含水 45.5%。正丁醚和正丁醇形成二元恒沸物，b.p. 117.6℃，含正丁醇 82.5%。此外，正丁醚还能和正丁醇、水形成三元恒沸物，b.p. 90.6℃，含正丁醇 34.6%，含水 29.9%。这些含水的恒沸物冷凝后，在分水器中分层，上层主要是正丁醇和正丁醚，下层主要是水。利用分水器可以使分水器中上层的有机物回流到反应器中。

④ 反应开始回流时，因为有恒沸物的存在，温度不可能马上达到 135℃。但随着水被蒸出，温度逐渐升高，最后达到 135℃以上，即应停止加热。如果温度升得太高，反应溶液会炭化变黑，并有大量副产物丁烯生成。

⑤ 50%硫酸的配制方法：20mL 浓硫酸缓慢加入到 34mL 水中。

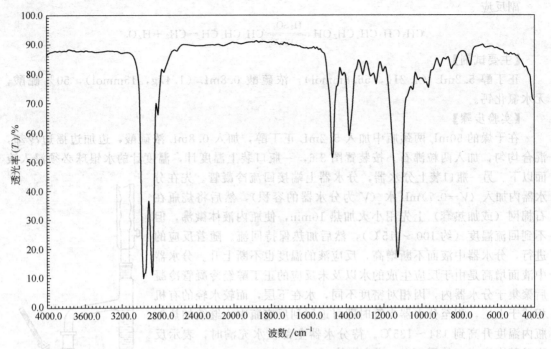

图 3-7　正丁醚的红外光谱图

⑥ 正丁醇能溶于 50%硫酸，而正丁醚溶解很少。因此，用 50%硫酸可以除去粗制正丁醚中的正丁醇。

思 考 题

1. 计算理论上分出的水量。若实验中分出的水量超过理论数值，试分析其原因。
2. 怎样得知反应已经比较完全了？
3. 反应结束后为什么要将混合物倒入 8mL 水中？各步洗涤的目的是什么？

实验 10　苯乙醚的合成

【反应式】

$$PhOH + NaOH \longrightarrow PhONa + H_2O$$
$$PhONa + CH_3CH_2Br \longrightarrow PhOCH_2CH_3 + NaBr$$

【主要试剂】

苯酚 3.75g（39.8mmol），氢氧化钠 2.00g（50mmol），溴乙烷 4.25mL（6.21g，57mmol），乙醚，无水氯化钙，饱和食盐水。

【实验步骤】

在 50mL 三口瓶上，装上电动搅拌器、回流冷凝管和恒压滴液漏斗。将 3.75g 苯酚、2g 氢氧化钠和 2mL 水加入瓶中，开动搅拌器，使固体全部溶解。用油浴加热反应瓶，控制油浴温度在 80~90℃之间，并开始慢慢滴加 4.25mL 溴乙烷，大约 40min 可滴加完毕，然后继续保温搅拌 1h，并降至室温。加入适量水（5~10mL）使固体全部溶解。将液体转入到分液漏斗中，分出水相，水相用 4mL 乙醚萃取一次，合并有机相。有机相用等体积饱和食盐水洗两次，分出有机相，并将水相每次都用 3mL 乙醚萃取一次，合并有机相。有机相用无水氯化钙干燥。先用水浴蒸出乙醚，然后常压蒸馏，收集 168~171℃的馏分。

纯苯乙醚为无色透明液体，b.p. 170℃，d_4^{15} 0.9666，n_D^{20} 1.5076。

【注意事项】
① 溴乙烷沸点低，实验时回流冷却水流量要大，以便保证有足够量的溴乙烷参与反应。
② 蒸去乙醚时不能用明火加热，将尾气通入下水道，以防乙醚蒸气外漏引起着火。

思 考 题

1. 制备苯乙醚时，萃取时用饱和食盐水洗涤的目的是什么？
2. 反应过程中，回流的液体是什么？出现的固体又是什么？为什么恒温到后期回流就不明显了？

实验 11　β-萘乙醚的合成

β-萘乙醚又称橙花醚，是一种合成香料，其稀溶液具有类似橙花和洋槐花的香味，并伴有甜味和草莓、菠萝的芳香。β-萘乙醚呈无色片状结晶，若将其加入到一些易挥发的香料中，便会减慢这些香料的挥发速度（具有这种性质的化合物称为定香剂），因而它较广泛地用于肥皂中作为香料和用作其他香料（如玫瑰香、薰衣草香、柠檬香等）的定香剂。β-萘乙醚属芳基烷基混合醚，若以硫酸脱水法制 β-萘乙醚，反应中的副产物有乙醚，若反应温度偏高，还会产生乙烯。由于这些副产物都是低沸点化合物，易于分离。因此，β-萘乙醚类芳基烷基混合醚，既可以用威廉姆逊（Williamson）反应来制备，也可采用硫酸脱水法来合成。

【反应式】

方法 1　硫酸脱水法合成

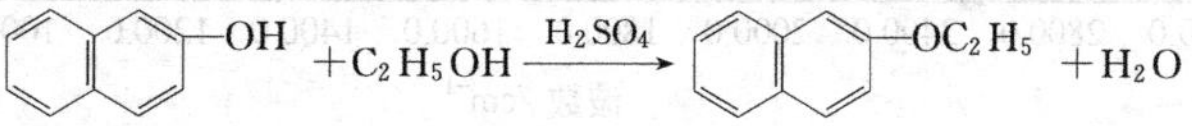

萘-OH + C_2H_5OH $\xrightarrow{H_2SO_4}$ 萘-OC_2H_5 + H_2O

方法 2　Williamson 法合成

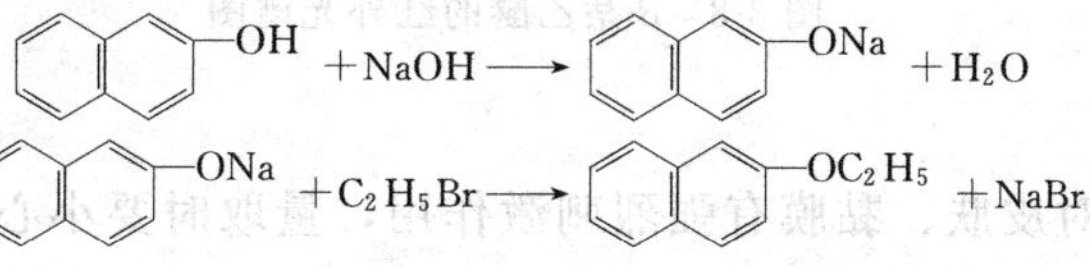

萘-OH + NaOH ⟶ 萘-ONa + H_2O

萘-ONa + C_2H_5Br ⟶ 萘-OC_2H_5 + NaBr

【主要试剂】

方法 1　β-萘酚 2.9g（20mmol），无水乙醇 6mL（4.74g，102.8mmol），浓硫酸 1mL（1.84g，18.4mmol），5%氢氧化钠水溶液 20mL，95%乙醇 10mL。

方法 2　β-萘酚 2.9g（20mmol），无水乙醇 20mL，氢氧化钠 0.9g（23mmol），溴乙烷 2.1g（1.44mL，19.3mmol）。

【实验步骤】

方法 1　在 25mL 圆底烧瓶中，加入 2.9g β-萘酚和 6mL 无水乙醇，然后在振摇下小心地加入 1mL 浓硫酸，使其混合均匀。加入两粒沸石，装上回流冷凝管，在 120℃ 的油浴中加热回流 2h。反应结束后，将反应液倒入盛有 30mL 冰水的烧杯中，有晶体析出。倾去水层，将固体物研细，再分别用 7～8mL 5%氢氧化钠水溶液和水洗涤（10mL×2），每次洗涤都要用玻璃棒充分搅拌，用倾倒法分除洗涤液，即得 β-萘乙醚粗品。干燥后称量、测熔点并计算产率。粗产物可用 95%乙醇重结晶。

方法 2　在 50mL 干燥的圆底烧瓶中，加入 2.9g β-萘酚、20mL 无水乙醇和 0.9g 氢氧化钠，在室温下超声波辐射振荡使固体溶解（约 5min）。然后加入 1.5mL 溴乙烷，摇匀，加入两粒沸石，并配置回流冷凝管，采用加热套加热，回流反应 2h。反应结束后，改回流装置为蒸馏装置，蒸馏回收大部分乙醇。将瓶内残余物趁热倒入 30mL 冰水混合物中（边倒边搅拌），彻底冷却后，倾去水层，粗产物用水洗涤两次，抽滤，用少量冷水洗涤固体，得到粗产品。粗产物可用 95%乙醇重结晶，得到白色片状晶体。干燥后称量、测熔点并计算产率,测定红外光谱。

纯品 β-萘乙醚为白色片状结晶,m.p. 37～38℃,b.p. 281～282℃。

纯 β-萘乙醚的红外光谱见图 3-8。

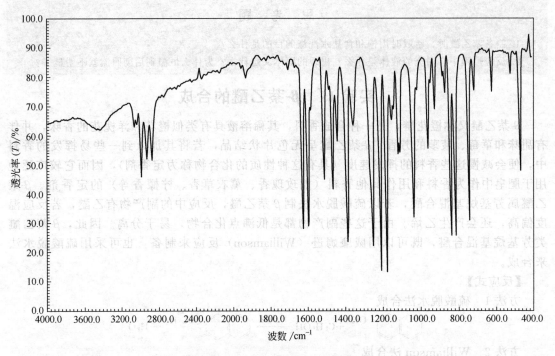

图 3-8　β-萘乙醚的红外光谱图

【注意事项】

① β-萘酚有毒,对皮肤、黏膜有强烈刺激作用,量取时要小心。若触及皮肤,应立即用肥皂清洗。

② 浓硫酸有强腐蚀性,若不慎溅到皮肤上,马上用水冲洗。

③ 氢氧化钠有强腐蚀性,若沾到皮肤上,应用水冲洗。

④ 用 Williamson 法合成时,使用溴乙烷应小心,因溴乙烷蒸气具有麻醉性,能刺激眼睛和呼吸系统。溴乙烷沸点低,实验时回流冷却水流量要大,以保证有足够量的溴乙烷参与反应。回流时烧瓶内有固体析出。

> **思　考　题**
>
> 1. 以硫酸脱水法制取 β-萘乙醚会产生哪些副产物?这些副产物对产物的精制有无影响?为什么要用稀的氢氧化钠水溶液对粗产物进行处理?
> 2. 用 Williamson 法合成 β-萘乙醚时,为什么不以 2-溴萘和乙醇作为原料?
> 3. 在制取 β-萘酚盐时,为什么用氢氧化钠的乙醇溶液,而不用氢氧化钠水溶液?
> 4. 在后处理时,如果不先蒸除乙醇而直接将反应混合物倒入水中,对实验结果会有何影响?

3.4　傅-克反应

傅-克反应(Friedel-Crafts Reaction)是指某些芳香族化合物在无水三氯化铝等 Lewis 酸催化剂存在下,同酰氯、酸酐或卤代烷等作用,芳环上的氢原子被酰基或烷基取代的反应。前者称为傅-克酰基化反应，后者称为傅-克烷基化反应。

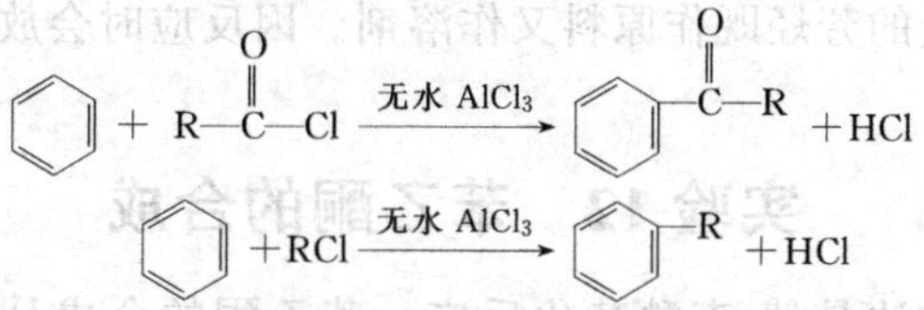

$$\text{C}_6\text{H}_5\text{H} + \text{R-CO-Cl} \xrightarrow{\text{无水 AlCl}_3} \text{C}_6\text{H}_5\text{-CO-R} + \text{HCl}$$

$$\text{C}_6\text{H}_5\text{H} + \text{RCl} \xrightarrow{\text{无水 AlCl}_3} \text{C}_6\text{H}_5\text{-R} + \text{HCl}$$

芳烃的烷基化反应和酰基化反应被用来制备烷基芳烃和芳酮。

在烷基化反应中，除了卤代烃外，烯烃以及醇类也可作烷基化试剂。若以卤代烃作为烷基化试剂，使用的催化剂除常用的无水三氯化铝外，还有无水氯化锌、三氟化硼等 Lewis 酸，其中以无水三氯化铝催化效果最好，并且仅用催化剂量的催化剂。若以烯烃或醇作为烷基化试剂，一般用质子酸作催化剂，如氟化氢、硫酸及磷酸。

对于烷基化反应，由于烷基对芳烃具有活化作用，反应并不停止在一烷基化阶段，因为生成的烷基芳烃比原料芳烃更容易发生烷基化，因而还可以生成多烷基取代的芳烃。此外，芳烃的烷基化反应是通过烷基正碳离子的形成与进攻而发生的，因而会生成重排产物。这些导致烷基化反应应用在合成多于 2 个碳以上的直链烷基芳烃时，受到一定的限制。

在酰基化反应中，由于酰基对芳环具有钝化作用，因而反应可以停止在一酰基化阶段，这对于选择性地制备单取代芳烃是十分有利的。由于在酰基化反应中不会发生重排，因此该反应在合成直链烷基芳烃或带其他支链结构的烷基芳烃具有特殊的应用价值。常用的酰化剂是酰氯和酸酐。催化剂除了常用的无水三氯化铝外，还可用其他的 Lewis 酸，但催化效果以无水三氯化铝最佳。其反应过程如下：

第一步

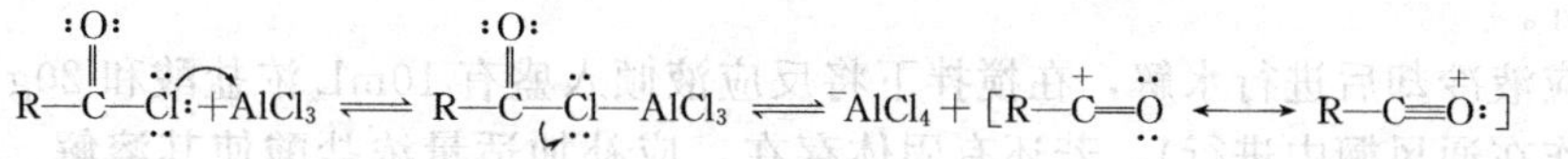

$$\text{R-CO-Cl} + \text{AlCl}_3 \rightleftharpoons \text{R-CO-Cl-AlCl}_3 \rightleftharpoons \text{AlCl}_4^- + [\text{R-C}^+\text{=O} \leftrightarrow \text{R-C}\equiv\text{O}^+]$$

第二步

$$\text{C}_6\text{H}_6 + \text{R-C}^+\text{=O} \longrightarrow [\text{中间体}] \xrightarrow{\text{Cl-AlCl}_3^-}$$

$$\text{C}_6\text{H}_5\text{-CO-R} + \text{AlCl}_3 + \text{HCl} \longrightarrow \text{C}_6\text{H}_5\text{-CO-R} \cdot \text{AlCl}_3$$

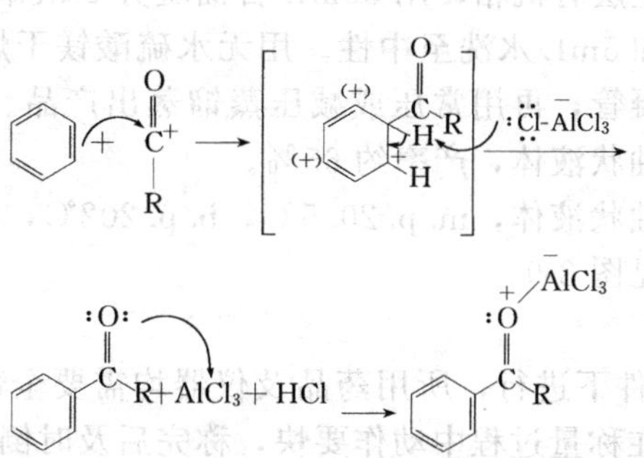

在烷基化反应和酰基化反应中，三氯化铝的用量有所不同。在烷基化反应中，三氯化铝的投入量仅需催化剂量，但在酰基化反应中因三氯化铝可与酰氯及产物芳酮生成配合物，所以 1mol 酰氯需用多于 1mol 的无水三氯化铝，一般过量 10%。当用酸酐作酰基化试剂时，因为酸酐先要和三氯化铝作用，故比用酰卤需多消耗 1mol AlCl$_3$，因此 1mol 酸酐需要 2mol 三氯化铝，在实际制备时一般还要需过量 10%～20%。

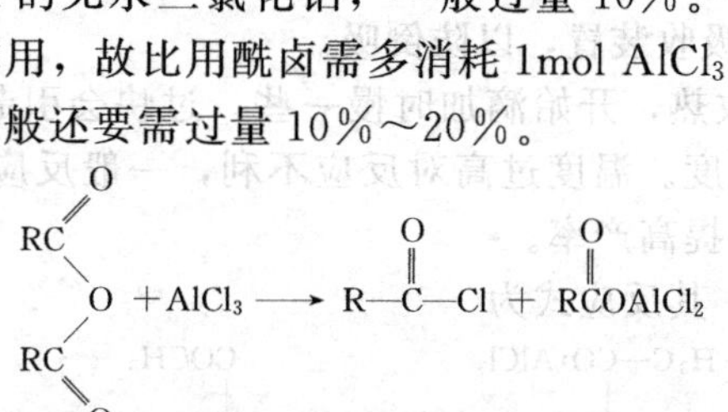

$$(\text{RCO})_2\text{O} + \text{AlCl}_3 \longrightarrow \text{R-CO-Cl} + \text{RCOOAlCl}_2$$

傅-克反应是放热反应，故常将酰基化试剂配成溶液后慢慢加入盛有芳香族化合物溶液的反应瓶中。反应常用的溶剂有二硫化碳、硝基苯、硝基甲烷等。若原料为液态芳烃，如苯、甲苯等，则常用过量的芳烃既作原料又作溶剂。因反应时会放出氯化氢气体，故需连接气体吸收装置。

实验 12 苯乙酮的合成

制备芳酮最重要的方法是傅-克酰基化反应。苯乙酮的合成是利用苯与乙酸酐在 Lewis 酸催化剂（无水三氯化铝）作用下发生的反应。乙酸酐为酰化剂，虽然其酰化能力较弱，但较便宜。

【反应式】

$$C_6H_6 + (CH_3CO)_2O \xrightarrow{AlCl_3} C_6H_5COCH_3 + CH_3COOH$$

【主要试剂】

无水苯 8mL（约 7g，90mmol），乙酸酐 2mL（2.15g，21mmol），无水三氯化铝 6.5g（48.7mmol），浓盐酸，氢氧化钠，石油醚，无水硫酸镁。

【实验步骤】

在装有电动搅拌器、恒压滴液漏斗和回流冷凝管（上口接一个装有无水氯化钙的干燥管并与 HCl 气体吸收装置相连）的 50mL 的三口瓶中，迅速加入 6.5g 无水三氯化铝和 8mL 无水苯，开动搅拌器，边搅拌边滴加 2mL 乙酸酐，开始先少加几滴，待反应开始后再继续滴加。此反应为放热反应，应注意控制乙酐的滴加速度，以使三口瓶稍热为宜，切勿使反应过于激烈，必要时用冷水冷却反应瓶，此过程约需 10min 左右。加料完毕，待反应稍缓和后，用沸水浴加热回流并搅拌，直至无 HCl 气体逸出为止，此过程约需 50min 左右。

待反应液冷却后进行水解，在搅拌下将反应液倾入盛有 10mL 浓盐酸和 20g 碎冰的烧杯中（此操作在通风橱中进行），若还有固体存在，应补加适量浓盐酸使其溶解。然后，将反应液转入分液漏斗中，分出上层有机相，用 30mL 石油醚分 2 次萃取下层水相，合并有机相，依次用 5mL 10% NaOH 和 5mL 水洗至中性。用无水硫酸镁干燥。在水浴上蒸出石油醚和苯后，稍冷后改用空气冷凝管，再用常压或减压蒸馏蒸出产品。常压蒸馏收集 198～202℃ 的馏分。产品为无色透明油状液体，产率约 65%。

纯苯乙酮为无色透明油状液体，m.p. 20.5℃，b.p. 202℃，d_4^{15} 1.0281，n_D^{20} 1.53718。

纯苯乙酮的红外光谱见图 3-9。

【注意事项】

① 本实验应在无水条件下进行，所用药品及仪器均需要全部干燥无水。无水三氯化铝在空气中容易吸潮分解，在称量过程中动作要快，称完后及时倒入烧瓶中，并将烧瓶及药品瓶盖子及时盖好。苯用无水氯化钙干燥过夜后再用。乙酸酐必须在临用前重新蒸馏，收集 137～140℃ 的馏分使用。

② 使用无水三氯化铝时，应避免与皮肤接触，以免被灼烧。

③ 正确安装 HCl 气体吸收装置，以防倒吸。

④ 滴加醋酸酐，反应放热，开始滴加时慢一些，过快会引起爆沸，反应高峰过后可以加快速度，注意控制反应温度。温度过高对反应不利，一般反应液温度控制在 60℃ 以下为宜。反应时间长一些，可以提高产率。

⑤ 加酸使苯乙酮析出，其反应式为

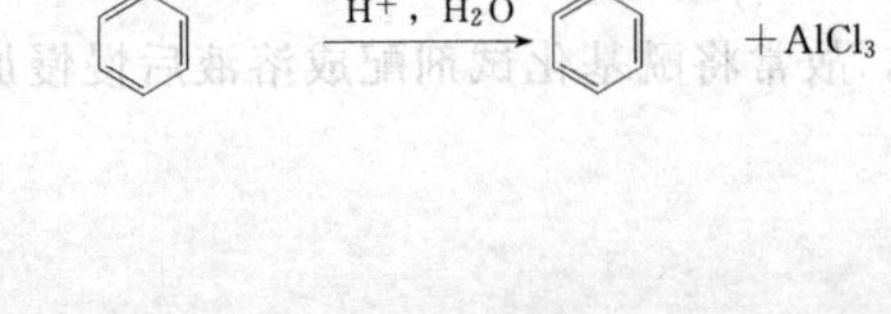

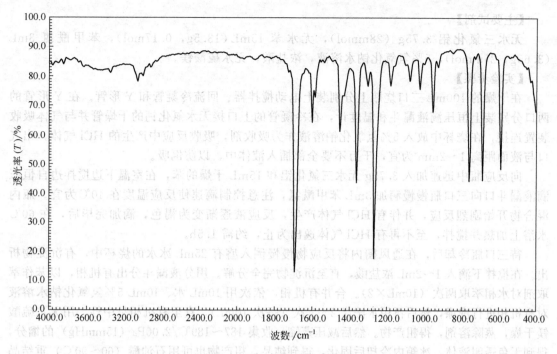

图 3-9 苯乙酮的红外光谱图

⑥ 也可用减压蒸馏收集苯乙酮。苯乙酮在不同压力下的沸点如下。

压力/Pa	533	667	800	933	1067	1120	1333	3333
沸点/℃	60	64	68	71	73	76	78	98
压力/Pa	4000	5333	6666	7999	13332	19998	26664	
沸点/℃	102	110	115.5	120	134	146	155	

思 考 题

1. Friedel-Crafts 酰基化反应和烷基化反应各有何特点？在两个反应中，无水三氯化铝和芳烃的用量有何不同？为什么？
2. 为什么要用过量的苯和无水三氯化铝？
3. 反应完后，为什么要用含盐酸的冰水来分解产物？
4. 试解析苯乙酮的红外光谱。

实验 13 二苯甲酮（酰基化法）的合成

二苯甲酮是无色带光泽晶体，具有甜味及玫瑰香味，故用作香料，它能赋予香精以甜的气息，用在许多香水和香皂香精中。还可用于合成有机颜料、杀虫剂等，在医药工业上可用于生产苯甲托品氢溴酸盐以及苯海拉明盐酸盐等。

二苯甲酮的合成方法有很多，既可以用苄氯作原料经烷基化、氧化等反应制得，也可以由苯作起始原料通过烷基化、水解等步骤制备。本实验采取由苯甲酰氯和苯进行酰基化反应一步法制取二苯甲酮。

【反应式】

【主要试剂】

无水三氯化铝 3.75g（28mmol），无水苯 15mL（13.5g，0.17mol），苯甲酰氯 3mL（3.65g，25mmol），5%氢氧化钠水溶液，浓盐酸，无水硫酸镁。

【实验步骤】

在干燥的 100mL 三口烧瓶上分别装上电动搅拌器、回流冷凝管和 Y 形管。在 Y 形管的两口分别装上恒压滴液漏斗和温度计，在冷凝管的上口接无水氯化钙的干燥管并与气体吸收装置连接，在烧杯中放入 5%氢氧化钠溶液作为吸收剂，吸收反应中产生的 HCl 气体，出气口与液面距离 1~2mm 为宜，千万不要全部插入液体中，以防倒吸。

向反应瓶中迅速加入 3.75g 无水三氯化铝和 15mL 干燥的苯，在室温下边搅拌边自恒压滴液漏斗口向三口瓶慢慢滴加 3mL 苯甲酰氯，注意控制滴速使反应温度在 40℃为宜。瓶内混合物开始剧烈反应，并伴有 HCl 气体产生，反应液逐渐变为褐色，滴加完毕后，在 60℃水浴上加热并搅拌，至不再有 HCl 气体逸出为止，约需 1.5h。

待三口瓶冷却后，在通风橱内将反应物慢慢倒入盛有 25mL 冰水的烧杯中，有沉淀物析出。在搅拌下滴入 1~2mL 浓盐酸，直至沉淀物完全分解。用分液漏斗分出有机相，以苯作萃取剂对水相萃取两次（10mL×2）。合并有机相，依次用 10mL 水、10mL 5%氢氧化钠水溶液对有机相进行洗涤，然后再用水洗涤 2~3 次（每次 10mL），直至有机相呈中性。用无水硫酸镁干燥，蒸除溶剂，得粗产物。然后减压蒸馏，收集 187~189℃/2.00Pa（15mmHg）的馏分，得到无色透明液体，冰箱内冷却后固化，得到纯品，粗产物也可用石油醚（60~90℃）重结晶代替减压蒸馏得到纯品。干燥后称量、测定熔点及红外光谱，计算产率。

二苯甲酮为无色晶体，m.p. 47~48℃，b.p. 305.4℃。

纯二苯甲酮的红外光谱见图 3-10。

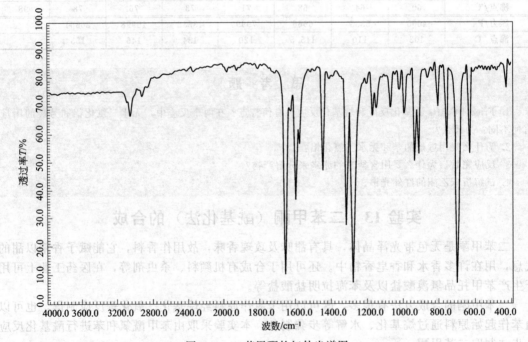

图 3-10 二苯甲酮的红外光谱图

【注意事项】

① 参见苯乙酮合成的注意事项。

② 二苯甲酮有多种晶型，它们的熔点各不相同：α 型为 49℃；β 型为 26℃；γ 型为 45~

48℃；δ 型为 51℃。其中 α 晶型较稳定。

> **思 考 题**
>
> 1. 在酰基化反应中，是否容易产生多酰基取代芳烃？
> 2. 与脂肪酮相比，芳酮分子中的羰基红外吸收峰是向高波数移动还是向低波数移动？为什么？
> 3. 为什么硝基苯可作为傅-克反应溶剂？芳环上有 OH、OR、NH_2 等基团存在时对傅-克反应不利甚至不发生反应，为什么？
> 4. 为什么在减压蒸馏时，即使沸点较高的化合物最好也不用石棉网直接加热，而采用热浴加热？

实验 14 二苯甲酮（烷基化法）的合成

【反应式】

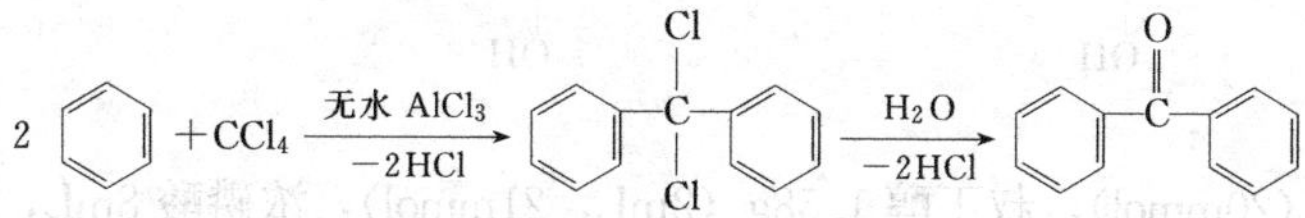

$$2\,C_6H_6 + CCl_4 \xrightarrow[-2HCl]{\text{无水 } AlCl_3} (C_6H_5)_2CCl_2 \xrightarrow[-2HCl]{H_2O} (C_6H_5)_2C=O$$

【主要试剂】

无水苯 4.5mL（3.95g，51mmol），四氯化碳 11mL（17.5g，114mmol），无水三氯化铝 3.35g(25mmol)，无水硫酸镁。

【实验步骤】

在干燥的 250mL 三口烧瓶上分别装上电动搅拌器、回流冷凝管和 Y 形管。在 Y 形管的两口分别装上恒压滴液漏斗和温度计，在冷凝管的上口接无水氯化钙的干燥管并与气体吸收装置连接，在烧杯中放入 5% 氢氧化钠溶液作为吸收剂，吸收反应中产生的 HCl 气体，出气口与液面距离 1～2mm 为宜，千万不要全部插入液体中。

将 3.35g 无水三氯化铝和 7.5mL 四氯化碳迅速投入三口烧瓶中，将三口瓶置于冰水浴中，待瓶内温度降至 10～15℃ 时，在搅拌下慢慢滴加 2mL 由 4.5mL 无水苯和 3.5mL 四氯化碳配成的溶液。反应开始后，有 HCl 气体产生，反应混合物温度逐渐升高，此期间应注意用冰水浴控制反应温度在 5～10℃ 之间。当反应变温和后，将余下的苯溶液逐渐加入反应瓶中，滴速以保持反应瓶内温度在 5～10℃ 之间为宜。滴加完毕（约需 10min 左右），继续搅拌 1h，反应温度保持在 10℃ 左右。然后改用冰水浴，在搅拌下，通过滴液漏斗慢慢滴加 50mL 水使其水解。改用蒸馏装置，在水浴上蒸出过量的四氯化碳，再在石棉网上加热 0.5h，以除去残余的四氯化碳，并促使二氯二苯甲烷水解完全。

冷却反应混合物，将其转移到分液漏斗中，分出有机相。水相用蒸出的四氯化碳（约 10mL）萃取一次，萃取液与有机相合并，用无水硫酸镁干燥。常压蒸除四氯化碳后，改用减压蒸馏，蒸出产品，收集 144～148℃/280Pa（2.1mmHg）或 187～189℃/2.00kPa（15mmHg）的馏分。产品为无色透明油状液体，冷却后固化，得到白色晶体，用石油醚（60～90℃）重结晶得到纯品，m.p. 45～46℃，产率约 70%。

【注意事项】

① 参见苯乙酮合成的注意事项。
② 反应温度低于 5℃ 时，反应缓慢；高于 10℃ 时，则易产生焦油状树脂产物。

> **思 考 题**
>
> 1. 傅-克反应有几种类型，各自的特点是什么，本实验的反应属于哪一类？
> 2. 哪些物质可用作傅-克反应的催化剂？
> 3. 本实验为什么要在无水条件下进行？
> 4. 本实验为什么是四氯化碳过量而不是苯过量？如苯过量会有什么结果？

实验 15　2-叔丁基对苯二酚的合成

抗氧剂有多种类型，酚类抗氧剂是其中的一大类。2-叔丁基对苯二酚，又称叔丁基氢醌（ter-butyl hydroquinone，简记 TBHQ），就是一种酚类抗氧剂，具有抗氧、阻聚等性能，并且低毒、廉价。因此，2-叔丁基对苯二酚广泛被用作橡胶、塑料的抗氧剂以及食品的添加剂。2-叔丁基对苯二酚可以由叔丁醇作烷基化试剂，在酸催化下与对苯二酚发生烷基化反应而制得，也可由卤代烃或烯烃作烷基化试剂进行合成。

【反应式】

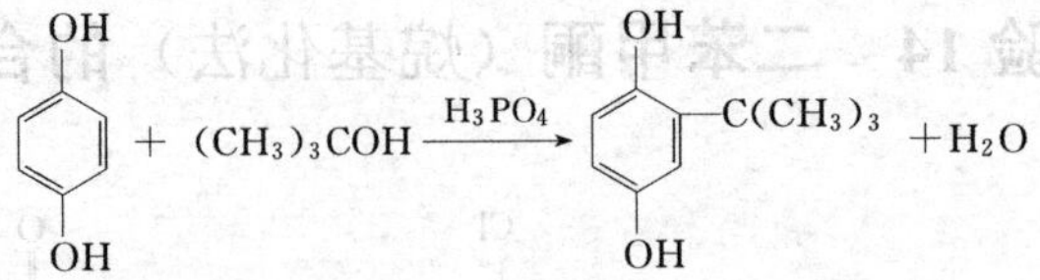

$$\text{对苯二酚} + (CH_3)_3COH \xrightarrow{H_3PO_4} \text{2-叔丁基对苯二酚} + H_2O$$

【主要试剂】

对苯二酚 2.2g（20mmol），叔丁醇 1.58g（2mL，21mmol），浓磷酸 8mL，甲苯 10mL。

【实验步骤】

在 100mL 的三口烧瓶上，配置磁力搅拌器、恒压滴液漏斗、温度计和回流冷凝管。依次将 2.2g 对苯二酚、8mL 浓磷酸和 10mL 甲苯加入到三口烧瓶中。在搅拌下，用水浴加热使反应瓶中的混合物升温至 90℃，然后自滴液漏斗向反应瓶慢慢滴加 2mL 叔丁醇，控制反应温度在 90~95℃。滴加完毕，在 90℃ 条件下继续搅拌 0.5h 左右，直到混合物中的固体物全部溶解。

反应完毕，趁热将反应物倒入分液漏斗并分出磷酸层。再将有机相转入三口瓶中，并加入 20mL 水，直接进行水蒸气蒸馏（见图 2-34），蒸除溶剂。蒸馏完毕，将烧瓶中的残余水溶液趁热过滤，弃去不溶物，滤液转入烧杯中。如果残余液体积不足 20mL，应补加热水，使产物尽可能被热水所提取。让滤液静置冷却至室温，有白色晶体析出，过滤，滤饼用冷水洗涤两次，抽滤后干燥，即得 2-叔丁基对苯二酚，称量，测熔点并计算产率。

2-叔丁基对苯二酚可用水重结晶，纯品为白色针状晶体，m.p. 127~129℃。

【注意事项】

① 磷酸具有腐蚀性，量取时要小心。

② 甲苯作为反应溶剂。羟基和甲基相比较，羟基对苯的活化效应更强，故在对苯二酚的烷基化反应中，只要烷基化试剂不过量，甲苯可以作为惰性溶剂来使用。

③ 对苯二酚不溶于甲苯，而 2-叔丁基对苯二酚溶于甲苯。因此，当固体物质对苯二酚完全溶解时，可认为反应结束。

④ 过滤用的漏斗要事先预热，以防漏斗堵塞。

⑤ 2-叔丁基对苯二酚溶于热水中，微溶于冷水；二取代物 2,5-叔丁基对苯二酚不溶于热水。

思 考 题

1. 本实验以甲苯作溶剂是否会产生甲苯烷基化产物？
2. 本实验可能有哪些副反应？为了减少副反应，实验中采取了什么措施？
3. 水蒸气蒸馏时，如何判断终点？水蒸气蒸馏后为什么要趁热对残液过滤？

3.5　酯化反应

在人类的日常生活中，大部分酯具有广泛的用途。有些酯可作为食用油、脂肪、塑料以及油漆的溶剂。许多酯具有令人愉快的香味，是廉价的香料。

羧酸与醇或酚在无机或有机强酸催化下发生反应生成酯和水,这个过程称为酯化反应(Esterification Reaction)。常用的催化剂有浓硫酸、干燥的氯化氢、有机强酸或阳离子交换树脂。在酯化反应中,如果参与反应的羧酸本身就具有足够强的酸性,如甲酸、草酸等,就可以不另加催化剂。

$$RCOOH + R'OH \underset{}{\overset{H^+}{\rightleftharpoons}} RCOOR' + H_2O$$

该反应是一个可逆反应。在酯化反应中,如用等物质的量的有机酸和醇,反应达到平衡后,只能得到理论产量的67%。为了得到较高产量的酯,通常使用过量的酸或醇,促使平衡向产物方向移动。至于使用过量的酸还是过量的醇,取决于哪一种原料易得和价廉。另外,也可采用把反应中生成的酯或水及时地从体系中除去的方法来促使反应趋于完成。这可通过向反应体系中加入一些能与水形成低沸点共沸物的有机溶剂,如苯、甲苯、氯仿等,通过蒸馏共沸物带出生成的水。此外,还可用酰氯或酸酐与醇反应制取相应的酯。

实验16 乙酸乙酯的合成

【反应式】

$$CH_3COOH + CH_3CH_2OH \underset{}{\overset{H_2SO_4}{\rightleftharpoons}} CH_3COOCH_2CH_3 + H_2O$$

【主要试剂】

冰醋酸6mL(6.0g,0.10mol),无水乙醇9.5mL(9.2g,0.20mol),浓硫酸2.5mL,饱和碳酸钠溶液,饱和氯化钠溶液,饱和氯化钙溶液,无水硫酸镁。

【实验步骤】

在50mL圆底烧瓶中加入9.5mL(0.2mol)无水乙醇和6mL(0.1mol)冰醋酸,再小心加入2.5mL浓硫酸,混匀后,加入沸石,然后装上冷凝管。

小火加热反应瓶,保持缓慢回流30min,待瓶内反应物冷却后,将回流装置改成蒸馏装置,接收瓶用冷水冷却。加热蒸出生成的乙酸乙酯,直到流出液体积约为反应物总体积的1/2为止。

在馏出液中慢慢加入饱和碳酸钠溶液,并不断振荡,直至不再有二氧化碳气体产生(或调节至石蕊试纸不再显酸性),然后将混合液转入分液漏斗,分去下层水溶液,有机层用5mL饱和食盐水洗涤,再用5mL饱和氯化钙洗涤,最后用水洗一次,分去下层液体。有机层倒入一干燥的锥形瓶中,用无水硫酸镁干燥,粗产物约6.8g(产率约77%)。将干燥后的有机层进行蒸馏,收集73~78℃的馏分,产量约4.2g(产率约48%)。

纯乙酸乙酯为无色而有香味的液体,b.p.为77.06℃,d_4^{20}为0.9003,n_D^{20}为1.3723。

【注意事项】

① 在馏出液中除了酯和水外,还含有少量未反应的乙醇和乙酸,也含有副产物乙醚。故必须用碱除去其中的酸,并用饱和氯化钙除去未反应的醇,否则会影响到酯的收率。

② 当有机层用碳酸钠洗过后,若紧接着就用氯化钙溶液洗涤,有可能会产生絮状碳酸钙沉淀,使进一步分离变得困难,故在两步操作间必须用水洗一下。由于乙酸乙酯在水中有一定的溶解度,为了尽可能减少由此而造成的损失,所以实际上用饱和食盐水来进行水洗。

③ 乙酸乙酯与水或乙醇可分别生成共沸混合物,若三者共存则生成三元共沸混合物。因此,有机层中的乙醇不除净或干燥不够时,由于形成低沸点共沸混合物,从而影响酯的产率。

沸点/℃	质量分数/%		
	酯	乙醇	水
70.2	82.6	8.4	9.0
70.4	91.6	—	8.1
71.8	69.0	31.0	—

思 考 题

1. 酯化反应有什么特点？在试验中如何创造条件促使酯化反应尽量向生成物方向进行？
2. 本实验若采用醋酸过量的做法是否合适？为什么？
3. 蒸出的粗乙酸乙酯中主要有哪些杂质？如何除去？

实验17　乙酸正丁酯的合成

【反应式】

$$CH_3COOH + n\text{-}C_4H_9OH \xrightleftharpoons{H_2SO_4} CH_3COOC_4H_9 + H_2O$$

【主要试剂】

正丁醇5mL（4.05g，54.5mmol），冰醋酸 3.5mL（3.67g，61mmol），浓硫酸，10%碳酸钠溶液，无水硫酸镁。

【实验步骤】

在50mL两口烧瓶上安装分水器、回流冷凝器及温度计，在分水器内预先加入（$V-1.2$）mL水（V为分水器的容积）。

向反应瓶中加入5mL正丁醇和3.5mL冰醋酸，再滴入2滴浓硫酸，混合均匀，加入2粒沸石。在80℃左右加热15min后提高温度使反应处于回流状态约15min，当看不到水珠穿行时，表示反应完毕。

冷却后卸下回流冷凝管，将分水器中的液体及两口瓶中的反应液倒入分液漏斗中，分去水层，用5mL 10%碳酸钠水溶液洗涤有机层，使有机层pH=7，再用5mL水洗涤一次，分去水层，有机层倒入一个干燥的锥形瓶中，用无水硫酸镁干燥。将干燥后的粗产物倒入10mL干燥的圆底烧瓶中，常压蒸馏产物，收集124～126℃的馏分。称量，测折射率并计算产率。

纯乙酸正丁酯为无色透明液体，b.p.126.3℃，d_4^{18} 0.8824，n_D^{20} 1.3947。

【注意事项】

① 本实验利用形成的共沸混合物除去反应中生成的水。正丁醇、乙酸正丁酯和水形成以下几种恒沸混合物。含水的恒沸混合物冷凝为液体时，分为两层，上层为含少量水的酯和醇，下层主要是水。

	恒沸混合物	沸点/℃	组成的质量分数/%		
			乙酸正丁酯	正丁醇	水
二元	乙酸正丁酯-水	90.7	72.9		27.1
	正丁醇-水	93.0		55.5	44.5
	乙酸正丁酯-正丁醇	117.6	32.8	67.2	
三元	乙酸正丁酯-正丁醇-水	90.7	63.0	8.0	29.0

② 滴加浓硫酸时，要边加边摇边冷却，以免局部炭化。浓硫酸在反应中起催化作用，故只需少量。

思 考 题

1. 试计算反应完全时应分出多少水？
2. 本实验根据什么原理提高乙酸正丁酯的产率？浓硫酸在此起什么作用？

实验 18　乙酸正戊酯的合成

【反应式】

主反应

$$CH_3COOH + CH_3(CH_2)_3CH_2OH \xrightleftharpoons{H^+} CH_3COOCH_2(CH_2)_3CH_3 + H_2O$$

副反应

$$CH_3(CH_2)_3CH_2OH \xrightarrow[\Delta]{H^+} CH_3(CH_2)_2CH=CH_2 + H_2O$$

$$2CH_3(CH_2)_3CH_2OH \xrightarrow[\Delta]{H^+} CH_3(CH_2)_3CH_2OCH_2(CH_2)_3CH_3 + H_2O$$

【主要试剂】

正戊醇 5.5g（6.75mL，62.5mmol），冰醋酸 4.5g（4.3mL，75mmol），浓硫酸 0.3mL，5%碳酸钠水溶液，饱和食盐水，无水硫酸镁，苯。

【实验步骤】

将 6.75mL 正戊醇、4.3mL 冰醋酸和 7mL 苯加入到 30mL 干燥的圆底烧瓶中，在振摇下缓缓加入 0.3mL 浓硫酸，再投入两粒沸石。在反应瓶上连一个分水器，分水器中用苯充满，再在分水器上连一个冷凝管。加热回流至无水分出为止（约 2h），此期间分水器中不断有水分出。

回流结束后，冷却至室温，将反应混合物从烧瓶中倒入分液漏斗。用 7mL 冷水洗涤反应瓶，洗涤液也倒入分液漏斗。振摇分液漏斗，静止分层，分出水层。有机层经 6.0mL 5% 碳酸钠水溶液洗涤后，再用 7.0mL 饱和食盐水洗涤至中性。分出水层后将酯层转入干燥的锥形瓶中，用无水硫酸镁干燥。粗产物经过滤转入干燥的蒸馏瓶中，加入沸石进行分馏，收集 144～150℃ 的馏分。称量产品并计算产率。测定产物的折射率和红外光谱。

乙酸正戊酯为无色透明液体，b.p. 148.4℃，d_4^{20} 0.8970，n_D^{20} 1.4000。

思　考　题

1. 在实验中，使用分水器的目的是什么？
2. 溶剂中的苯是在后处理中的哪一步被除去的？

实验 19　乙酸异戊酯（香蕉油）的合成

【反应式】

$$CH_3COOH + HOCH_2CH_2\underset{CH_3}{CH}CH_3 \xrightleftharpoons{H^+} CH_3COOCH_2CH_2\underset{CH_3}{CH}CH_3 + H_2O$$

【主要试剂】

异戊醇 4.05g（5mL，46.5mmol），冰醋酸 7.35g（7mL，122.5mol），浓硫酸 1.0mL，5%碳酸钠水溶液，饱和食盐水，无水硫酸镁。

【实验步骤】

将 5mL 异戊醇和 7mL 冰醋酸加入到 25mL 干燥的圆底烧瓶中，在振摇下缓缓加入 1.0mL 浓硫酸，再投入几粒沸石，并配置回流冷凝管，加热回流 1h。回流结束后，冷却至室温，再将反应混合物从烧瓶中倒入分液漏斗。用 10mL 冷水洗涤反应瓶，洗涤液也倒入分液漏斗。振摇分液漏斗，静止分层，分出水层。有机层经 5.0mL 5%碳酸钠水溶液洗涤后，再用 5.0mL 饱和食盐水洗涤至中性。分出水层后将酯层转入一个干燥的 25mL 锥形瓶中，用无水硫酸镁干燥。粗产物经过滤转入蒸馏瓶，蒸馏收集 138～142℃ 馏分，接收瓶应插入冰浴中以减少气味。称量产品并计算产率，测定折射率。

乙酸异戊酯为无色透明液体，b.p.138～142℃，d_4^{20} 0.876，n_D^{20} 1.4000。

【注意事项】

① 加入浓硫酸时，反应液会放热，应小心振荡，使热量迅速扩散。

② 碳酸钠水溶液洗涤粗产品时会产生二氧化碳气体，振摇时不易剧烈，并留意放气。

③ 一定要将有机相洗涤至中性，否则在蒸馏过程中产物易发生分解。

思 考 题

1. 有利于乙酸异戊酯生成的一种方法是使冰醋酸过量，请提出另一种有利于酯生成平衡向右方向进行的方法。

2. 过量的冰醋酸为什么比过量的异戊醇易从产物中除去？

实验20 苯甲酸乙酯的合成

【反应式】

$$\text{C}_6\text{H}_5\text{COOH} + \text{C}_2\text{H}_5\text{OH} \xrightleftharpoons{\text{H}^+} \text{C}_6\text{H}_5\text{COOC}_2\text{H}_5 + \text{H}_2\text{O}$$

【主要试剂】

苯甲酸 1.53g（12.5mmol），无水乙醇 6.5mL（5.13g，0.114mol），浓硫酸 0.5mL（0.92g，9.2mmol），乙醚，10% Na_2CO_3 溶液，无水硫酸镁。

【实验步骤】

在30mL干燥的圆底烧瓶中，加入1.53g苯甲酸，6.5mL无水乙醇，0.5mL浓硫酸和两粒沸石，在石棉网上小火加热回流1.5h，稍冷后，将回流装置改为蒸馏装置，补加沸石，在沸水浴上蒸出过量的乙醇。并将蒸出的乙醇倒入指定的回收瓶内。

残余物冷却后倒入盛有10mL冷水的分液漏斗，用乙醚萃取两次（5mL×2），合并有机

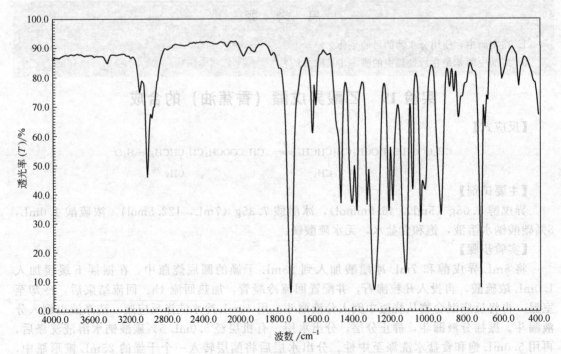

图3-11 苯甲酸乙酯的红外光谱图

相。分别用 10mL 10% Na_2CO_3 溶液、水（5mL×2）洗涤有机相，分出有机相，用无水硫酸镁干燥。水浴蒸出乙醚后，进行减压蒸馏，收集 b.p.73~74℃/533Pa(4mmHg) 的馏分，即得到无色透明的液体纯品。称量，测定折射率及红外光谱，计算产率（产品约 1.3g，产率 69.3%）。

纯苯甲酸乙酯为无色透明液体，b.p.213℃，d_4^{20} 1.0468，n_D^{20} 1.5007。

纯苯甲酸乙酯的红外光谱见图 3-11。

【注意事项】
① 水浴蒸出乙醚时，要无明火，并控制水浴温度。
② 减压蒸馏操作时，要按操作规程进行。

思 考 题

1. 为什么要加入过量的乙醇？
2. 在本实验中还可以采用什么原理和措施来提高产率？

实验 21　乙酰水杨酸（阿司匹林）的合成

乙酰水杨酸（又称阿司匹林，Aspirin）是一种非常普遍的治疗感冒的药物，具有镇痛、退热及抗风湿等功效，同时还有软化血管的作用。近年来的研究结果表明，阿司匹林会使肠癌的发生率减少 30%~50%。

水杨酸是一个具有双官能团的化合物，酚羟基和羧基都可以发生酯化反应。水杨酸经乙酸酐酰化后生成乙酰水杨酸（阿司匹林）。

【反应式】
主反应

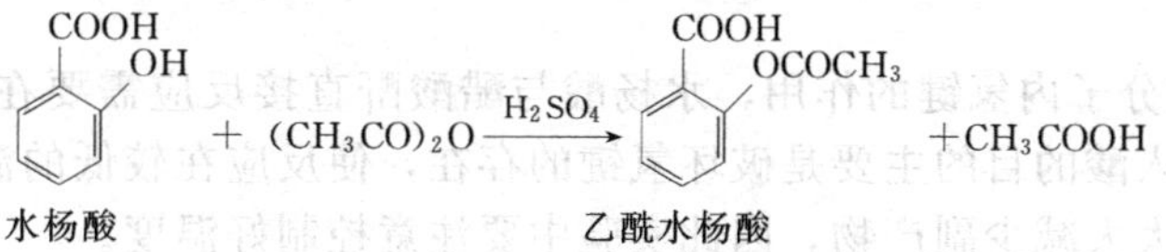

水杨酸 + $(CH_3CO)_2O$ $\xrightarrow{H_2SO_4}$ 乙酰水杨酸 + CH_3COOH

副反应

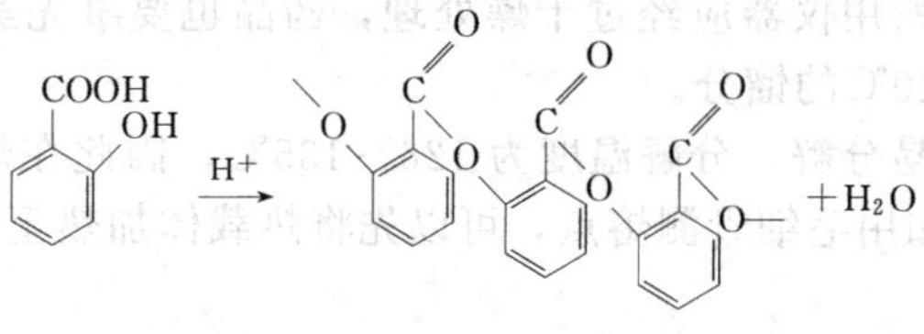

【主要试剂】
水杨酸 0.5g(3.6mmol)，乙酸酐 1.25mL(13.25mmol)，浓硫酸。

【实验步骤】
称取 0.5g 水杨酸放入 30mL 干燥的锥形瓶中，慢慢加入 1.25mL 新蒸馏的乙酸酐，使用滴管滴加浓硫酸（或 85% 磷酸）2 滴，摇动反应瓶使水杨酸溶解，水浴加热（温度 90℃）10min 后冷却至室温，再用冰水浴冷却片刻即有乙酰水杨酸晶体析出。若无晶体析出，可用玻璃棒摩擦瓶壁促使结晶，再放入冰水中冷却。晶体析出后再加入 13mL 水，继续在冰水浴中冷却，使晶体完全析出。抽滤，用少量冷水洗涤晶体，完全抽干后放置于空气中自然晾干。粗产品可用 1% 三氯化铁溶液检验是否有酚羟基存在。取少量粗产品，加入几滴乙醇使晶体溶解，滴加 1~2 滴 1% 三氯化铁溶液。如果发生颜色反应，产物可用乙醇-水混合溶剂重结晶：先将粗产品溶于少量的沸乙醇中，再向乙醇溶液中添加热水直到溶液中出现混浊，再加热至溶液澄清透明，静置慢慢冷却，过滤、干燥、称量，测定熔点并计算产率。

乙酰水杨酸为白色针状晶体，m.p. 132～135℃。

纯乙酰水杨酸的红外光谱见图 3-12。

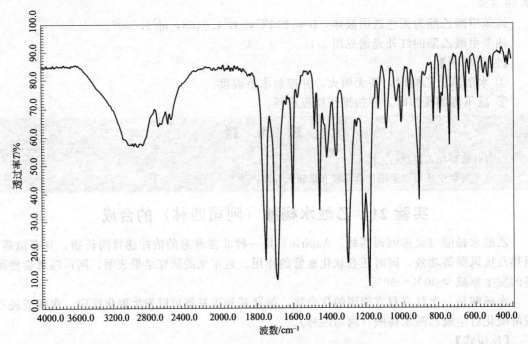

图 3-12 乙酰水杨酸的红外光谱图（KBr 压片）

【注意事项】

① 由于水杨酸分子内氢键的作用，水杨酸与醋酸酐直接反应需要在 150～160℃才能生成乙酰水杨酸。加入酸的目的主要是破坏氢键的存在，使反应在较低的温度下（90℃）就可以进行，而且可以大大减少副产物，因此实验中要注意控制好温度。

② 此反应开始时，所用仪器应经过干燥处理，药品也要事先经过干燥处理。乙酸酐需重新蒸馏，收集 139～140℃的馏分。

③ 乙酰水杨酸受热易分解，分解温度为 126～135℃，因此在烘干、重结晶、熔点测定时均不宜长时间加热。如用毛细管测熔点，可以先将热载体加热至 120℃左右，再放入毛细管测定。

④ 如粗产品中混有未反应的水杨酸，用 1% 三氯化铁检验时会显紫色。

⑤ 粗产品也可用 1:1(体积比) 的稀盐酸，或苯和石油醚（30～60℃）的混合溶剂进行重结晶。

思 考 题

1. 在水杨酸与醋酸酐的反应过程中，浓硫酸起什么作用？
2. 在硫酸存在下，水杨酸与乙醇作用将会得到什么产物？写出反应方程式。
3. 试解析乙酰水杨酸的红外光谱图。

实验 22 水杨酸甲酯（冬青油）的合成

水杨酸甲酯是在 1843 年从冬青植物中提取出来的，故称冬青油，是一种天然酯，存在于依兰油、月下香油、丁香油中，具有冬青树叶的香气。常作香料，用于食品、牙膏、化妆品等。后来发现，水杨酸甲酯还具有止痛和退热的特性。水杨酸甲酯可由水杨酸和甲醇作原料在硫酸催化下酯化而得。

【反应式】

水杨酸 + CH_3OH $\xrightarrow{H_2SO_4}$ 水杨酸甲酯 + H_2O

【主要试剂】

水杨酸 3.45g（25mmol），甲醇 12g（15mL，370.5mmol），浓硫酸 1mL，10%碳酸氢钠 10～15mL。

【实验步骤】

在 30mL 干燥的圆底烧瓶中，加入 3.45g 水杨酸和 15mL 甲醇。振摇溶解后，冷却下慢慢加入 1mL 浓硫酸，振摇反应瓶，使反应物混合均匀，再加入 2 粒沸石。并配置回流冷凝管。

加热，回流 1.5h，然后将回流装置改为蒸馏装置，水浴加热，蒸除过量的甲醇。剩余反应混合物经冷却后加入 10mL 水，振摇后转入分液漏斗静置分层。分出有机层，水层用 10mL 乙醚萃取，合并有机相。有机相依次用 10mL 水、10mL 10%碳酸氢钠水溶液洗涤，然后再加水洗涤数次，使有机相呈中性。分出有机相，并用无水硫酸镁干燥。除去干燥剂，水浴蒸除乙醚。将粗产物进行减压蒸馏，收集 115～117℃/2.7kPa（20mmHg）的馏分。称量，测折射率并计算产率。

纯水杨酸甲酯为无色透明液体，m.p. $-8 \sim -7℃$，b.p. 222℃，d_4^{25} 1.1787，n_D^{20} 1.5360。

纯水杨酸甲酯的红外光谱见图 3-13。

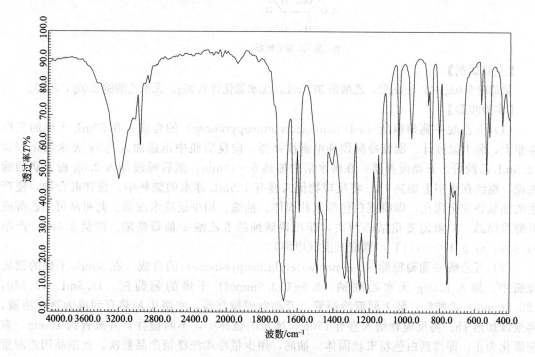

图 3-13 水杨酸甲酯的红外光谱图

【注意事项】
① 反应容器需干燥，如含水会使酯化收率降低。
② 滴加浓硫酸时，如果没及时振摇反应瓶，有时会出现部分原料炭化现象。
③ 要彻底蒸除甲醇，否则加水后产物溶解度增大，产率降低。

思 考 题

1. 水杨酸与甲醇的酸催化酯化反应属于可逆反应，为了使反应正向进行，本实验采用了什么措施？还可采取其他什么方法吗？
2. 酯化反应结束后，如果不先蒸除甲醇而直接用水洗涤，将会对实验结果有何影响？
3. 在后处理中，为什么要用碳酸氢钠水溶液对粗产物进行洗涤？若用氢氧化钠水溶液来洗涤，将会产生怎样的情况？
4. 硫酸在此反应中的作用是什么？试写出水杨酸和甲醇的酸催化酯化反应机理。
5. 试解析水杨酸甲酯的红外光谱。

实验 23　五乙酸葡萄糖酯的合成

【反应式】

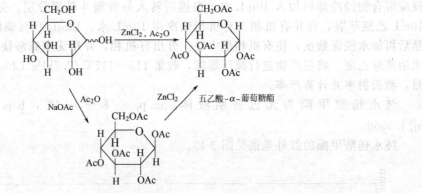

【主要试剂】
葡萄糖 5.00g（27.8mmol），乙酸酐 37.5mL，无水氯化锌 0.95g，无水乙酸钠 2.00g，乙醇。

【实验步骤】
(1) 五乙酸-α-葡萄糖酯（α-D-pentaacetylglucopyranose）的合成　在 50mL 干燥的三口烧瓶上，装上温度计、回流冷凝器和电磁搅拌器。向反应瓶中迅速加入 0.7g 无水氯化锌和 12.5mL 乙酸酐。开动搅拌器，在沸水浴中加热 5~10min，然后慢慢加入 2.5g 粉末状的葡萄糖，继续在水浴上加热 1h。将反应物倒入盛有 125mL 冰水的烧杯中，搅拌混合物，使产生的油状物完全固化，即得到白色粉末状固体。抽滤，用少量冷水洗涤。此粗品可用乙醇或甲醇重结晶，一般需要重结晶两次，即可得到纯品五乙酸-α-葡萄糖酯。产量 3.50g，产率 64.6%，m.p. 110~111℃，测旋光度（甲醇）。

(2) 五乙酸-β-葡萄糖酯（β-D-pentaacetylglucopyranose）的合成　在 50mL 干燥的圆底烧瓶中，加入 2.00g 无水乙酸钠、2.5g（13.9mmol）干燥的葡萄糖、12.5mL（13.52g，122.6mmol）乙酸酐，装上回流冷凝管，开动电磁搅拌器，水浴中加热直到成为透明溶液，再继续加热 1h。将反应物倒入盛有 150mL 冰水的烧杯中，不断搅拌，并放置约 10min，直至固化为止，即得到白色粉末状固体。抽滤，用少量冷水洗涤粗产物数次。此粗品用乙醇重结晶两次（25mL×2）后得到纯品五乙酸-β-葡萄糖酯。产量 3.50g，产率 64.6%，m.p.

131~132℃，测旋光度（甲醇）。

（3）五乙酸-β-葡萄糖酯转化为五乙酸-α-葡萄糖酯的反应　在50mL干燥的三口烧瓶中，加入12.5mL乙酸酐，迅速加入0.25g无水氯化锌，装上回流冷凝管，开动搅拌器，在沸水浴上加热5~10min至固体全部溶解。然后迅速加入2.5g纯五乙酸-β-葡萄糖酯，在水浴上加热30min。将热溶液倒入盛有125mL冰水的烧杯中，剧烈搅拌，以诱导油滴尽快结晶。抽滤，用少量冰水洗涤粗产品，抽干后，用乙醇重结晶即得到纯品五乙酸-α-葡萄糖酯，称量，计算产率，m.p. 110~111℃。

【注意事项】

① 自然界中D-(+)-葡萄糖是以环形半缩醛形式存在，有α和β两种端基差向异构体。将葡萄糖与过量的乙酸或乙酸酐在催化剂存在下加热，可使葡萄糖的五个羟基都被乙酰化，相应地，生成α和β五乙酸葡萄糖酯。但使用的催化剂不同时，所生成的主要产物也不同。用无水氯化锌作催化剂时，α-葡萄糖酯为主要产物；用无水乙酸钠作催化剂时，大部分为β-葡萄糖酯。从立体构型来看β异构体比α-异构体稳定，但在无水氯化锌催化作用下，β异构体也能转换为α-异构体。

② 市售的无水氯化锌使用前必须除水，乙酸酐应使用重新蒸馏的。无水氯化锌极易潮解，称取研碎时，操作要迅速。

③ 反应物倒入冰水中后，要强烈搅拌使块状固体成粉末，防止固体中包藏溶剂，使产物在重结晶时部分水解。

3.6　坎尼扎罗反应

坎尼扎罗（Cannizzaro）反应是指不含α-氢的醛，在强碱作用下，进行的自身氧化还原反应，一个分子醛被氧化成酸，另一个分子醛被还原为醇。

坎尼扎罗反应的实质是羰基的亲核加成反应。反应时，首先是OH^-对一个醛分子的羰基进行亲核进攻，生成负离子Ⅰ，然后由Ⅰ转移一个负氢到另一个醛分子的羰基碳上，接着进行质子的转移产生相应的醇和羧酸盐，反应机理如下。

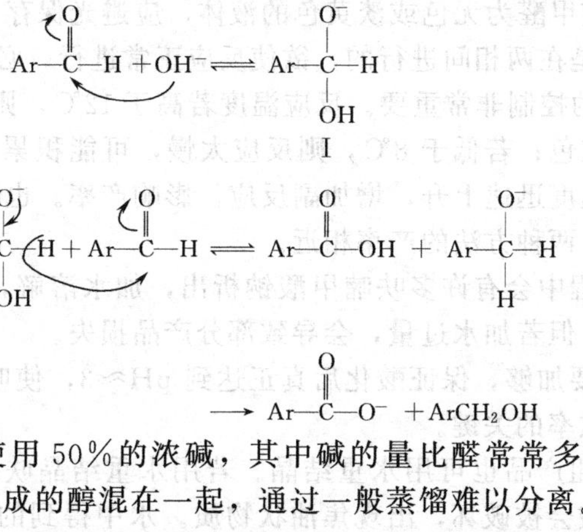

$$Ar-\overset{O}{\underset{}{C}}-H + OH^- \rightleftharpoons Ar-\overset{O^-}{\underset{OH}{C}}-H$$
$$\text{I}$$

$$Ar-\overset{O^-}{\underset{OH}{C}}-H + Ar-\overset{O}{\underset{}{C}}-H \rightleftharpoons Ar-\overset{O}{\underset{}{C}}-OH + Ar-\overset{O^-}{\underset{H}{C}}-H$$

$$\longrightarrow Ar-\overset{O}{\underset{}{C}}-O^- + ArCH_2OH$$

在坎尼扎罗反应中，通常使用50%的浓碱，其中碱的量比醛常常多一倍以上。否则，反应不易完全，未反应的醛与生成的醇混在一起，通过一般蒸馏难以分离。

实验24　呋喃甲酸和呋喃甲醇的合成

【反应式】

$$2\ \text{furan-CHO} \xrightarrow{OH^-} \text{furan-CH}_2OH + \text{furan-COO}^-$$

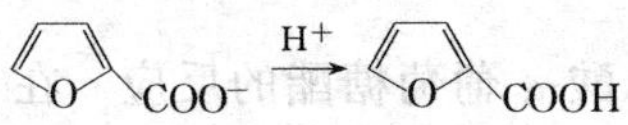

$$\text{furan-CHO} \xrightarrow{H^+} \text{furan-COO}^- + \text{furan-COOH}$$

【主要试剂】

呋喃甲醛（新蒸的）10mL（11.6g，120.7mmol），42％氢氧化钠 9mL。

【实验步骤】

准确量取 9mL 42％氢氧化钠溶液，置于 100mL 的小烧杯中，充分搅匀，将烧杯置于冰盐浴中冷却内容物至 5℃。在不断搅拌下由毛细滴管慢慢滴入 10mL 新蒸馏过的呋喃甲醛，把反应温度保持在 8～12℃，加完后于室温下继续搅拌约 25min，反应即可完全，得到淡黄色浆状物。

在搅拌下加入适量（约 15mL）的水，至沉淀恰好完全溶解，此时溶液呈暗红色。将溶液转入分液漏斗中，用乙醚（每次用 10mL）萃取溶液，重复 4 次，合并乙醚萃取液，加入约 2g 无水碳酸钠或无水硫酸镁干燥，过滤除去干燥剂之后，先在水浴上蒸去乙醚（注意：蒸去乙醚时须无明火），再蒸去呋喃甲醇，收集 169～172℃ 的馏分（或减压蒸馏收集 88℃/4666Pa 的馏分），产量 4～5g，产率 68％～84％。

纯呋喃甲醇为无色液体，b.p. 171℃/100kPa(750mmHg)，d_4^{20} 1.1296，n_D^{20} 1.4868。

经乙醚萃取后的水溶液内主要含呋喃甲酸钠，在搅拌下加入 25％盐酸酸化至 pH＝3，得到白色沉淀，充分冷却，使呋喃甲酸完全析出后，加入 30mL 乙醚（用制呋喃甲醇时蒸出的回收乙醚）于上述烧杯中，使白色沉淀完全溶解于醚层后转入分液漏斗中。烧杯再用少量乙醚清洗，将清洗液并入分液漏斗中，分出有机相，水相再用乙醚萃取一次，合并有机相。水浴蒸去乙醚后得到淡黄色呋喃甲酸晶体。将此粗产品用苯重结晶，冷却后析出晶体，抽滤，并用少量苯淋洗晶体，彻底抽干，得到白色晶体。放置空气中自然晾干，产量 5.5g，产率 81.5％。

纯呋喃甲酸的 m.p. 133～134℃。

【注意事项】

① 呋喃甲醛存放过久会变成棕褐色甚至黑色，同时往往含有水分。因此，使用前需蒸馏提纯，收集 160～162℃ 的馏分，最好用减压蒸馏收集 54～55℃/2.27kPa(17mmHg) 的馏分。新蒸馏的呋喃甲醛为无色或淡黄色的液体，应避光保存，最好蒸后就使用。

② 这个反应是在两相间进行的，欲使反应正常进行，必须充分搅拌。

③ 反应温度的控制非常重要。反应温度若高于 12℃，则反应温度极易上升而难于控制，反应物会变成深红色；若低于 8℃，则反应太慢，可能积累一些氢氧化钠，一旦反应发生，则过于猛烈，使温度迅速上升，增加副反应，影响产率。也可采用将氢氧化钠溶液滴加到呋喃甲醛中的方法。两种方法的产率相近。

④ 在反应过程中会有许多呋喃甲酸钠析出，加水溶解，可使奶油色的浆状物转为酒红色透明状的液体。但若加水过量，会导致部分产品损失。

⑤ 酸量一定要加够，保证酸化后真正达到 pH≈3，使呋喃甲酸充分游离出来，这一步是影响呋喃甲酸收率的关键。

⑥ 呋喃甲酸粗产品也可用水重结晶。若用水重结晶呋喃甲酸粗品，不要长时间加热，否则部分呋喃甲酸会被破坏，出现焦油状物质。水中得到的呋喃甲酸呈叶状晶体，100℃ 有部分升华，最好自然晾干。

思 考 题

1. 本实验根据什么原理来分离提纯呋喃甲醇和呋喃甲酸？
2. 在反应过程中析出的黄色浆状物是什么？

实验 25　苯甲酸和苯甲醇的合成

【反应式】
$$2C_6H_5CHO + NaOH \longrightarrow C_6H_5COONa + C_6H_5CH_2OH$$
$$C_6H_5COONa + HCl \longrightarrow C_6H_5COOH + NaCl$$

【主要试剂】

苯甲醛 6.3mL（6.6g，62.2mmol），氢氧化钠 5.5g（137.5mmol），浓盐酸，乙醚，饱和亚硫酸氢钠溶液，10%碳酸钠溶液，无水硫酸镁。

【实验步骤】

在 100mL 锥形瓶中加入 5.5g 氢氧化钠和 5.5mL 水振荡成溶液。冷却至室温后，在振荡下，分次将 6.3mL 新蒸馏过的苯甲醛加入到锥形瓶中，每次约加 1.5mL，每次加完后都应盖紧瓶塞，用力振摇，使反应物充分混合。若温度过高，可适时地把锥形瓶放入冷水浴中冷却。最后反应物变成白色糊状物，塞紧瓶塞，放置 24h 以上。

向反应混合物中逐渐加入足够量的水（约 20~25mL），微热，不断搅拌使其中的苯甲酸盐全部溶解。冷却后将溶液倒入分液漏斗中，用 15mL 乙醚分 3 次萃取苯甲醇。将乙醚萃取过的水溶液保存好。合并乙醚萃取液，依次用 3mL 饱和亚硫酸氢钠溶液、5mL 10%碳酸钠溶液和 5mL 冷水洗涤。分离出乙醚溶液，用无水硫酸镁干燥。

将干燥后的乙醚溶液倒入 25mL 圆底烧瓶中，用热水浴加热，蒸出乙醚（无明火蒸馏，乙醚回收，倒入指定的回收瓶中）。蒸完乙醚后，改用空气冷凝管，在石棉网上加热，蒸馏苯甲醇，收集 198~204℃的馏分，产量约 2.2g。

纯苯甲醇为无色液体，b.p. 205.4℃，d_4^{20} 1.045，n_D^{20} 1.5396。

在不断搅拌下，向乙醚萃取过的水溶液、20mL 水和 12.5g 碎冰的混合物，用盐酸调 pH=2。充分冷却使苯甲酸完全析出，抽滤，用少量冷水洗涤，挤压去水分，取出产物，晾干。粗苯甲酸可用水重结晶。产量 3.5g。

纯苯甲酸为无色针状晶体，m.p. 122.4℃。

【注意事项】

① 充分振摇是反应成功的关键。如果混合充分，放置 24h 后混合物通常在瓶内固化，苯甲醛气味消失。

② 此反应是放热反应。但反应温度不宜过高，而要适时冷却，以免过量的苯甲酸生成。

思 考 题

1. 为什么要用新蒸过的苯甲醛？苯甲醛长期放置后含有什么杂质？若不除去，对实验有何影响？
2. 本实验是根据什么原理分离提纯苯甲醇和苯甲酸的？用饱和亚硫酸氢钠溶液及 10%碳酸钠溶液洗涤乙醚萃取液的目的是什么？

3.7　缩合反应

实验 26　肉桂酸的合成

肉桂酸是生产冠心病药物"心可安"的重要中间体，其酯类衍生物是配制香精或食品香料的重要原料，在农药、塑料、感光树脂等精细化工产品生产中也有着广泛应用。

利用珀金（Perkin）反应将芳醛与酸酐混合后，在相应的羧酸盐存在下加热，可以制得 α,β-不饱和酸。应用此反应可以合成肉桂酸及其同系物。在本实验中，按照 Kalnin 所提出的方法，用碳酸钾代替 Perkin 反应中的醋酸钾，可缩短反应时间，提高产率。

【反应式】

$$PhCHO + (CH_3CO)_2O \xrightarrow[140\sim180℃]{CH_3COOK} PhCH=CHCOOH + CH_3COOH$$

【主要试剂】

苯甲醛 1.5mL（1.56g，15mmol），乙酸酐 4mL（4.33g，42.4mmol），无水碳酸钾 2.2g（16mmol），10%氢氧化钠，浓盐酸，乙醇，活性炭。

【实验步骤】

在 100mL 两口烧瓶中放入 1.5mL 新蒸馏过的苯甲醛、4mL 新蒸馏过的酸酐和 2.2g 无水碳酸钾。在两口烧瓶上安装空气冷凝管和温度计。在石棉网上加热回流 40min。开始小火加热，然后控制反应温度在 150～180℃。由于有二氧化碳放出，反应初期有泡沫产生。

待反应物冷却后，加入 35mL 温水，将装置改为水蒸气蒸馏装置［见图 2-36(f)］，蒸出未反应完的苯甲醛。再将烧瓶冷却，加入 10mL 10%氢氧化钠溶液，以保证所有的肉桂酸形成钠盐而溶解。抽滤，除去不溶物，用少量水洗涤两次（10mL×2）。将滤液倒入 250mL 烧杯中，加入少量活性炭煮沸 10min，以便脱色。稍冷却，抽滤。滤液倾入 250mL 烧杯中，冷却至室温后，在搅拌下用浓盐酸酸化至刚果红试纸变蓝。冷却，待晶体完全析出后进行抽滤，用少量冷水洗涤晶体，抽干。粗产品可用 5:1 的水-乙醇重结晶。产率约 68%。

纯肉桂酸为无色晶体，m.p. 135～136℃，b.p. 300℃，d_4^{20} 1.2475。

纯肉桂酸的红外光谱见图 3-14。

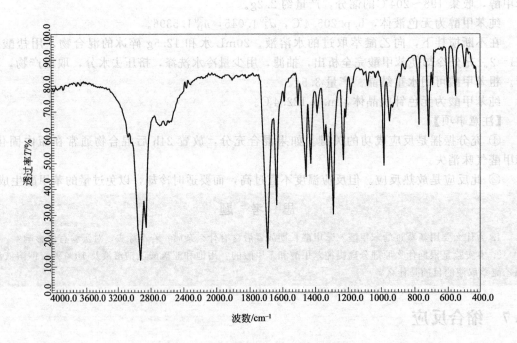

图 3-14 肉桂酸的红外光谱图（研糊法）

【注意事项】

① 久置的苯甲醛，由于自动氧化而生成较多的苯甲酸。这不但影响反应的进行，而且苯甲酸混在产品中不易除净，将影响产品的纯度。故本实验应使用重新蒸馏的苯甲醛。

② 醋酐放久了因吸潮和水解转变为乙酸，故本实验所需的醋酐必须在实验前进行重新蒸馏。

③ 肉桂酸有顺反异构体，通常制得的是其反式异构体，m.p.135.6℃。
④ 开始加热不要过猛，以防醋酐受热分解而挥发。

思 考 题

1. 用苯甲醛和丙酸酐在无水丙酸钾的存在下相互作用后得到什么产物？
2. 本实验用水蒸气蒸馏的目的是什么？如何判断蒸馏终点？
3. 在反应中，为什么可用碳酸钾代替醋酸钾？
4. 具有何种结构的醛能进行 Perkin 反应？试写出 Perkin 反应的反应机理。
5. 试解析肉桂酸的红外光谱中的主要吸收峰。

实验 27 乙酰乙酸乙酯的合成

含有 α-氢的酯在碱性催化剂存在下，能和另一分子酯起缩合反应，生成 β-酮酸酯，这就是 Claisen 酯缩合。如果所用的碱是乙醇钠，Claisen 酯缩合反应过程可表示如下。

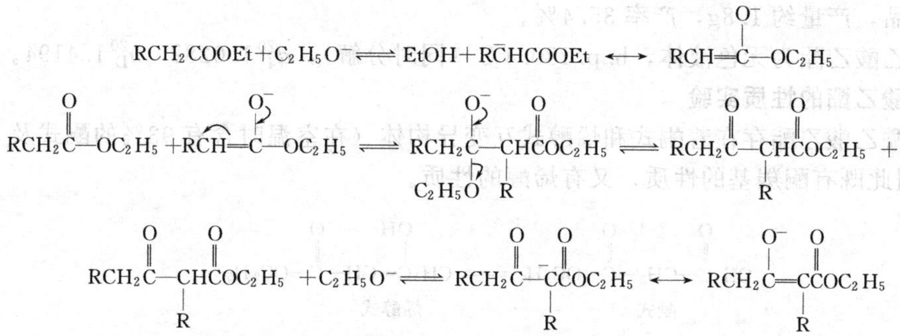

$$RCH_2COOEt + C_2H_5O^- \rightleftharpoons EtOH + R\overset{-}{C}HCOOEt \leftrightarrow RCH=\overset{O^-}{\underset{}{C}}-OC_2H_5 \tag{1}$$

$$RCH_2\overset{O}{\underset{}{C}}-OC_2H_5 + RCH=\overset{O^-}{\underset{}{C}}-OC_2H_5 \rightleftharpoons RCH_2\overset{O}{\underset{}{C}}-\underset{R}{\underset{|}{C}H}COC_2H_5 \rightleftharpoons RCH_2\overset{O}{\underset{}{C}}-\underset{R}{\underset{|}{C}H}COC_2H_5 + C_2H_5O^- \tag{2}$$

$$RCH_2\overset{O}{\underset{}{C}}-\underset{R}{\underset{|}{C}H}COC_2H_5 + C_2H_5O^- \rightleftharpoons RCH_2\overset{O}{\underset{}{C}}-\underset{R}{\underset{|}{\overset{-}{C}}}COC_2H_5 \leftrightarrow RCH_2\overset{O^-}{\underset{}{C}}=\underset{R}{\underset{|}{C}}COC_2H_5 \tag{3}$$

其中（1）的平衡偏向左方，因酯的 α-氢酸性（$pK_a=25$）较乙醇酸性（$pK_a=17$）弱。（2）也是平衡反应。酯缩合反应之所以能向右边进行是由于（3）的平衡偏向于右方。这是因为生成的 β-酮酸酯的两个羰基中的氢的酸性（$pK_a=11$）比乙醇的大。

β-酮酸酯的钠盐可以用乙酸中和，得到 β-酮酸酯。乙酰乙酸乙酯就是通过这个反应来制备的。这里使用的催化剂是乙醇钠，也可以是金属钠。本实验用乙酸乙酯及金属钠为原料，并以过量的乙酸乙酯作为溶剂。这是因为乙酸乙酯中含有少量乙醇（少于 2%），钠与乙醇作用生成乙醇钠催化该反应，在进行缩合反应的同时产生乙醇，后者继续与金属钠反应，从而可使反应不断进行下去，直至金属钠消耗完毕。但若乙酸乙酯中含有较多乙醇和水，会使产量显著降低。

【反应式】

$$2CH_3COOEt \xrightarrow{C_2H_5ONa} [CH_3COCHCOOEt]^-Na^+ \xrightarrow{H^+} CH_3COCH_2COOEt$$

【主要试剂】

乙酸乙酯 10mL(9g, 102.2mmol)，钠 0.9g(39.1mmol)，甲苯，50% 乙酸，饱和食盐水，无水硫酸镁。

【实验步骤】

在干燥的 50mL 圆底烧瓶内，加入 10mL 干燥的甲苯，0.9g 已除掉表面氧化膜的金属钠，安装回流冷凝管、无水氯化钙干燥管，加热回流使金属钠全部熔融，约需 10~15min，停止加热。待回流停止后，拆去冷凝管，用橡皮塞子塞紧圆底烧瓶的瓶口，按住瓶塞，用力振荡几下，使钠分散成尽可能小而均匀的小珠（小米粒大小）。随着甲苯逐渐冷却，钠珠迅速固化。待甲苯冷却至室温后，将甲苯倾倒入甲苯的回收瓶中（切记不可将甲苯倒入水池中，以免着火）。然后，立即加入10mL乙酸乙酯，迅速装上带有氯化钙干燥管的回流冷凝管，反应立即开始。反应液处于微沸状态。若反应不立即开始，可用小火直接加热，促使反应开始后即移去热源。若反应过于剧烈则用冷水稍微冷却一下。

待剧烈反应阶段过后，利用小火保持反应体系一直处于微沸状态，至金属钠全部溶解（约需2h）。如果长时间不溶，可补加2mL乙酸乙酯。反应结束时，反应瓶内生成的乙酰乙酸乙酯钠盐为棕红色透明溶液（但有时也可能夹带有少量黄白色沉淀）。

待反应液冷却后，将圆底烧瓶取下，边摇边加入50%乙酸（等体积的冰醋酸和水配成），直至整个体系呈弱酸性（pH=5~6）为止（约需8mL）。此时，固体应全部溶解（若还有固体，可加水使其溶解）。将反应液转入分液漏斗中，加入等体积的饱和氯化钠水溶液，用力振荡后放置，分出有机层，水层用8mL苯萃取，萃取液和酯层合并后，用无水硫酸镁或无水硫酸钠干燥。将干燥过的有机层转入蒸馏瓶中，水浴蒸去苯和未作用的乙酸乙酯。当温度升至95℃时停止蒸馏。将剩余液体进行减压蒸馏，收集54~55℃/931Pa(7mmHg)的馏分即为产品，产量约1.8g，产率35.4%。

纯乙酰乙酸乙酯为无色液体，b.p.180.4℃（同时分解），d_4^{20} 1.0282，n_D^{20} 1.4194。

乙酰乙酸乙酯的性质实验

由于乙酰乙酸乙酯存在着酮式和烯醇式互变异构体（在室温时含有93%的酮式及7%的烯醇式），因此既有酮羰基的性质，又有烯醇的性质。

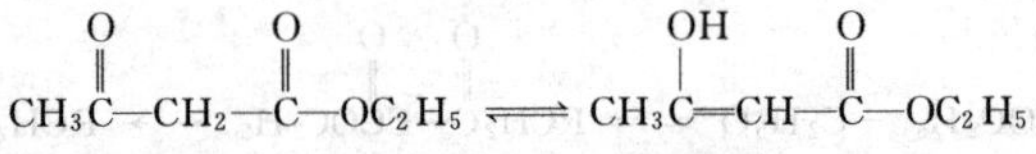

$$CH_3\text{—}\underset{O}{\overset{\|}{C}}\text{—}CH_2\text{—}\underset{O}{\overset{\|}{C}}\text{—}OC_2H_5 \rightleftharpoons CH_3\text{—}\underset{OH}{\overset{|}{C}}\text{=}CH\text{—}\underset{O}{\overset{\|}{C}}\text{—}OC_2H_5$$

酮式　　　　　　　　　　　烯醇式

在这种结构中存在着两个配位中心，可以与一些金属离子形成螯合物，利用这一性质可以进行定性检测。

(1) 与2,4-二硝基苯肼的反应　在试管中加入3滴新配制的2,4-二硝基苯肼溶液，然后加入2滴乙酰乙酸乙酯，微热后冷却可见黄色沉淀物。

(2) 与溴水和三氯化铁的反应　在试管中加入2滴乙酰乙酸乙酯和1滴1%三氯化铁溶液，观察溶液的颜色有何变化。然后再加入几滴溴水，振荡，观察溶液的颜色变化，放置片刻再观察颜色变化。记录这些现象并解释。

【注意事项】

① 本实验所使用的仪器均应是干燥的。金属钠在处理过程中应严格防止与水接触，切钠过程要迅速，以免被空气中水汽侵蚀或被空气氧化。

② 乙酸乙酯必须绝对干燥，但其中应含有1%~2%的乙醇。其提纯方法为：将普通乙酸乙酯用饱和氯化钙溶液洗涤数次，以洗去其中所含的部分乙醇，再用高温烘焙过的无水碳酸钾干燥，最后在水浴上蒸馏，收集76~78℃的馏分，即能得到符合要求（含醇量1%~2%）的乙酸乙酯。如果是分析纯的乙酸乙酯则可直接使用。

③ 一定要等钠反应完后再加入乙酸水溶液，以防着火。

④ 用乙酸溶液中和时，开始有固体析出，继续加酸并不断振摇，固体将逐渐消失，最后得到澄清的液体。当溶液已呈弱酸性，而尚有少量固体未溶解时，可加少许水使其溶解。但要注意避免加入过量的乙酸溶液，否则会增加酯在水中的溶解度。另外，酸度过高，会促使副产物"去水乙酸"的生成，从而降低产量。

⑤ 乙酰乙酸乙酯在常压蒸馏时易发生分解，其分解产物为"去水乙酸"，这将影响产率，故应采用减压蒸馏收集产品，且压力越低越好。

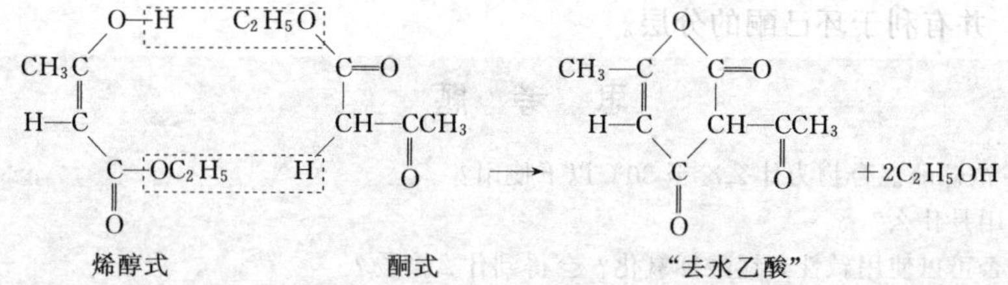

$$\text{烯醇式} \quad \text{酮式} \quad \longrightarrow \quad \text{"去水乙酸"} \quad +2C_2H_5OH$$

"去水乙酸"通常溶解于酯层内，随着过量的乙酸乙酯的蒸出，特别是最后减压蒸馏时，随着部分乙酰乙酸乙酯的蒸出，"去水乙酸"就呈棕黄色固体析出。

⑥ 产率以钠计算，乙酸乙酯是过量的。

思 考 题

1. 本实验所用仪器未经干燥处理，对反应有何影响？
2. 在本实验中加入50%乙酸及氯化钠饱和溶液的目的是什么？
3. 什么是互变异构现象？如何用实验证明乙酰乙酸乙酯是两种互变异构体的平衡混合物？
4. 本实验所用的缩合剂是什么？它与反应物的摩尔比如何？应以哪种原料为基准计算产率？

3.8 氧化还原反应

实验28 环己酮的合成

【反应式】

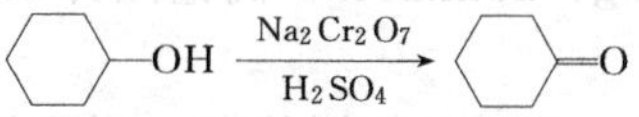

$$\text{C}_6\text{H}_{11}\text{OH} \xrightarrow[H_2SO_4]{Na_2Cr_2O_7} \text{C}_6\text{H}_{10}\text{O}$$

【主要试剂】

环己醇 5.00g（5.2mL，50mmol），重铬酸钠 5.25g（$Na_2Cr_2O_7 \cdot 2H_2O$，17.6mmol），氯化钠，浓硫酸，无水硫酸镁。

【实验步骤】

在100mL的烧杯中加入30mL水和5.25g重铬酸钠，搅拌使之溶解，然后在冷却和搅拌下慢慢加入4.3mL浓硫酸，冷至30℃以下备用。

在100mL的两口瓶中加入5.00g环己醇，然后将上述已溶解的铬酸溶液加入其中，振荡使之混合。观察温度变化，当温度上升至55℃时，立即用冷水浴控制反应温度在55~60℃之间。大约0.5h后，温度开始下降，撤去冷水浴，将反应瓶放置1h，其间不断振荡，反应溶液变成墨绿色。

在反应瓶中加入30mL水及沸石，安装成蒸馏装置。将环己酮与水一起蒸出，环己酮与水能形成沸点为95℃的共沸混合物。直至馏出液不再浑浊，收集约25mL馏出液。向馏出液中加入氯化钠使溶液饱和后，分出有机相，水相用30mL乙醚分两次萃取，萃取液与有机相合并，用无水硫酸镁干燥。在水浴上蒸去乙醚后，改用空气冷凝管进行常压蒸馏，收集151~155℃的馏分，产品3.0~3.50g，产率61%~71%。

纯环己酮为无色液体，b.p.155.7℃，d_4^{20} 0.9478，n_D^{20} 1.4507。

【注意事项】

① 加水蒸馏产品，实质上是一种简化了的水蒸气蒸馏。环己酮和水形成恒沸混合物，b.p.95℃，含环己酮38.4%。

② 水的馏出量不易过多，否则即使使用盐析仍不可避免有少量环己酮溶于水中而损失。环己酮在水中的溶解度：31℃时为2.4g/100mL水。馏出液中加入食盐是为了降低环己酮在

水中的溶解度，并有利于环己酮的分层。

思 考 题

1. 重铬酸钠-浓硫酸混合物为什么冷至30℃以下使用？
2. 盐析的作用是什么？
3. 该反应是否可以使用碱性高锰酸钾氧化？会得到什么产物？

实验29 己二酸的合成

己二酸是合成尼龙-66的主要原料之一，可以用高锰酸钾氧化环己酮制得，也可以用硝酸或高锰酸钾氧化环己醇制得。

【反应式】

方法1 环己酮 $\xrightarrow{KMnO_4}$ HOOC(CH$_2$)$_4$COOH

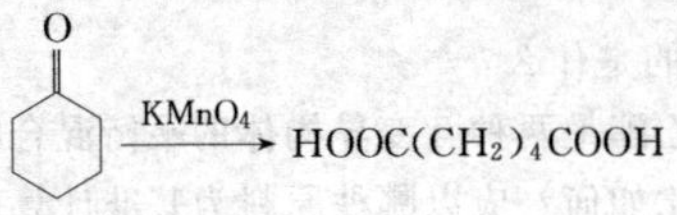

方法2 环己醇 $\xrightarrow{[O]}$ 环己酮 $\xrightarrow{[O]}$ HOOC(CH$_2$)$_4$COOH

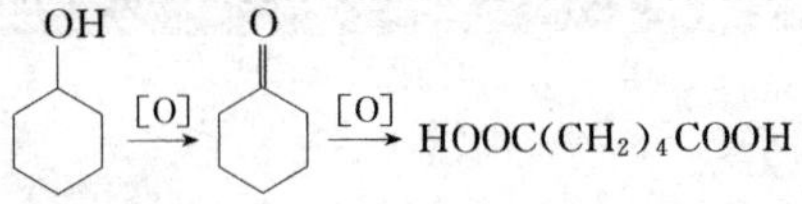

【主要试剂】

方法1 环己酮 2mL（1.9g，19.3mmol），高锰酸钾 6.3g（40mmol），氢氧化钠，亚硫酸氢钠，浓盐酸。

方法2 环己醇 2.6mL（2.5g，25mmol），高锰酸钾 12g（76mmol），碳酸钠，浓硫酸。

【实验步骤】

方法1 在100mL三口烧瓶上分别安装搅拌器、温度计和回流冷凝管。瓶内放入6.3g高锰酸钾，50mL 0.3mol/L氢氧化钠溶液和2mL环己酮。注意反应温度，如反应温度超过45℃时，应用冷水浴适当冷却，然后保持温度45℃反应25min，再在石棉网上加热至沸腾5min使反应完全。取1滴反应混合物放在滤纸上检查高锰酸钾是否还存在，若有未反应的高锰酸钾存在，会在棕色二氧化锰周围出现紫色环。假如有未反应的高锰酸钾存在则可加少量的固体亚硫酸氢钠直至点滴试验呈负性。抽气过滤反应混合物，用水充分洗涤滤饼。滤液置于烧杯中，在石棉网上加热浓缩到10mL左右，用浓盐酸酸化使溶液pH=1~2后再多加2mL浓盐酸，冷却后过滤，即得到粗产物。粗制的己二酸可用水重结晶，并用活性炭脱色，得到白色晶体1.5g，产率53%，熔点为151~152℃。

纯己二酸为白色棱状晶体，熔点为153℃。

方法2 在250mL三口烧瓶上安装好搅拌器和温度计。向反应瓶内加入2.6mL环己醇和碳酸钠水溶液（3.8g碳酸钠溶于35mL温水）。在搅拌下，分批加入研细的12g高锰酸钾，约需2.5h。加入时控制反应温度始终大于30℃。加完后继续搅拌，直至反应温度不再上升为止，然后在50℃水浴中加热并不断搅拌30min。反应过程中有大量二氧化锰沉淀产生。

将反应混合物抽滤，用10mL 10%碳酸钠溶液洗涤滤渣。搅拌下慢慢滴加浓硫酸，直至溶液呈强酸性，己二酸沉淀析出，冷却，抽滤，晾干。产量约2.2g，产率约60.2%，m.p.153℃。

【注意事项】

方法1 此反应是放热反应，反应开始后会使混合物超过45℃。假如在室温下反应开始5min后，混合物温度尚不能上升至45℃，则可小心温热至40℃，使反应开始。

方法2

① 配制碳酸钠水溶液时水太少将影响搅拌效果，使高锰酸钾不能充分反应。

② 加入高锰酸钾后，反应可能不立即开始，可用水浴温热，当温度升到 30℃时，必须立刻撤去温水浴，该放热反应自动进行。

思 考 题

1. 为什么必须严格控制氧化反应的温度？
2. 方法 2 的反应体系中加入碳酸钠有何作用？

实验 30　苯甲酸的合成

苯甲酸俗称安息香酸。苯甲酸及其钠盐是食品的重要防腐剂，苯甲酸可用作制药和染料的中间体，还用于制造增塑剂、聚酯聚合用引发剂、香料等，此外还可用作钢铁设备的防锈剂。

【反应式】

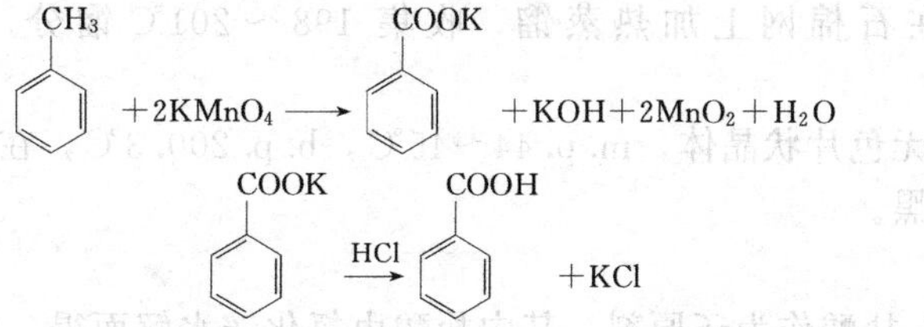

$$C_6H_5CH_3 + 2KMnO_4 \longrightarrow C_6H_5COOK + KOH + 2MnO_2 + H_2O$$

$$C_6H_5COOK \xrightarrow{HCl} C_6H_5COOH + KCl$$

【主要试剂】

甲苯 2.7mL（2.3g，25mmol），高锰酸钾 8.5g（53.8mmol），浓盐酸。

【实验步骤】

在 250mL 圆底烧瓶中加入 2.7mL 甲苯和 100mL 水，安装上回流冷凝管，在石棉网上加热至沸腾。从冷凝管上口分批加入 8.5g 高锰酸钾，黏附在冷凝管内壁的高锰酸钾最后用 25mL 水冲洗入瓶内。继续加热煮沸并间歇振摇烧瓶，直到甲苯层几乎近于消失、回流液不再出现油珠（约需 4~5h）。

将反应混合物趁热减压过滤，用少量的热水（苯甲酸溶于热水，难溶于冷水）洗涤滤渣二氧化锰。合并滤液和洗涤液，放在冰水浴中冷却，然后用浓盐酸酸化至刚果红试纸变蓝，苯甲酸晶体析出。

待溶液彻底冷却后，减压过滤出苯甲酸，用少量冷水洗涤，彻底抽干后，即得到粗产品，干燥后产量约 1.7g。粗产品可在水中重结晶得到纯品。

纯苯甲酸为无色针状晶体，m.p. 122.4℃。

【注意事项】

① 滤液如果呈紫色，可加入少量亚硫酸氢钠使紫色退去，重新减压过滤。

$$KMnO_4 + NaHSO_3 \longrightarrow MnO_2 \downarrow + Na_2SO_4 + K_2SO_4 + H_2O$$

② 若苯甲酸的颜色不纯，可在适量的热水中重结晶，并用活性炭脱色。苯甲酸在 100g 水中的溶解度为：4℃ 0.18g；18℃ 0.27g；75℃ 2.2g。

思 考 题

1. 还可以用什么方法来制备苯甲酸？
2. 反应完毕后，若过滤液呈紫色，为什么要加亚硫酸氢钠？

实验 31　对甲苯胺的合成

【反应式】

$$p\text{-}CH_3C_6H_4NO_2 \xrightarrow[H^+]{Fe} p\text{-}CH_3C_6H_4NH_2$$

【主要试剂】

对硝基甲苯 4.50g（32.8mmol），还原铁粉 7.00g（125.4mmol），氯化铵 0.90g，碳酸氢钠。

【实验步骤】

在 50mL 的三口烧瓶上安装搅拌器和回流冷凝管，向烧瓶中加入 7.00g 还原铁粉，0.90g 氯化铵及 20mL 水，边搅拌边加热，小火煮沸 15min。稍冷，加入 4.50g 对硝基甲苯，在搅拌下加热回流 1h。反应结束后，冷却至室温，用 5％碳酸氢钠溶液中和。搅拌下将适量苯加入反应混合物内，抽滤，除去铁粉残渣，用少量苯洗涤残渣。滤液倒入分液漏斗中，分出苯层，水相用苯萃取三次，合并苯萃取液。再用 5％盐酸对上述苯萃取液提取三次，合并盐酸提取液，搅拌下往盐酸提取液中加入 20％氢氧化钠溶液，析出粗产品。抽滤，并用少量水洗涤，再用少量苯萃取水相，苯萃取液与粗产品合并。水浴蒸馏除去苯，然后向残留物中加少量锌粉，在石棉网上加热蒸馏，收集 198～201℃馏分。产品约 2.50g，m.p. 44～45℃。

纯对甲苯胺为无色片状晶体，m.p. 44～45℃，b.p. 200.3℃，在空气及光的作用下因发生氧化作用而易变黑。

【注意事项】

① 本实验以铁-盐酸作为还原剂，其中盐酸由氯化铵水解而得。

$$NH_4Cl + H_2O \rightleftharpoons NH_4OH + HCl$$

② 加入碳酸氢钠要控制 pH 在 7～8 之间，避免因碱性过强产生胶状氢氧化铁使分离发生困难。

③ 铁残渣为活性铁泥，内含二价铁 44.7％（以 FeO 计算），呈黑色颗粒状，在空气中会剧烈发热，故应及时倒入盛水的废物缸中。

④ 除蒸馏法外，还可用乙醇-水混合溶剂重结晶方法纯化对甲苯胺。

思 考 题

1. 在还原反应开始前，为什么要对铁粉作预处理？
2. 后处理时，为什么先加碳酸氢钠水溶液和苯，再用 5％盐酸对苯层进行萃取？

实验 32　二苯甲醇的合成

【反应式】

方法 1　$C_6H_5COC_6H_5 \xrightarrow[CH_3OH]{NaBH_4} C_6H_5CH(OH)C_6H_5$

方法 2　$C_6H_5COC_6H_5 \xrightarrow{Zn, NaOH} C_6H_5CH(OH)C_6H_5$

【主要试剂】

方法 1　二苯甲酮 0.92g（5.05mmol），硼氢化钠 0.12g（3.2mmol），甲醇 4mL，石油醚（沸程 60～90℃）。

方法 2　二苯甲酮 1.83g（10.04mmol），锌粉 1.97g（30.1mmol），氢氧化钠 1.97g（49.3mmol），95％乙醇，盐酸，石油醚（沸程 60～90℃）。

【实验步骤】

方法 1　在 30mL 干燥磨口的锥形瓶中，加入 0.92g 二苯甲酮和 4mL 甲醇，摇动使其溶解。迅速称取 0.12g 硼氢化钠加入反应瓶中，并迅速安装上回流冷凝管。反应物自然升温至沸腾，室温下放置 20min，不时振荡。加入 1.5mL 水，在水浴上加热至沸腾，保持 5min。稍冷后，将反应瓶置于冰水浴中冷却，析出结晶。减压过滤，粗品干燥后用石油醚（沸程

60～90℃) 或环己烷重结晶。产率70%～80%，m.p.67～68℃。

纯品为白色针状晶体，m.p.69℃。

方法2　在50mL干燥的圆底烧瓶中，依次加入1.97g氢氧化钠、1.83g二苯甲酮、1.97g锌粉和20mL 95%乙醇，振摇，使氢氧化钠和二苯甲酮逐渐溶解。装上回流冷凝管，置于80℃水浴中，电磁搅拌2h。然后停止搅拌，冷却。用布氏漏斗过滤，残渣用少量95%乙醇洗涤。将滤液倒入盛有90mL冰水和4mL浓盐酸的烧杯中，立即出现白色沉淀，减压过滤。干燥后粗产品用石油醚（沸程60～90℃）重结晶，得到白色针状晶体1.4～1.6g，产率约80%，m.p.67～68℃。

【注意事项】

① 硼氢化钠是强碱性物质，易吸潮，具腐蚀性。称量时要迅速，放置潮解，操作要小心，勿与皮肤接触。

② 方法2中，酸化时酸性不宜太强，一般pH为5～6为宜，否则难于析出晶体。

纯二苯甲醇的红外光谱见图3-15。

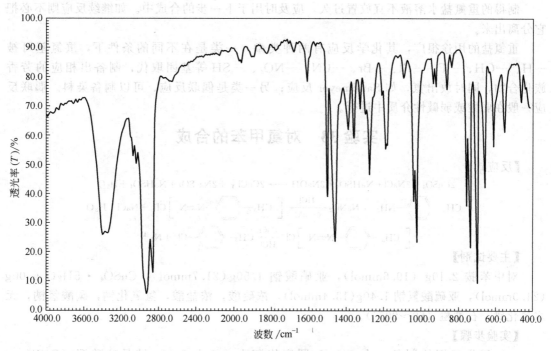

图3-15　二苯甲醇的红外光谱图（研糊法）

思　考　题

1. 比较LiAlH$_4$和NaBH$_4$的还原特性有什么区别？
2. 在方法1中，为什么反应后加入1.5mL水，并加热至沸腾，然后再冷却结晶？

3.9　重氮化反应

芳香族伯胺与亚硝酸钠在过量的、冷的无机酸（常用盐酸和硫酸）水溶液中反应生成重氮盐的反应称为重氮化反应。

$$ArNH_2 + 2HX + NaNO_2 \xrightarrow{0～5℃} ArN_2^+X^- + NaX + 2H_2O$$

通常重氮盐的制备方法是：将1mol芳香族伯胺溶于2.5～3mol的盐酸水溶液中，制成盐酸盐水溶液，然后将溶液冷却至0～5℃，在此温度下不断搅拌慢慢滴加亚硝酸钠水溶液，直到反应液使淀粉-碘化钾试纸变蓝为止，即得到重氮盐水溶液。

在制备重氮盐时应注意以下几点。

① 要控制温度。重氮化反应是一个放热反应，并且大多数重氮盐很不稳定，在室温时易分解，所以必须严格控制反应温度，一般保持在0～5℃进行反应。芳环上有强的间位定位基的伯芳胺，如—NO_2、—SO_3H时，它们的重氮盐比较稳定，往往可在较高的温度下进行重氮化反应。

② 要注意维持溶液的酸度。在制备重氮盐时，酸的用量比理论量多0.5～1mol。若酸性不够，生成的重氮盐会和尚未反应的芳香胺偶联。

③ 要控制亚硝酸钠的用量。若亚硝酸钠过量，则生成多余的亚硝酸，使重氮盐氧化而降低产率。因而，在滴入适量的亚硝酸钠溶液后，要不断用碘化钾-淀粉试纸检验至刚变蓝为止。过量的亚硝酸钠可以加入尿素来除去。

制得的重氮盐水溶液不宜放置过久，应及时用于下一步的合成中。如继续反应则不必把它分离出来。

重氮盐的用途很广，其化学反应有两种类型：一类是在不同的条件下，重氮基可被—H、—OH、—F、—Cl、—Br、—CN、—NO_2、—SH等基团取代，制备出相应的芳香族化合物，同时放出氮，如Sandmeyer反应；另一类是偶联反应，可以制备染料。偶联反应一般在弱酸或弱碱性介质中进行。

实验33 对氯甲苯的合成

【反应式】

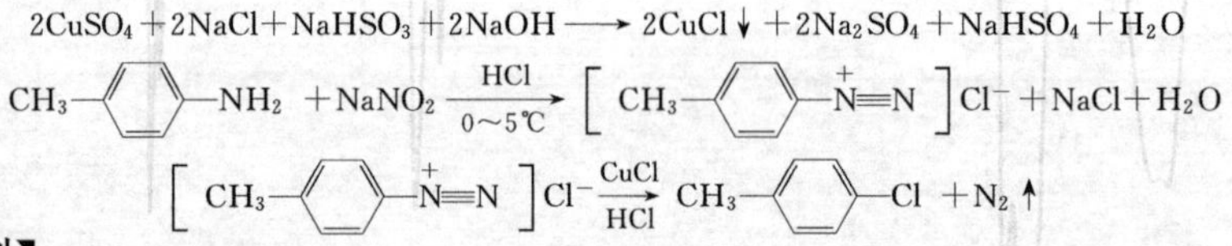

$$2CuSO_4 + 2NaCl + NaHSO_3 + 2NaOH \longrightarrow 2CuCl\downarrow + 2Na_2SO_4 + NaHSO_4 + H_2O$$

$$CH_3-\underset{}{\bigcirc}-NH_2 + NaNO_2 \xrightarrow[0\sim5℃]{HCl} \left[CH_3-\underset{}{\bigcirc}-\overset{+}{N}\equiv N\right]Cl^- + NaCl + H_2O$$

$$\left[CH_3-\underset{}{\bigcirc}-\overset{+}{N}\equiv N\right]Cl^- \xrightarrow[HCl]{CuCl} CH_3-\underset{}{\bigcirc}-Cl + N_2\uparrow$$

【主要试剂】

对甲苯胺2.10g（19.6mmol），亚硝酸钠1.50g（21.7mmol），$CuSO_4 \cdot 5H_2O$ 6.00g（24.0mmol），亚硫酸氢钠1.40g（13.4mmol），浓硫酸，浓盐酸，氢氧化钠，碳酸氢钠，无水氯化钙，石油醚（60～90℃）。

【实验步骤】

(1) 氯化亚铜的制备 在100mL圆底烧瓶中，加入6.00g结晶硫酸铜（$CuSO_4 \cdot 5H_2O$），1.80g精盐和20mL水，加热使固体溶解。趁热（60～70℃）在振摇下加入由1.40g亚硫酸氢钠和0.90g氢氧化钠及10mL水配成的溶液，反应液由原来的蓝绿色变成浅绿色或无色，并析出白色粉末状固体。将烧瓶置于冰水浴中冷却，用倾滗法尽量倒去上层溶液，再用水洗涤固体两次，得到白色粉末状氯化亚铜。再加入10mL冷的浓盐酸，使沉淀溶解，塞紧瓶塞，置于冰水浴中备用。

(2) 重氮盐溶液的制备 在100mL的小烧杯中加入6.0mL浓盐酸、6.0mL水和2.10g对甲苯胺，加热使对甲苯胺溶解。稍冷后，置于冰盐浴中不断搅拌使成糊状，控制温度在5℃下。在搅拌下，用滴管逐滴加入由1.50g亚硝酸钠溶于4.0mL水配成的溶液。控制滴加速度，使反应温度始终保持在5℃以下，必要时可在反应液中加入一小块冰，防止温度上升。85%～90%亚硝酸钠溶液加入后，用淀粉-碘化钾试纸检验。若立即出现深蓝色，表示亚硝酸钠已适量，不必再加，继续搅拌片刻使反应完全。将其置于冰浴中备用。

(3) 对氯甲苯的制备　将制备好的对甲苯胺重氮盐溶液慢慢倒入已冷至 0℃ 的氯化亚铜盐酸溶液中，边加边摇动烧瓶，不久可见瓶中有橙红色重氮盐-氯化亚铜复合物析出。加完后，在室温下放置 30min。然后用水浴慢慢加热到 50～60℃，以分解复合物，直至不再有氮气逸出。

将产物进行水蒸气蒸馏，蒸出对氯甲苯。分出有机层，水层用石油醚（60～90℃）萃取两次（10mL×2），萃取液与有机相合并，并依次用 5% 氢氧化钠溶液、水、浓硫酸各 2mL 洗涤，再用水、5% 碳酸氢钠溶液、水（每次 2mL）洗涤，直到溶液呈中性。有机层用无水氯化钙干燥后，常压下蒸去石油醚，然后再收集 158～162℃ 的馏分，产品 1.50g，产率 60%。

纯对氯甲苯为无色透明液体，b.p. 162℃，d_4^{20} 1.072，n_D^{20} 1.521。

【注意事项】

① 亚硫酸氢钠容易变质，必须使用优质品（纯度在 90% 以上），否则还原反应不完全，将影响产率。

② 在 60～70℃ 制得的氯化亚铜质量较好，颗粒较粗，易于漂洗处理。

③ 若实验中发现溶液仍呈蓝绿色，则表明还原不完全，应酌情多加亚硫酸氢钠溶液。若发现沉淀呈黄褐色，应立即滴入几滴盐酸并稍加振摇，以使其中氢氧化亚铜转化成氯化亚铜，但应控制好加酸的量，因为氯化亚铜会溶于酸中。

④ 用水洗涤氯化亚铜时，要轻轻摇动，静止，小心倾去水层。勿剧烈振摇，否则因其颗粒较细，沉降较慢。

⑤ 氯化亚铜在空气中遇热或见光易被氧化，重氮盐久置易分解，因此二者的制备应同时进行，并且制备好后应立即混合进行反应。因重氮盐生成接近终点时反应较慢，建议先进行重氮盐的制备反应，后制备氯化亚铜较为适宜。

⑥ 因重氮化反应越接近终点时反应越慢，所以每加一滴亚硝酸钠溶液后，搅拌 1～2min 再用淀粉-碘化钾试纸检验。若试纸显蓝色，则表明亚硝酸过量（析出的碘遇淀粉显蓝色）。

$$2KI + 2HNO_2 + 2HCl \longrightarrow I_2 + 2NO + 2H_2O + 2KCl$$

若加入过量的亚硝酸钠溶液，可用尿素分解。

$$H_2NCONH_2 + 2HNO_2 \longrightarrow CO_2\uparrow + 2N_2\uparrow + 3H_2O$$

⑦ 在重氮化操作中，应始终保持溶液呈酸性，使刚果红试纸变蓝。

⑧ 在制备对氯甲苯时，为避免副产物偶氮苯的生成，倒入重氮盐的速度不易太快。

⑨ 重氮盐-氯化亚铜复合物不稳定，15℃ 即会分解出对氯甲苯。稍加热可使分解加速。但若升温过速，温度过高会产生焦油状物与对甲苯酚，使产率降低。若时间许可，可室温放置过夜，再加热分解。在水浴加热分解时，有大量氮气逸出，应不断搅拌，以免反应液外溢。

⑩ 浓硫酸可去除副产物偶氮苯。

思　考　题

1. 什么是重氮化反应？它在有机合成中有何用途？
2. 为什么重氮化反应必须在低温下进行？温度过高或溶液酸度不够会产生什么副产物？
3. 能否用甲苯直接氯化制备对氯甲苯？
4. 氯化亚铜在盐酸存在下，被亚硝酸氧化，可以观察到一种红棕色的气体放出，试解释这一现象，并用反应式来表示。此气体对人体有何害处？
5. 淀粉-碘化钾试纸为什么能检验亚硝酸的存在？发生了哪些反应？如加入过量的亚硝酸钠对此反应有什么不利？

实验34 甲基橙的合成

【反应式】

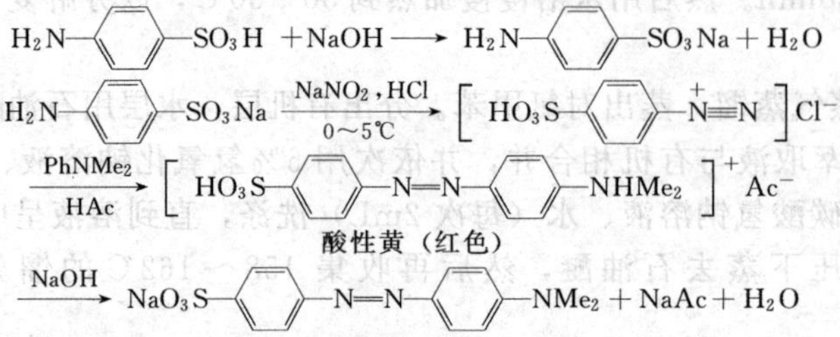

$$H_2N-C_6H_4-SO_3H + NaOH \longrightarrow H_2N-C_6H_4-SO_3Na + H_2O$$

$$H_2N-C_6H_4-SO_3Na \xrightarrow[0\sim5℃]{NaNO_2, HCl} [HO_3S-C_6H_4-N\equiv N]^+ Cl^-$$

$$\xrightarrow[HAc]{PhNMe_2} [HO_3S-C_6H_4-N=N-C_6H_4-NHMe_2]^+ Ac^-$$

酸性黄（红色）

$$\xrightarrow{NaOH} NaO_3S-C_6H_4-N=N-C_6H_4-NMe_2 + NaAc + H_2O$$

甲基橙（橙色）

【主要试剂】

对氨基苯磺酸（含两个结晶水）2.00g（9.56mmol），5％氢氧化钠10mL（263.4mmol），亚硝酸钠0.8g（11.6mmol），浓盐酸，N,N-二甲基苯胺1.3mL（1.24g，10.3mmol），冰醋酸，10％氢氧化钠，氯化钠，乙醇。

【实验步骤】

（1）对氨基苯磺酸重氮盐的制备　将2.00g对氨基苯磺酸晶体放入100mL的烧杯中，再加入10mL 5％氢氧化钠溶液，热水浴中温热使之溶解。然后放在冰水浴中冷却，待冷至室温后，在搅拌下加入0.8g亚硝酸钠使其溶解。然后，搅拌下将该混合物溶液分批滴入装有13mL冷水和2.5mL浓盐酸的烧杯中，使温度保持在5℃以下，很快就有对氨基苯磺酸重氮盐的细粒状白色沉淀析出。滴加完后用淀粉-碘化钾试纸检验。为了保证反应完全，将反应液继续在冰浴中放置15min。

（2）偶联　将新蒸馏的1.3mL N,N-二甲基苯胺和1mL冰醋酸加入一支试管中，振荡使之混合。在搅拌下将此溶液慢慢加到上述冷却的对氨基苯磺酸重氮盐溶液中，加完后，继续搅拌10min，以保证偶联反应完全，此时有红色的酸性黄沉淀生成。然后再在冷却下搅拌，慢慢地加入15mL 10％氢氧化钠溶液，直至产物变为橙色，此时反应液呈碱性。粗的甲基橙细粒状沉淀析出。

将反应物在沸水浴上加热5min，使沉淀溶解后，稍冷，置于冰浴中冷却，待甲基橙全部重新结晶析出后，抽滤收集结晶。依次用少量的饱和氯化钠水溶液、乙醇和乙醚洗涤产品，压干。若要制得较纯的产品，可用溶有少量氢氧化钠（约0.15g）的沸水（每克粗产物约需25mL）进行重结晶，得到橙红色片状晶体约2.5g，产率80％。

产品没有确定的熔点，因此不必测定。

取少许产品溶于水中，加几滴稀盐酸，然后用氢氧化钠溶液中和，观察溶液颜色有何变化。

【注意事项】

① 对氨基苯磺酸是一种两性化合物，其酸性比碱性强，以酸性内盐的形式存在，能与碱作用生成盐，难与酸作用生成盐，所以不溶于酸。但重氮化反应又要在酸性溶液中进行，因此进行重氮化反应时，首先将对氨基苯磺酸与碱作用，生成水溶性较大的对氨基苯磺酸钠。

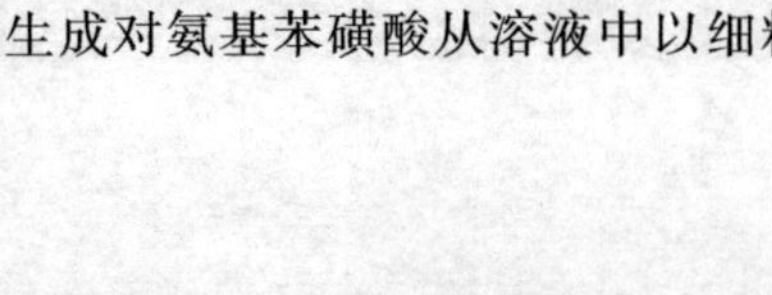

$$C_6H_4(NH_3^+)(SO_3^-) + NaOH \longrightarrow C_6H_4(NH_2)(SO_3^-Na^+) + H_2O$$

② 在重氮化反应中，当溶液酸化时生成亚硝酸。

$$NaNO_2 + HCl \longrightarrow HNO_2 + NaCl$$

同时，对氨基苯磺酸钠则生成对氨基苯磺酸从溶液中以细粒沉淀析出，并立即与亚硝酸作用，发生重氮化反应，生成粉末状的重氮盐。为了使对氨基苯磺酸完全重氮化，反应过程中必须不断搅拌。

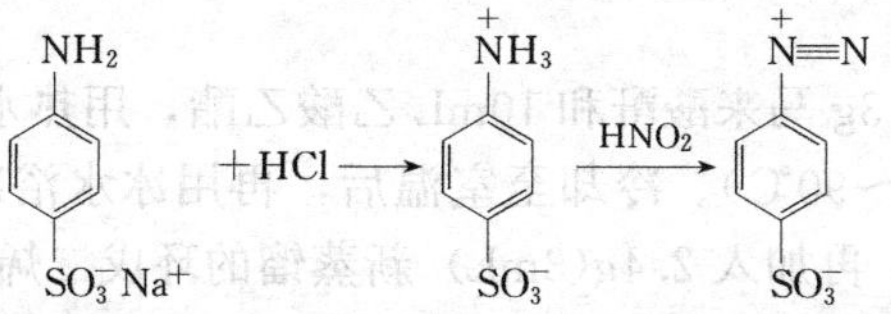

$$\underset{SO_3Na}{\underset{|}{C_6H_4}}-NH_2 \xrightarrow{+HCl} \underset{SO_3^-}{\underset{|}{C_6H_4}}-NH_3^+ \xrightarrow{HNO_2} \underset{SO_3^-}{\underset{|}{C_6H_4}}-N\equiv N$$

③ 重氮化反应过程中要严格控制反应温度，一般控制在 0~5℃。若反应温度高于 5℃，则生成的重氮盐易水解生成相应的酚类，从而降低产率。因此制备好重氮盐后仍要保存在冰水浴中备用。本实验中制得的对氨基苯磺酸重氮盐，因重氮基的对位有强吸电子的磺酸基，因此比较稳定。

④ 用淀粉-碘化钾试纸检验时，若试纸不显蓝色，则需补充亚硝酸钠溶液，并充分搅拌，直到使淀粉-碘化钾试纸刚好呈蓝色；若亚硝酸钠已过量，可加少量尿素分解除去过量的亚硝酸钠。因为亚硝酸能起氧化和亚硝基化作用，所以亚硝酸的过量将会引起一系列副反应。

$$2HNO_2 + 2KI + 2HCl \longrightarrow I_2 + 2NO + 2H_2O + 2KCl$$
$$H_2NCONH_2 + 2HNO_2 \longrightarrow CO_2\uparrow + 2N_2\uparrow + 3H_2O$$

⑤ N,N-二甲基苯胺久置易被氧化，故需要用新蒸馏的。

⑥ 在偶联反应步骤中，10%氢氧化钠溶液一直滴加到当它接触到混合物的表面时，不再产生黄色为止，加碱期间反应混合物的温度始终保持在 0~5℃。必须用试纸测试反应物是否呈碱性，否则粗甲基橙的色泽不佳。

⑦ 粗产物呈碱性，温度稍高时易使产物变质，颜色变深。所以加热溶解时温度不易过高，一般约在 60℃。湿的甲基橙受日光照射亦会使颜色变深。用乙醇和乙醚洗涤产品的目的是使产品迅速干燥。

思 考 题

1. 什么是偶联反应？结合本实验讨论偶联反应在什么介质中进行？
2. 在本实验中，制备重氮盐时为什么要把对氨基苯磺酸变成钠盐？如果直接与盐酸混合，是否可行？
3. N,N-二甲基苯胺与重氮盐偶联时为什么总在氨基的对位上发生？
4. 试解释甲基橙在酸性介质中变色的原因，并用反应式表示。

3.10 Diels-Alder 反应

共轭二烯烃和亲双烯试剂发生 1,4-加成反应，生成环己烯型化合物，这种反应称为 Diels-Alder 反应。该反应不仅是合成六元环有机化合物的重要方法，而且在理论上占有非常重要的位置。当双烯体上含有给电子基团（如烷基、烷氧基等），亲双烯体上含有吸电子基团（如羰基、酯基、氰基等）时反应速率加快，容易进行。此反应是一步发生的协同反应，不存在活泼的反应中间体，具有可逆性和立体定向的顺式加成两大特点。

实验 35 环戊二烯与马来酸酐的反应

【反应式】

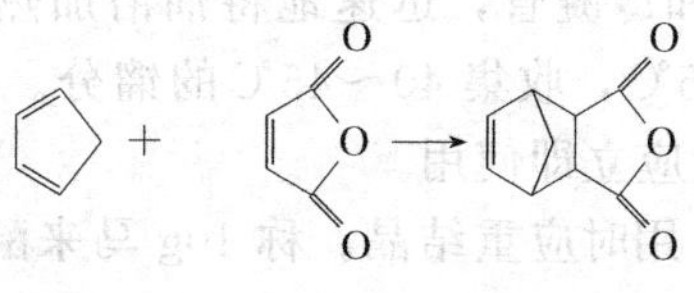

【主要试剂】

环戊二烯 2.4g（3mL，36mmol），马来酸酐 3g（31mmol），石油醚（60~90℃），乙酸乙酯。

【实验步骤】

在50mL锥形瓶中加入3g马来酸酐和10mL乙酸乙酯，用热水浴加热使固体物全部溶解，然后加入10mL石油醚（60~90℃）。冷却至室温后，再用冰水浴冷却（这时可能会有少量沉淀析出，但不会影响反应），再加入2.4g(3mL)新蒸馏的环戊二烯。将盛有反应液的锥形瓶置于冰水浴中并不断摇动，直到白色固体析出，放热停止。用水浴加热使析出的固体全部溶解，然后再静置让其缓缓地冷却，得到白色针状结晶，抽滤，用5mL乙酸乙酯和石油醚的混合液（体积比为1:1）淋洗，彻底抽干，干燥。称量，产量约4g，产率78.7%，m.p. 164~165℃。

此加成产物还具有双键，能使高锰酸钾溶液或溴的四氯化碳溶液褪色。

图3-16为环戊二烯与马来酸酐加成产物（双环[2.2.1]-2-庚烯-5,6-二酸酐，bicyclo[2.2.1] hept-2-en-5,6-dicarboxylic anhydride）的红外光谱。

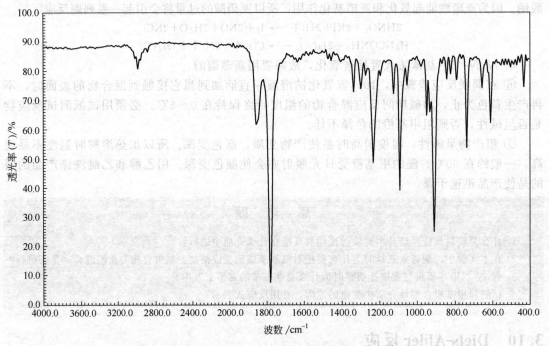

图3-16 环戊二烯与马来酸酐加成产物的红外光谱图

【注意事项】

① 环戊二烯在室温下容易二聚，生成环戊二烯的二聚体。但环戊二烯与其二聚体在二聚体的 b.p.170℃时能建立起一个平衡。

因此，纯净的环戊二烯可经二聚体的解聚、分馏而获得。在100mL圆底烧瓶中加入10g环戊二烯的二聚体，连接上一个长30cm的Vigrenx分馏柱（或一般的空气冷凝管），分馏柱头连上温度计（100℃）和冷凝管，迅速地将油浴加热到170℃以上使二聚体解聚。但要控制分馏头的温度不超过45℃，收集40~45℃的馏分。接收瓶中加入无水氯化钙，并用冰水冷却。新蒸出的环戊二烯应立即使用。

② 马来酸酐如放置过久，用时应重结晶。称10g马来酸酐加15mL三氯甲烷，煮沸数分

钟，趁热过滤，滤液放冷，即得到纯净的马来酸酐。抽滤，置于干燥器中晾干，m.p. 60℃。

思 考 题

二聚环戊二烯解聚应注意什么？为什么？

实验 36　环戊二烯与对苯醌的反应

【反应式】

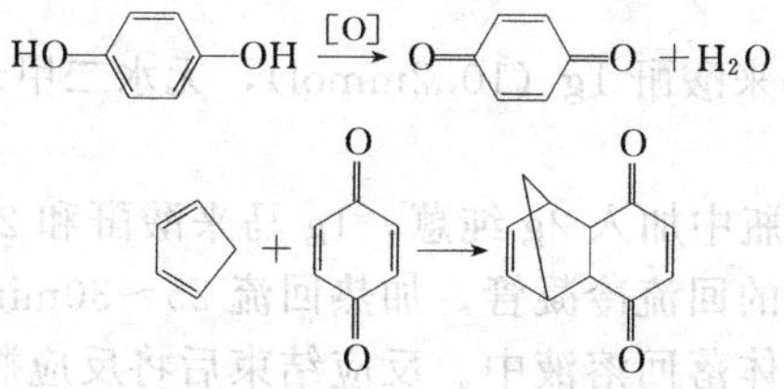

$$HO-\bigcirc-OH \xrightarrow{[O]} O=\bigcirc=O + H_2O$$

$$\bigcirc + \bigcirc\!\!=\!\!O \longrightarrow \text{（加成产物）}$$

【主要试剂】

对苯二酚 10g（90.8mmol），重铬酸钠 14g（47mmol），浓硫酸，苯，无水氯化钙，对苯醌 1.8g（17mmol），环戊二烯 1.2g（1.5mL，18mmol），石油醚（60～90℃），无水乙醇。

【实验步骤】

(1) 对苯醌的合成　在 250mL 烧杯中分别加入 200mL 水和 10g 对苯二酚，加热到 50℃ 并搅拌使对苯二酚溶解，然后冷却至室温。搅拌下逐滴加入 5.4mL 浓硫酸，此时可观察到溶液颜色变为黄色。把 14g 重铬酸钠用 6.5mL 水溶解，然后在搅拌下慢慢滴入上述反应液中，反应混合物的黏度逐渐增大，颜色也逐渐变深至黄绿色，并伴有少量绿黑色沉淀生成。冷却至 10℃，抽滤，并尽量除净所含水分。用 30mL 苯萃取水相两次，用 50mL 苯溶解滤出的粗产物，合并两溶液并用无水氯化钙干燥。数小时后得到橙黄色透明溶液，过滤除去氯化钙，减压旋转蒸掉大部分苯，直到有少量亮黄色结晶析出，然后用冰浴冷却剩余物直到晶体全部析出为止。抽滤，干燥，称量，产率约为 71%。产物为亮黄色晶体，m.p. 115～116℃。

(2) 环戊二烯与对苯醌的加成　在 100mL 锥形瓶中分别加入 1.8g 对苯醌和 7mL 无水乙醇，得到黄色悬浮液，将此锥形瓶置于冰水浴中冷却至 0～5℃。然后把新蒸馏的环戊二烯 1.2g 迅速加入到上述悬浮液中，摇匀后置于冰浴中，15min 后除去冰浴，在室温下放置 45min。在该过程中，溶液由浑浊变澄清呈橙黄色，并伴有大量淡黄色沉淀析出。然后，在水泵减压下蒸出乙醇，得淡黄色固体。在 70℃ 下用石油醚（60～90℃）溶解，回流，重结晶，抽滤，得淡黄色针状晶体。干燥后得到产品，m.p. 77～78℃。

【注意事项】

① 当对苯二酚不纯时，在硫酸存在下将变成黏稠状物质，所以应过滤除去不溶物。如使用市售的对苯醌，因其常常不纯，呈紫黑色，所以应将其提纯处理，可采用升华法提纯。

② 在制备对苯醌时由于体系逐渐变得黏稠，所以应快速搅拌以提高产率。

③ 对苯醌见光易变色，应避光保存。对苯醌易于挥发，因此用苯溶解醌时温度不宜过高。

思 考 题

1. 所得加成产物属于内型（endo-）结构还是外型（exo-）结构？
2. 本实验中对苯醌与环戊二烯的摩尔比是多少？若加大反应中环戊二烯的用量，将得到什么产物？写出其结构式。

实验 37　蒽与马来酸酐的反应

【反应式】

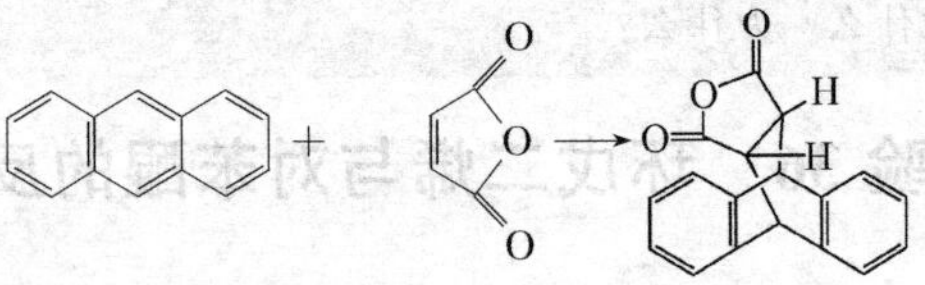

【主要试剂】

蒽 2g（11.2mmol），马来酸酐 1g（10.2mmol），无水二甲苯。

【实验步骤】

在干燥的 50mL 圆底烧瓶中加入 2g 纯蒽，1g 马来酸酐和 25mL 无水二甲苯，然后在烧瓶上连接配有氯化钙干燥管的回流冷凝管。加热回流 25～30min，并且在回流过程中间歇振摇烧瓶，使沾在瓶壁上的晶体落回溶液中。反应结束后将反应瓶静置冷却至室温，有大量白色固体析出。抽滤，用无水二甲苯重结晶粗产物，产物放在盛有石蜡片和硅胶的真空干燥器内干燥。产量约 2g，m.p. 262～263℃（分解）。

纯化合物 m.p. 263～264℃。

【注意事项】

① 马来酸酐和生成物遇水会水解成二元酸，因此反应仪器及所用的试剂必须干燥。
② 石蜡能吸收烃类气体，用它除去吸附在产物中的痕量二甲苯。
③ 产物应保存在干燥器内，否则产物会吸收空气中的水分，发生水解生成二元酸，给测熔点带来困难。

思 考 题

1. 若本实验中使用未干燥的二甲苯，对实验会产生什么影响？
2. 为什么蒽与马来酸酐可发生 Diels-Alder 反应？且反应一般发生在蒽的 9,10 位？

3.11　相转移催化反应

在有机合成中，通常均相反应容易进行，而非均相反应则难以发生。但有机合成中却常遇到两种反应物处于不同相的非均相反应，由于反应物不能彼此靠拢，其反应速率慢、产率低，甚至很难发生反应。此时可利用相转移催化法，即使用一种催化剂使得互不相溶的两相物质发生反应或者加速反应，这种反应就称为相转移催化反应。作为相转移催化剂（Phase Transfer Catalyst，PTC）应具备两个基本条件：①能够将一个试剂由一相转移到另一相中；②被转移的试剂处于较活泼的状态。

常用的相转移催化剂有三类：季盐类、冠醚类和非环多醚类。

（1）季盐类　包括季铵盐、季𬭸盐等。在这类化合物中，烃基是酯溶性的，而负离子亲水性较强。烃基的阳离子体积要适中，若体积太大，就会降低在水中的溶解度；若太小，在水中的溶解度会增大，在有机相中的溶解度则会降低，这将影响相转移催化作用。

（2）冠醚类　能与某些金属离子配合而使其溶于有机相。但由于其价格较贵，毒性较大，因而应用受到一定的限制。

（3）非环多醚类　非环多醚类相转移催化剂的作用机理与冠醚类似，对一些金属离子也具有一定的配合能力。如聚乙二醇（PEG），当其呈弯曲状时，形如冠醚。一般相对分子质量在 400～600 之间的聚乙二醇，其弯曲结构的孔径大小适中，对金属离子的配合能力较强，相转移催化效果较好。

实验 38　7,7-二氯二环［4.1.0］庚烷的合成

【反应式】

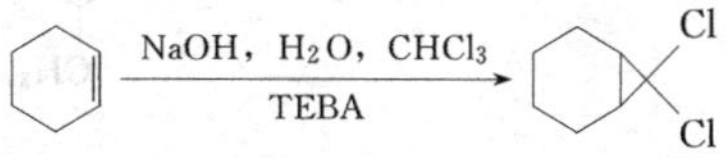

$$\text{环己烯} \xrightarrow[\text{TEBA}]{\text{NaOH, H}_2\text{O, CHCl}_3} \text{7,7-二氯二环[4.1.0]庚烷}$$

【主要试剂】

环己烯 1.62g（2.0mL，20.0mmol），TEBA（三乙基苄基氯化铵）0.3g，氯仿 20mL，氢氧化钠 4.00g（100mmol）。

【实验步骤】

在装有电磁搅拌器、回流冷凝管、温度计和恒压滴液漏斗的 50mL 三口烧瓶中，加入 2.0mL 新蒸馏的环己烯，0.3g 相转移催化剂 TEBA 和 10mL 氯仿，在快速搅拌下，由滴液漏斗滴加 4.00g 氢氧化钠溶于 4mL 水中的溶液，此时有放热现象。滴加完毕后，在剧烈搅拌下水浴加热回流 40min。反应液为黄色，并有固体析出。

待反应液冷却至室温，加入 10mL 水使固体溶解。将混合液转移到分液漏斗中，分出有机层，水层用 10mL 氯仿提取一次，将提取液与有机层合并，有机层用水洗涤 3 次（10mL×3）至中性。有机层用无水硫酸镁干燥，水浴蒸出氯仿后，进行减压蒸馏，收集 80～82℃/2.13kPa（16mmHg）的馏分。产品 1.80g，为无色透明液体。

纯 7,7-二氯二环［4.1.0］庚烷为无色透明液体，b.p. 197～198℃，n_D^{20} 1.5014。

【注意事项】

① 本实验属于卡宾反应，应用相转移催化反应来合成目标分子。所用相转移催化剂可有多种选择，如四丁基溴化铵、聚乙二醇-400（PEG-400）等，效果相同。本实验相转移循环式如下：

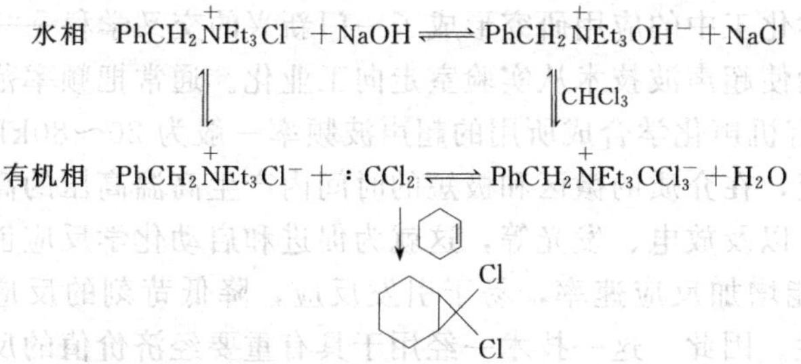

水相　　$PhCH_2NEt_3Cl^- + NaOH \rightleftharpoons PhCH_2NEt_3OH^- + NaCl$

$\Updownarrow$　　　　　　　　　$\Updownarrow CHCl_3$

有机相　$PhCH_2NEt_3Cl^- + :CCl_2 \rightleftharpoons PhCH_2NEt_3CCl_3^- + H_2O$

$\Downarrow$ 环己烯

7,7-二氯二环[4.1.0]庚烷

② 环己烯最好用新蒸馏过的。

③ 反应温度必须控制在 50～55℃，低于 50℃ 则反应不完全，高于 60℃ 反应液颜色加深，黏稠，产率低，原料或中间体卡宾均可能挥发损失。

④ 此反应是在两相中进行的，反应过程中必须剧烈搅拌反应物，否则影响产率。

⑤ 反应液在分层时，常出现较多絮状物，可用布氏漏斗过滤处理。

⑥ 分液时，水层要分尽，有机层干燥要彻底，才不会影响蒸馏。

思 考 题

1. 本实验中有水存在，为什么二氯卡宾还能与环己烯进行加成反应？
2. 相转移催化剂在本实验中起什么作用？
3. 本实验中，滴加氢氧化钠溶液时，剧烈搅拌的目的是什么？

实验39 对甲苯硫代乙酸的合成

【反应式】

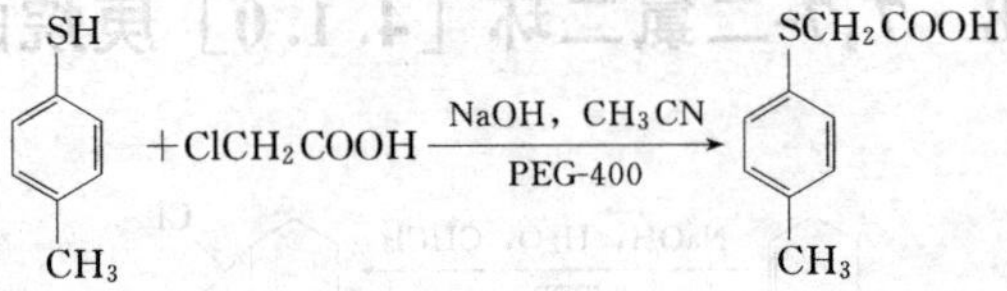

$$\text{对-CH}_3\text{-C}_6\text{H}_4\text{-SH} + \text{ClCH}_2\text{COOH} \xrightarrow[\text{PEG-400}]{\text{NaOH, CH}_3\text{CN}} \text{对-CH}_3\text{-C}_6\text{H}_4\text{-SCH}_2\text{COOH}$$

【主要试剂】

对甲苯硫酚 2.48g（20mmol），氯乙酸 1.89g（20mmol），氢氧化钠 2.0g（50mmol），聚乙二醇-400（PEG-400），乙腈，浓盐酸。

【实验步骤】

在配有电动搅拌器和回流冷凝管的 250mL 干燥三口烧瓶中，加入 2.48g 对甲苯硫酚、1.89g 氯乙酸、2.0g 氢氧化钠、100mL 乙腈和 0.4g PEG-400，在 125℃油浴温度下，加热搅拌回流反应 3h。然后，水浴加热彻底蒸出乙腈，将烧瓶冷却后加入约 40mL 水使瓶内固体钠盐溶解。将此溶液转入烧杯中，在搅拌下滴加盐酸酸化至刚果红试纸变蓝，得到固体沉淀。彻底冷却后，抽滤、并用少量冷水洗涤滤饼，得到粗产品。干燥后，粗产品可用石油醚（60～90℃）重结晶。干燥后称量，测熔点，计算产率。

纯对甲苯硫代乙酸 m.p. 93.5～94.5℃。

【注意事项】

① 对甲苯硫酚有毒并有刺激性臭味，应在通风橱内量取。

② 水浴蒸馏乙腈，且回收后放在指定回收瓶内。

3.12 超声波辐射反应

超声波（Ultrasonic Wave）作为活化和促进化学反应的高新技术是在 20 世纪 80 年代中期以后才发展起来的，这一技术在化学化工中的应用研究形成了一门新兴的交叉学科——超声化学，超声化学的研究和发展已开始使超声波技术从实验室走向工业化。通常把频率范围为 20～1000kHz 的声波称为超声波。有机声化学合成所用的超声波频率一般为 20～80kHz。合成化学主要利用超声波的声空化效应，在介质的微区和极短的时间内产生高温高压的高能环境，并伴有强大的冲击波和微射流，以及放电、发光等，这就为促进和启动化学反应创造了一个极端的物理环境。超声波不仅能增加反应速率，易于引发反应，降低苛刻的反应条件，而且可以改变反应的途径和选择性。因此，这一技术一经用于具有重要经济价值的反应即显示出巨大的应用前景。

3.12.1 格氏反应

在无水乙醚或四氢呋喃中，镁可以和许多脂肪族及芳香族的卤代烃反应生成有机镁化合物，这种有机镁化合物称为格林那（Grignard）试剂，简称格氏试剂。它是由法国化学家格林那（V. Grignard）发明的。格氏试剂易与不饱和化合物，特别是在碳原子和其他元素之间含有重键的物质发生加成反应。与镁有机化合物发生的反应称为格氏反应。格氏试剂是有机合成中用途极广的一种试剂，可用来合成烷烃、醇、醛、羧酸等各类化合物。

在 Grignard 试剂的合成中，传统的方法需使用绝对干燥的乙醚，且需要加入少量碘作诱导剂。而在超声波辐射下该反应可使用无需特殊处理的无水乙醚，而且反应速率快，没有诱导期，产率良好。

实验 40　三苯甲醇的合成

【反应式】

方法 1　二苯甲酮与苯基溴化镁的反应。

$$\text{Ph-Br} \xrightarrow{\text{Mg, Et}_2\text{O}} \text{Ph-MgBr} \xrightarrow{\text{Ph-CO-Ph}}_{\text{Et}_2\text{O,)))}} \text{Ph}_3\text{C-OMgBr} \xrightarrow{\text{H}^+, \text{H}_2\text{O}} \text{Ph}_3\text{C-OH}$$

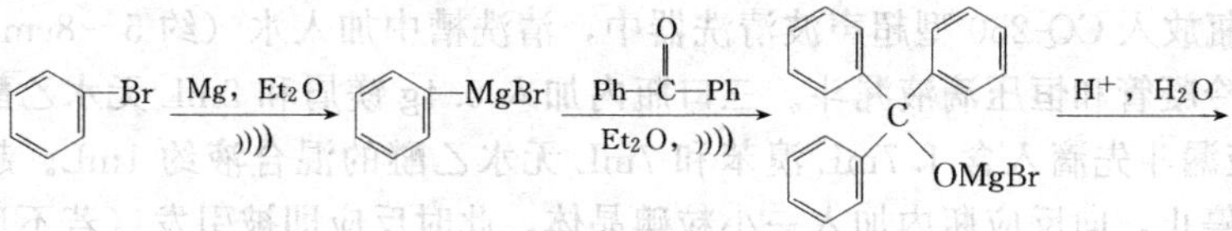

方法 2　苯甲酸乙酯与苯基溴化镁的反应。

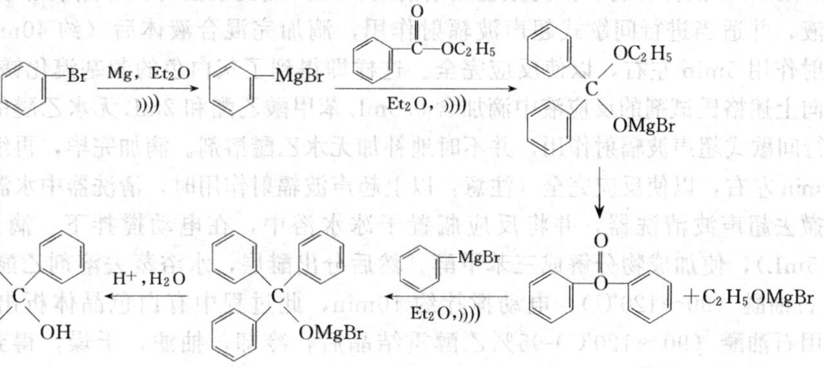

$$\text{Ph-Br} \xrightarrow{\text{Mg, Et}_2\text{O}} \text{Ph-MgBr} \xrightarrow{\text{Ph-COOC}_2\text{H}_5}_{\text{Et}_2\text{O,)))}} \text{Ph}_2\text{C(OC}_2\text{H}_5)(\text{OMgBr})$$

$$\longrightarrow \text{Ph-CO-Ph} + \text{C}_2\text{H}_5\text{OMgBr} \xrightarrow{\text{PhMgBr}}_{\text{Et}_2\text{O,)))}} \text{Ph}_3\text{C-OMgBr} \xrightarrow{\text{H}^+, \text{H}_2\text{O}} \text{Ph}_3\text{C-OH}$$

副反应

$$\text{Ph-MgBr} + \text{Ph-Br} \longrightarrow \text{Ph-Ph}$$

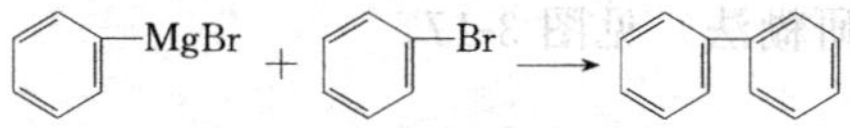

【主要试剂】

方法 1　溴苯 2.7mL（4.04g，25.7mmol），镁（表面明亮的镁屑）0.7g（28.8mmol），二苯甲酮 4.5g（25.0mmol），无水乙醚，20％硫酸，碘少许，95％乙醇，石油醚（90～120℃）。

方法 2　溴苯 1.7mL（2.54g，16.0mmol），镁（表面明亮的镁屑）0.4g（16.5mmol），苯甲酸乙酯 0.9mL(0.95g，6.3mmol)，无水乙醚，20％硫酸，碘少许，95％乙醇，石油醚（90～120℃）。

【实验步骤】

方法 1　二苯甲酮与苯基溴化镁的反应。

将 250mL 的三口瓶放入 CQ-250 型超声波清洗器中，清洗槽中加入水（约 5～8cm 高）。三口瓶上分别安装回流冷凝管和恒压滴液漏斗。三口瓶内加入 0.7g 镁屑和 5mL 无水乙醚（新开瓶的），再自恒压滴液漏斗先滴入含 2.7mL 溴苯和 10mL 无水乙醚的混合液约 1mL。超声波辐射作用约 1～2min 后停止，向反应瓶内加入一小粒碘晶体，此时反应即被引发（若不反应可用温水浴温热），液体沸腾，碘的颜色逐渐消失。当反应变缓慢时，开始滴加溴苯和无水乙醚的混合液，并适当进行间歇式超声波辐射作用，滴加完混合液体后（约 40min），再继续超声波辐射作用 5min 左右，以使反应完全。这样即得到了灰白色的苯基溴化镁格氏试剂。

向格氏试剂的反应液中缓慢滴加含 4.5g 二苯甲酮和 13mL 无水乙醚的混合液，在此期间进行间歇式超声波辐射作用，并不时地补加无水乙醚溶剂。滴加完毕，再继续超声波辐射作用 10min 左右，以使反应完全（注意：以上超声波辐射作用时，清洗器中水温不得超过 25℃）。

撤去超声波清洗器，并将反应瓶置于冰水浴中，在电动搅拌下，滴加 20％硫酸溶液（约 25mL），使加成物分解成三苯甲醇。然后分出醚层，水浴蒸去溶剂乙醚，剩余物中加入 10mL 石油醚（90～120℃），电动搅拌约 10min，此过程中有白色晶体析出。抽滤收集粗产品。用石油醚（90～120℃）-95%乙醇重结晶后，冷却，抽滤，干燥，得到白色片状晶体。称量，测熔点，计算产率。

纯三苯甲醇为白色片状晶体，m.p. 164.2℃。

方法2 苯甲酸乙酯与苯基溴化镁的反应。

将50mL的三口瓶放入CQ-250型超声波清洗器中，清洗槽中加入水（约5～8cm高）。三口瓶上分别安装回流冷凝管和恒压滴液漏斗。三口瓶内加入0.4g镁屑和2mL无水乙醚（新开瓶的），再自恒压滴液漏斗先滴入含1.7mL溴苯和7mL无水乙醚的混合液约1mL。超声波辐射作用约1～2min后停止，向反应瓶内加入一小粒碘晶体，此时反应即被引发（若不反应可用温水浴温热），液体沸腾，碘的颜色逐渐消失。当反应变缓慢时，开始滴加溴苯和无水乙醚的混合液，并适当进行间歇式超声波辐射作用，滴加完混合液体后（约40min），再继续超声波辐射作用5min左右，以使反应完全。这样即得到了灰白色的苯基溴化镁格氏试剂。

向上述格氏试剂的反应液中滴加含0.9mL苯甲酸乙酯和2mL无水乙醚的混合液，在此期间进行间歇式超声波辐射作用，并不时地补加无水乙醚溶剂。滴加完毕，再继续超声波辐射作用10min左右，以使反应完全（注意：以上超声波辐射作用时，清洗器中水温不得超过25℃）。

撤去超声波清洗器，并将反应瓶置于冰水浴中，在电动搅拌下，滴加20%硫酸溶液（约15mL），使加成物分解成三苯甲醇。然后分出醚层，水浴蒸去溶剂乙醚，剩余物中加入5mL石油醚（90～120℃），电动搅拌约10min，此过程中有白色晶体析出。抽滤收集粗产品。用石油醚（90～120℃）-95%乙醇重结晶后，冷却，抽滤，干燥，得到白色片状晶体。称量，计算产率。

三苯甲醇的红外光谱（研糊法）见图3-17。

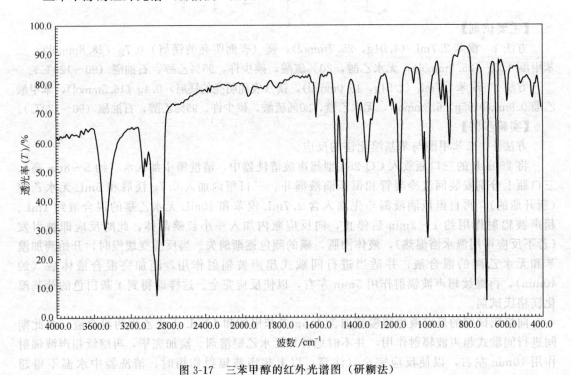

图3-17 三苯甲醇的红外光谱图（研糊法）

【注意事项】

① 超声波辐射作用过程中，清洗器中水温不得超过25℃，否则超声空化效应减弱，产率降低，并且乙醚也易挥发。

② 实验中所用的无水乙醚无需特殊处理，使用新开瓶的无水乙醚即可满足制备格氏试剂的要求。实验中所用仪器必须充分干燥。

③ 保持卤代烃在反应液中局部高浓度，有利于引发反应，因而在反应初期不用超声波辐射振荡。但是，如果整个反应过程中都保持高浓度卤代烃，则容易发生偶联副反应。因此，反应开始后要保持间歇式超声波辐射作用，卤代烃的滴加速度也不宜过快。

$$RMgBr + RBr \longrightarrow R-R + MgBr_2$$

④ 副产物易溶于石油醚中而被除去。

思 考 题

1. 格氏反应在有机合成中有哪些应用？
2. 本反应中可能发生什么副反应？
3. 如果实验中溴苯滴加速度太快或一次加入，对反应有何影响？
4. 合成化学利用超声波的什么特点来加速反应进程？
5. 试比较超声波辐射法合成三苯甲醇与经典法合成三苯甲醇有何优点？

实验 41 2-甲基-1-苯基-2-丙醇的合成

2-甲基-1-苯基-2-丙醇具有微甜、清香和温暖的药草花香味，有玫瑰等新鲜花香气息，广泛应用于日化和食用香精中。

【反应式】

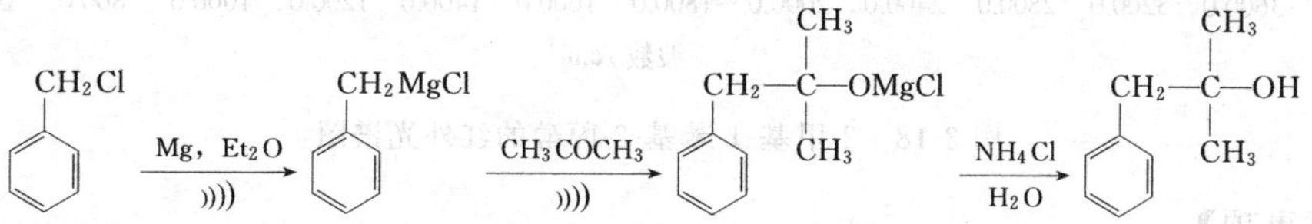

PhCH$_2$Cl $\xrightarrow{\text{Mg, Et}_2\text{O}}$ PhCH$_2$MgCl $\xrightarrow{\text{CH}_3\text{COCH}_3}$ PhCH$_2$C(CH$_3$)$_2$OMgCl $\xrightarrow{\text{NH}_4\text{Cl}, \text{H}_2\text{O}}$ PhCH$_2$C(CH$_3$)$_2$OH

【主要试剂】

丙酮（无水硫酸镁干燥过的）2.0mL（1.58g，27.2mmol），氯化苄 3.2mL（3.52g，27.8mmol），镁（表面明亮的镁屑）0.7g（28.8mmol），无水乙醚（无水氯化钙干燥过的），氯化铵溶液，碘少许。

【实验步骤】

将 250mL 的三口瓶放入超声波清洗器中（CQ-250 型，工作频率 33kHz，上海超声波仪器厂），清洗槽中加入水（约 5~8cm 高）。然后，在三口瓶上分别安装回流冷凝管和恒压滴液漏斗。瓶内加入 0.7g 镁屑和 5mL 无水乙醚，再自滴液漏斗滴入含 3.2mL 氯化苄和 10mL 无水乙醚的混合液约 1mL。超声波辐射作用约 1~2min 后停止，向反应瓶内加入一小粒碘晶体，此时反应即被引发，在镁表面有气泡产生，液体沸腾，碘的颜色逐渐消失。当反应变缓慢时，开始滴加氯化苄和无水乙醚的混合液，同时进行间歇式超声波辐射作用。滴加完混合液体后（约 40min），再继续超声波辐射作用 5min 左右，以使反应完全。这样即得到了灰白色的苄基氯化镁格氏试剂。

向格氏试剂的反应液中滴加含 2.0mL 丙酮和 6.0mL 无水乙醚的混合液，在此期间进行间歇式超声波辐射作用，并不时地补加无水乙醚溶剂。滴加完毕（约 20min），再继续超声波辐射作用 10min 左右，以使反应完全（注意：以上超声波辐射作用时，清洗器中水温不得超过 20℃）。

撤去超声波清洗器，并将反应瓶置于冰水浴中，电动搅拌下，滴加饱和氯化铵水溶液约 25mL，使白色加成物分解成 2-甲基-1-苯基-2-丙醇。然后分出醚层，水层再用乙醚萃取两次，合并醚层。醚层分别用饱和碳酸氢钠溶液和水洗涤，再用无水碳酸钾干燥，热水浴蒸去溶剂乙醚后进行减压蒸馏收集沸点在88～90℃/533Pa（4mmHg）的馏分。

纯2-甲基-1-苯基-2-丙醇为无色油状液体，b.p.88～90℃/400～533Pa（文献值）。

2-甲基-1-苯基-2-丙醇的红外光谱见图3-18。

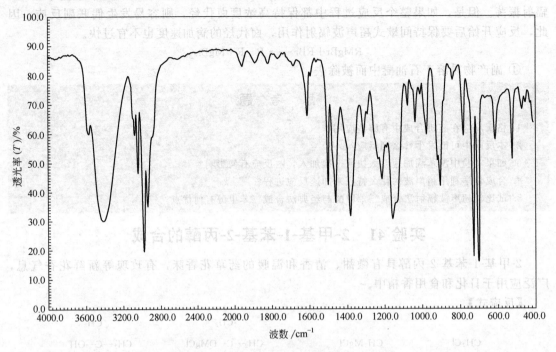

图 3-18　2-甲基-1-苯基-2-丙醇的红外光谱图

【注意事项】

参见实验40　三苯甲醇的合成。

3.12.2　羟醛缩合反应

在稀碱或稀酸催化下，两分子具有α-活泼氢的醛或酮发生分子间缩合反应，生成β-羟基醛或酮，若提高反应温度将进一步失水生成α,β-不饱和醛或酮，称为羟醛缩合反应。这是合成α,β-不饱和羰基化合物的重要方法，也是有机合成中增长碳链的重要反应。

羟醛缩合分为自身缩合和交叉缩合两种。不含α-氢的芳醛与含α-氢的醛或酮发生羟醛缩合，得到α,β-不饱和醛或酮，这种交叉的羟醛缩合反应称为克莱森-斯密特（Claisen-Schmidt）反应。

实验42　苯亚甲基苯乙酮的合成

【反应式】

$$PhCHO + CH_3COPh \xrightarrow[\text{)))), 25～30℃}]{\text{NaOH (10\%), EtOH}} PhCH=CHCOPh + H_2O$$

【主要试剂】

苯乙酮1.00mL（1.03g，8.57mmol），新蒸馏的苯甲醛0.8mL（0.83g，7.82mmol），10%NaOH，95%乙醇。

【实验步骤】

在50mL锥形瓶中，依次加入2.1mL 10% NaOH水溶液、2.5mL 95%乙醇、1.00mL苯乙酮，冷却至室温，再加入0.8mL新蒸馏过的苯甲醛。启动超声波清洗器（KQ-500B型，江苏昆山市超声仪器公司），将反应瓶置于超声波清洗槽中，并使清洗槽中水面略高于反应瓶中的液面，控制清洗槽中水温在 25～30℃，超声波辐射 30～35min，停止反应。然后将反应瓶置于冰水浴中冷却，使其结晶完全。抽滤，用少量冰水洗涤产品至滤液呈中性。粗产品用 95% 乙醇重结晶。产率约 85%，m.p. 55～57℃。

【注意事项】
① 反应温度高于 30℃ 或低于 15℃ 对反应均不利。
② 久置的苯甲醛，由于自动氧化而生成较多量的苯甲酸。故实验中所需的苯甲醛要重新蒸馏。
③ 由于产物熔点较低，重结晶回流时产品可能会出现熔融状态，这时应补加溶剂使其成均相。
④ 氢氧化钠的量不宜过多，以免产生大量聚合物，降低产率。
⑤ 关闭超声波清洗器之后，才能用温度计测试清洗槽内的水温。

思 考 题

通过查阅有关实验教材或文献，得出超声波辐射法合成苯亚甲基苯乙酮的优点。

3.13 鲁卡特反应

醛或酮与氨反应形成 α-氨基醇，因其不稳定继而脱水成亚胺，亚胺经催化加氢被还原生成胺，这是由羰基化合物合成胺的一个重要方法。

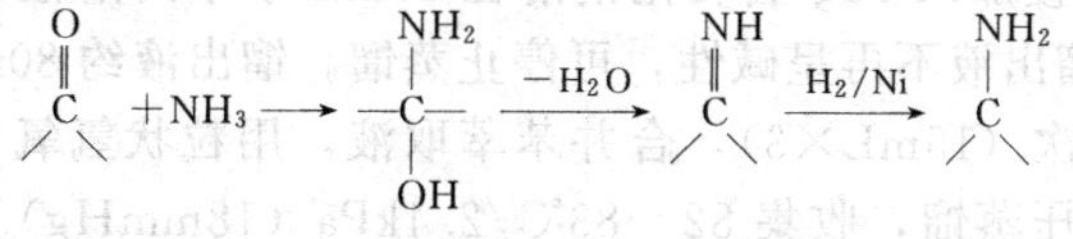

$$\underset{}{\overset{O}{\underset{}{C}}} + NH_3 \longrightarrow \underset{OH}{\overset{NH_2}{\underset{}{C}}} \xrightarrow{-H_2O} \underset{}{\overset{NH}{\underset{}{C}}} \xrightarrow{H_2/Ni} \underset{}{\overset{NH_2}{\underset{}{C}}}$$

α-氨基醇　　亚胺　　　胺

在此反应中，如果用甲酸作还原剂来代替 H_2/Ni，则此还原氨化的过程称为鲁卡特反应（Leuckart Reaction）。在鲁卡特反应中，甲酸或甲酸根离子在此起还原剂的作用，氢原子以氢负离子的形式转移至亚胺上。

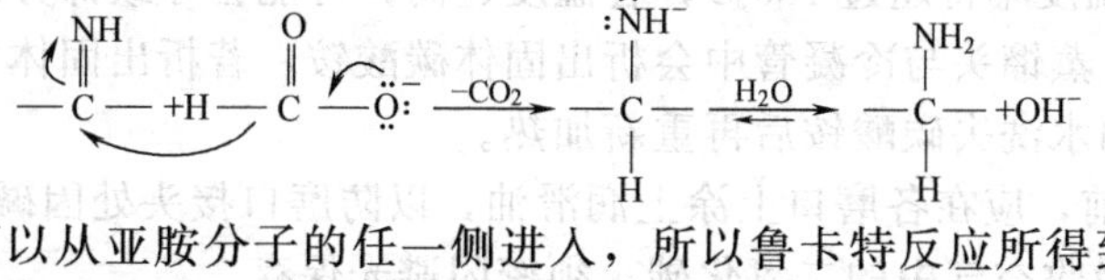

$$\underset{}{\overset{NH}{\underset{}{C}}} + H\underset{}{\overset{O}{\underset{}{C}}} \ddot{O}^- \xrightarrow{-CO_2} \underset{H}{\overset{\ddot{N}H_2}{\underset{}{C}}} \xrightarrow{H_2O} \underset{H}{\overset{NH_2}{\underset{}{C}}} + OH^-$$

由于氢负离子可以从亚胺分子的任一侧进入，所以鲁卡特反应所得到的还原产物是外消旋体。如果要得到具有旋光性的对映异构体，还需进行拆分。

事实上，由于存在着甲酸根离子，在上述反应中不直接生成游离的胺，而是先生成其 N-甲酰化物，后者需经酸解及碱中和后才能得到胺。

实验 43 (±)-α-苯乙胺的合成

利用鲁卡特反应，以苯乙酮和甲酰胺为原料制备苯乙胺。

【反应式】

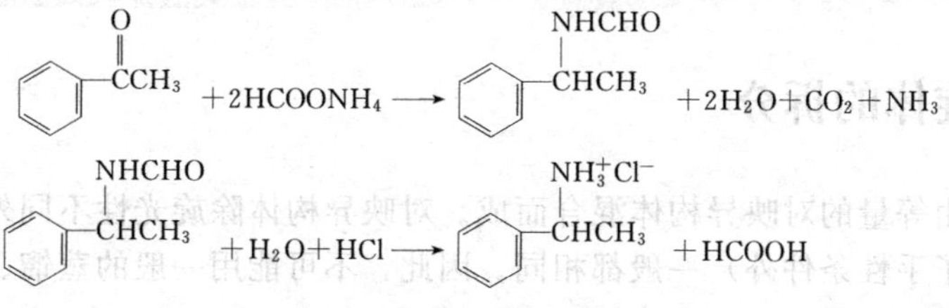

$$\text{Ph-}\underset{}{\overset{O}{\underset{}{C}}}\text{CH}_3 + 2\text{HCOONH}_4 \longrightarrow \text{Ph-}\underset{\text{CH}_3}{\overset{\text{NHCHO}}{\underset{}{C}}}\text{H} + 2\text{H}_2\text{O} + \text{CO}_2 + \text{NH}_3$$

$$\text{Ph-}\underset{\text{CH}_3}{\overset{\text{NHCHO}}{\underset{}{C}}}\text{H} + \text{H}_2\text{O} + \text{HCl} \longrightarrow \text{Ph-}\underset{\text{CH}_3}{\overset{\text{NH}_3^+\text{Cl}^-}{\underset{}{C}}}\text{H} + \text{HCOOH}$$

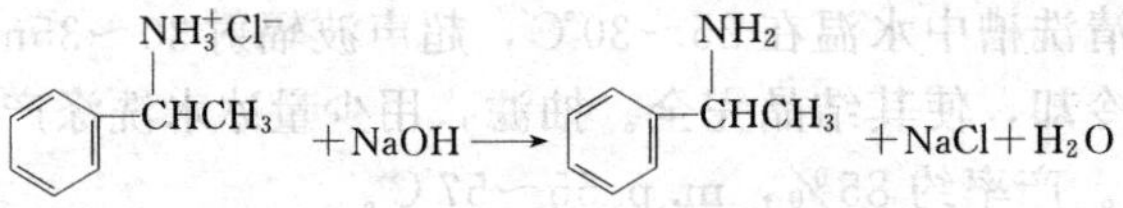

$$\underset{\text{苯乙酮}}{\underset{|}{\overset{NH_3^+Cl^-}{\underset{C_6H_5-CHCH_3}{}}}} + NaOH \longrightarrow \underset{(\pm)\text{-}\alpha\text{-苯乙胺}}{\underset{|}{\overset{NH_2}{\underset{C_6H_5-CHCH_3}{}}}} + NaCl + H_2O$$

【主要试剂】

苯乙酮 11.7mL(12.0g, 100mmol), 甲酰胺 20.3g(322mmol), 苯, 浓盐酸, 氢氧化钠。

【实验步骤】

在 100mL 三口瓶中, 加入 11.7mL 苯乙酮、20.3g 甲酰胺和几粒沸石。在三口瓶上安装温度计（插入溶液中）、蒸馏头并连接冷凝管, 装配成一个简单蒸馏装置。三口瓶下用空气浴（或电热套）小火缓缓加热, 瓶内混合物逐渐熔化, 分成两相, 同时有液体慢慢蒸出。当反应混合物温度升至 150~155℃时, 混合物变成均相。控制加热使混合物温度逐渐升高（小火加热, 以免有太多的泡沫泛出）, 到达 185℃时停止加热, 此过程约需 1~1.5h。馏出物中有少量苯乙酮, 用分液漏斗分出上层苯乙酮并倒回反应瓶中, 重新开始加热反应 1~1.5h, 并维持反应物温度在 180~185℃。

待反应混合物冷却后, 向烧瓶中加入 10mL 水, 振摇后转入分液漏斗中, 再用 10mL 水洗涤烧瓶, 洗涤液一并转入分液漏斗中（水洗去什么？）。将有机层（N-甲酰-α-苯乙胺）倒回三口瓶中, 水层用苯萃取两次（10mL×2）。苯萃取液也倒入三口瓶中, 加入 12mL 浓盐酸和 2 粒沸石。加热蒸馏除去所有的苯。然后再将蒸馏装置改成回流装置, 加热回流 0.5h。

反应混合物冷至室温后, 用苯萃取两次 10mL×2（洗去什么？）。将分出的水层置于 250mL 圆底烧瓶中, 慢慢加入 10g 氢氧化钠溶在 15mL 水中的溶液, 进行水蒸气蒸馏（见图 2-36）。直至蒸出的馏出液不再呈碱性, 可停止蒸馏。馏出液约 80mL。

馏出液用苯萃取三次（15mL×3）, 合并苯萃取液, 用粒状氢氧化钠干燥后, 先简单蒸馏, 蒸除溶剂, 然后减压蒸馏, 收集 82~83℃/2.4kPa（18mmHg）的馏分, 即得（±)-α-苯乙胺, 产量约 5g, 产率约 41%。

纯 (±)-α-苯乙胺 b.p. 187.4℃, n_D^{20} 1.5238。

【注意事项】

① 反应加热时温度不得超过 185℃, 若温度过高, 可能会导致部分碳酸铵凝固在冷凝管中。在蒸馏过程中, 蒸馏头与冷凝管中会析出固体碳酸铵, 若析出固体太多会导致冷凝管堵塞。应停止加热, 用水洗去碳酸铵后再重新加热。

② 水蒸气蒸馏前, 应在各磨口上涂上润滑油, 以防磨口接头处因碱腐蚀而被粘住。

③ α-苯乙胺易吸收空气中的二氧化碳, 须密闭避光保存。

④ 因 α-苯乙胺具有较强的腐蚀性, 为保护折光仪, 产品不必测折射率。

思 考 题

1. 利用鲁卡特反应制备 (±)-α-苯乙胺为什么只能得到其外消旋体？
2. 反应混合物进行水蒸气蒸馏之前, 为什么要将溶液变成碱性？
3. 本实验为什么要比较严格地控制反应温度不超过 185℃？
4. 试写出鲁卡特反应合成 α-苯乙胺的反应机理。

3.14 外消旋体的拆分

外消旋体是由等量的对映异构体混合而成。对映异构体除旋光性不同外, 其他物理性质和化学性质（除了手性条件外）一般都相同。因此, 不可能用一般的蒸馏、萃取、重结晶等方法分离外消旋体。

目前，对外消旋体的拆分方法很多，而实验室里通常用的一种方法是采用一个旋光的化合物（拆分剂）和给定的外消旋体进行化学反应，使原来的外消旋体转变为两个非对映体，再利用非对映体在某种选定的溶剂中具有不同的溶解度的特性，用分步结晶法将它们分离、精制，然后再去掉拆分剂，即可得到纯的旋光异构体。

拆分酸性外消旋体，常用旋光性生物碱，如（—）-麻黄碱，（—）-马钱子碱等；拆分碱性外消旋体，常用旋光性酸，如酒石酸、樟脑-β-磺酸等。

实验 44 （±）-α-苯乙胺的拆分

（±）-α-苯乙胺属于碱性外消旋体，可用酸性拆分剂（如酒石酸）进行拆分。本实验通过 L-(+)-酒石酸与（±）-α-苯乙胺反应形成非对映异构体盐（—）-α-苯乙胺-(+)-酒石酸盐和（+）-α-苯乙胺-(+)-酒石酸盐。这两种盐在甲醇中的溶解度有显著差异，前者在甲醇中的溶解度比后者小。因此，可以用分步结晶法，使（—）-α-苯乙胺-(+)-酒石酸盐从溶液中先结晶析出，经纯化、碱化处理，即可得到纯的（—）-α-苯乙胺。母液中所含的（+）-α-苯乙胺-(+)-酒石酸盐经类似的处理也可得到纯的（+）-α-苯乙胺。

【反应式】

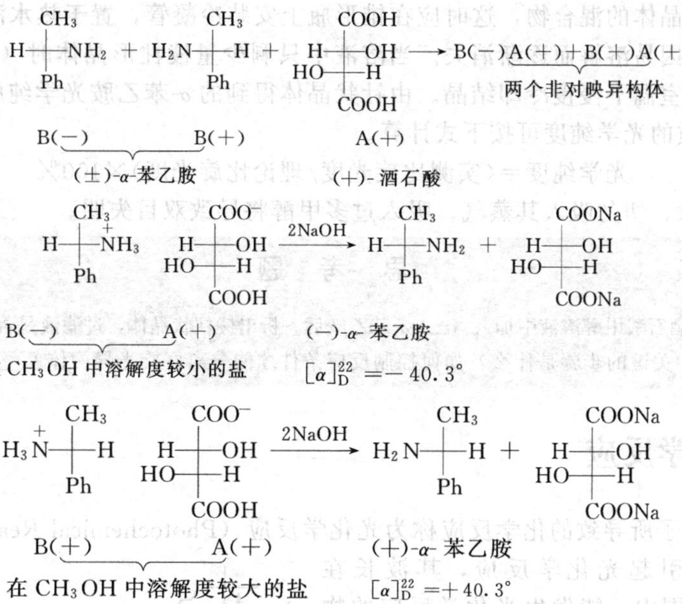

（结构式及反应图，略）

B(—)-α-苯乙胺   A(+)-酒石酸

在 CH_3OH 中溶解度较小的盐   （—）-α-苯乙胺 $[\alpha]_D^{22} = -40.3°$

在 CH_3OH 中溶解度较大的盐   （+）-α-苯乙胺 $[\alpha]_D^{22} = +40.3°$

【主要试剂】

（±）-α-苯乙胺 3g（3.25mL，25mmol），L-(+)-酒石酸 3.8g（25mmol），甲醇，乙醚，50%氢氧化钠。

【实验步骤】

(1) 分步结晶   在 100mL 锥形瓶中，加入 3.8g L-(+)-酒石酸、50mL 甲醇和 2 粒沸石，配置回流冷凝管，水浴加热使其溶解。稍冷，用滴管慢慢滴加 3g（±）-α-苯乙胺，边滴加边振摇（滴加速度不宜太快，否则会起泡），使之混合均匀。滴加完毕后，于室温下放置 24h，即可生成白色棱柱状晶体。过滤，所得溶液供分出（+）-α-苯乙胺用。晶体用少量甲醇洗涤，干燥后得（—）-α-苯乙胺-(+)-酒石酸盐，称量并计算产率。

（—）-α-苯乙胺-(+)-酒石酸盐为白色棱柱状晶体，m.p. 179～182℃（分解），$[\alpha]_D^{22} = 13°(H_2O, 8\%)$。

(2) 胺的分离

① (-)-α-苯乙胺的获得。将上述所得 (-)-α-苯乙胺-(+)-酒石酸盐晶体溶于 10mL 水中，加入 2mL 50%氢氧化钠溶液，搅拌至固体全部溶解，然后用乙醚萃取三次 (10mL×3)，合并萃取液并用粒状氢氧化钠 (或无水硫酸镁) 干燥。将干燥后的乙醚溶液转入蒸馏瓶，在水浴上蒸去乙醚后，即得到 (-)-α-苯乙胺粗产品。减压蒸馏 (因产量较小，可将几组的粗产品合并)，收集 81~81.5℃/2.4kPa (18mmHg) 的馏分。称量并计算产率，测定 (-)-α-苯乙胺的比旋光度，计算产物的光学纯度。

纯 (-)-α-苯乙胺 $[\alpha]_D^{22} = -40.3°$。

② (+)-α-苯乙胺的获得。上述析出 (-)-α-苯乙胺-(+)-酒石酸盐的母液中含有 CH_3OH，因此须先在水浴中将 CH_3OH 蒸尽，残留物呈白色固体，即 (+)-α-苯乙胺-(+)-酒石酸盐。可用同①一样的操作方法，用水、氢氧化钠来处理该盐，用乙醚萃取，粒状氢氧化钠干燥，水浴上蒸去乙醚，然后进行减压蒸馏，收集 85~86℃/2.8kPa(21mmHg) 的馏分。称量并计算产率，测定 (+)-α-苯乙胺的比旋光度，计算产物光学纯度。

纯 (+)-α-苯乙胺 $[\alpha]_D^{22} = +40.3°$。

【注意事项】

① 在制备 (-)-α-苯乙胺-(+)-酒石酸盐时，若得到的晶体不是棱柱状而是针状晶体，或棱柱状与针状晶体的混合物，这时应在锥形瓶上安装冷凝管，置于热水浴中加热，并不时振摇，针状晶体因易溶解而逐渐消失。当溶液中只剩少量棱柱形晶体时 (留作晶种)，停止加热，将溶液在室温下慢慢冷却结晶。由针状晶体得到的 α-苯乙胺光学纯度较差。

② 旋光物质的光学纯度可按下式计算。

$$光学纯度 = (实测比旋光度/理论比旋光度) \times 100\%$$

③ 甲醇有毒，切勿吸入其蒸气。吸入过多甲醇将导致双目失明。

思 考 题

1. 在 L-(+)-酒石酸甲醇溶液中加入 (±)-α-苯乙胺后，析出棱柱状晶体，过滤液是否有旋光性？为什么？
2. 拆分实验中关键的步骤是什么？如何控制反应条件才能分离好旋光异构体？

3.15 光化学反应

由光激发分子所导致的化学反应称为光化学反应 (Photochemical Reaction)。通常紫外光和可见光能引起光化学反应，其波长在 200~700nm 范围内，能发生光化学反应的物质一般具有不饱和键，如烯烃、醛、酮等。

紫外光和可见光对有机分子的照射可以引起分子中电子的跃迁，即将原来在成键轨道或非键轨道上的电子激发到能级更高的反键轨道上，从而使原来的基态分子变成激发态分子。

绝大多数有机分子在基态时是单线态 (基态单线态记作 S_0，此时电子自旋配对排列)，当吸收一定波长的光而受激发时，由于电子跃迁过程中电子自旋方向不变 (因为电子跃迁过程中自旋的改变是个量子力学上禁阻的过程)，所以总是产生单线激发态 (分子的第一激发态记作 S_1)。但

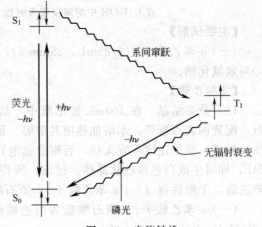

图 3-19 光能转换

是激发单线态很不稳定，很快会发生激发电子自旋方向的倒转，变成热力学上比较稳定的三线态（激发三线态记作 T_1，此时电子自旋平行排列），由激发单线态向三线态转化的过程称为系间窜跃（Intersystem Crossing，ISC）。激发单线态 S_1 可通过发出荧光释放出原来所吸收的光子能量，从而恢复到基态 S_0（见图 3-19）。三线态 T_1 可通过发出磷光（波长较荧光要长）恢复至基态，也可以通过无辐射跃迁，放出热能，返回基态。由 $T_1 \rightarrow S_0$ 这两种途径的跃迁都涉及电子自旋方向的转变，因而比较困难，需要一定的时间，故三线态比单线态的寿命要长。许多光化学反应都是当反应物分子处于激发三线态时发生的。因此，三线态在光化学反应中特别重要。

实验 45　苯频哪醇的合成

二苯甲酮的光化学还原是研究最为彻底的光化学反应。当二苯甲酮溶于氢给予体的溶剂（如异丙醇）中，在紫外光照射下，生成二聚体苯频哪醇。

实验证明，二苯甲酮的光化学还原是二苯甲酮的 n→π* 三线态（T_1）的反应。二苯甲酮的异丙醇溶液用 300～350nm 的紫外光照射时，异丙醇不吸收光能，只有二苯甲酮由于羰基接受光能后，外层的非键电子发生 n→π* 跃迁，经单线态（S_1）、系间窜跃成三线态（T_1），由于三线态（T_1）有较长的半衰期和相当的能量（314～334.7kJ/mol），可以从异丙醇的 C2 上夺取氢，使 C2 上的 C—H 键均裂，各自形成自由基，再经自由基的转移、偶合形成苯频哪醇。其反应机理如下。

【反应式】

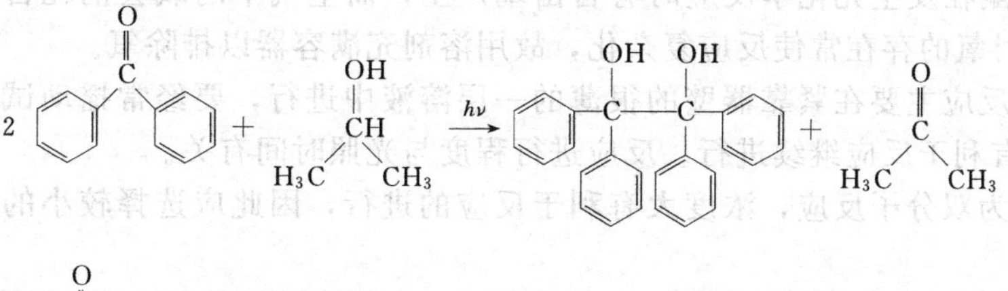

$$2 \; Ph_2C{=}O + (CH_3)_2CHOH \xrightarrow{h\nu} Ph_2C(OH){-}C(OH)Ph_2 + (CH_3)_2C{=}O$$

反应机理

$$Ph_2C{=}O \xrightarrow{h\nu} [Ph_2\dot{C}{-}\dot{O}]^{*(S_1)} \xrightarrow{(S_1)\rightarrow(T_1)} [Ph_2\dot{C}{-}\dot{O}]^{*(T_1)}$$

$$[Ph_2\dot{C}{-}\dot{O}]^{*(T_1)} + (CH_3)_2CHOH \longrightarrow Ph_2C(OH)\cdot + \cdot C(CH_3)_2OH$$

$$[Ph_2\dot{C}{-}\dot{O}]^{*(T_1)} + \cdot C(CH_3)_2OH \longrightarrow Ph_2C(OH)\cdot + (CH_3)_2C{=}O$$

$$2 \; Ph_2\dot{C}(OH) \longrightarrow Ph_2C(OH){-}C(OH)Ph_2$$

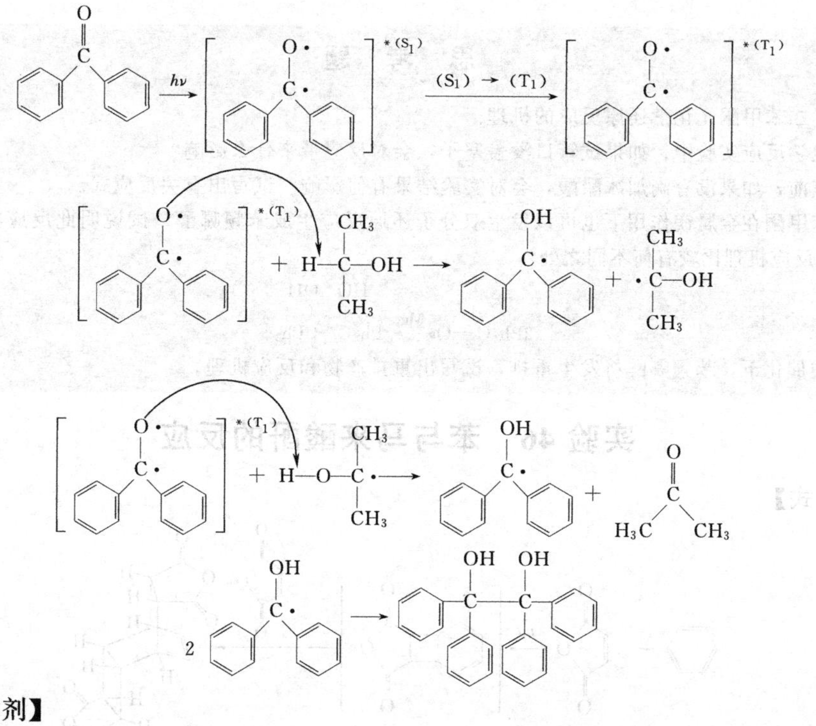

【主要试剂】

二苯甲酮 0.5g（2.7mmol），异丙醇，冰醋酸。

【实验步骤】
在一支 10mL 试管中，加入 0.5g 二苯甲酮和 3mL 异丙醇，在温水浴中加热，使二苯甲酮溶解。向试管中加入一滴冰醋酸，充分振摇后再补加异丙醇至试管口，以使反应尽量在无空气条件下进行。用塞子将试管塞住，置试管于烧杯中，并放在光照良好的窗台上，光照一周，试管内有大量无色晶体析出。经过滤、干燥后即得苯频哪醇粗品。粗产物可用冰醋酸作溶剂进行重结晶。干燥后，称量、测熔点并计算产率。

纯苯频哪醇为无色针状晶体，m.p. 188～190℃（分解）。

【注意事项】
① 玻璃具有微弱的碱性，加一滴冰醋酸的目的在于消除玻璃碱性的影响。因为痕量碱的存在将使苯频哪醇分解成二苯甲醇和二苯甲酮，反应式如下。

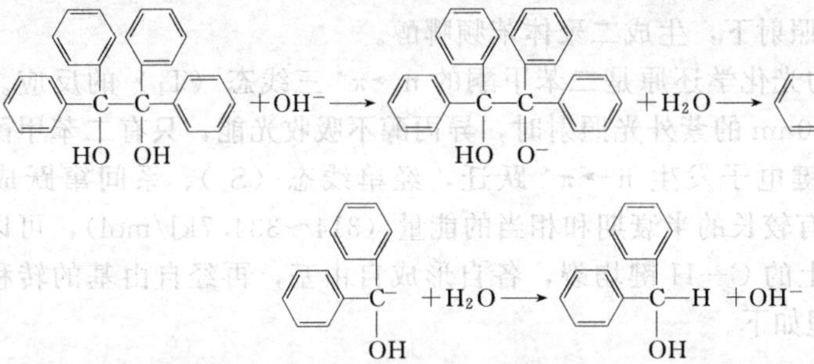

② 二苯甲酮在发生光化学反应时有自由基产生，而空气中的氧会消耗自由基，使反应速率减慢，同时氧的存在常使反应复杂化，故用溶剂充满容器以排除氧。

③ 光化学反应主要在紧靠器壁的很薄的一层溶液中进行，要经常摇动试管，防止晶体结在管壁上，有利于反应继续进行。反应进行程度与光照时间有关。

④ 该反应为双分子反应，浓度大有利于反应的进行，因此应选择较小的反应容器，产率较高。

思 考 题

1. 试述二苯甲酮光化学还原反应的机理。
2. 光化学反应实验中，如果试管口没盖塞子，会对反应带来什么影响？
3. 反应前，如果没有滴加冰醋酸，会对实验结果有何影响？试写出有关反应式。
4. 二苯甲酮在金属镁作用下也可以发生双分子还原反应生成苯频哪醇，试说明此反应的机理，并与光化学还原反应机理比较有何不同之处？

$$2Ph_2C=O \xrightarrow{Mg} Ph_2C(OH)-C(OH)Ph_2$$

5. 在酸催化下，苯频哪醇将发生重排，试写出重排产物和反应机理。

实验 46　苯与马来酸酐的反应

【反应式】

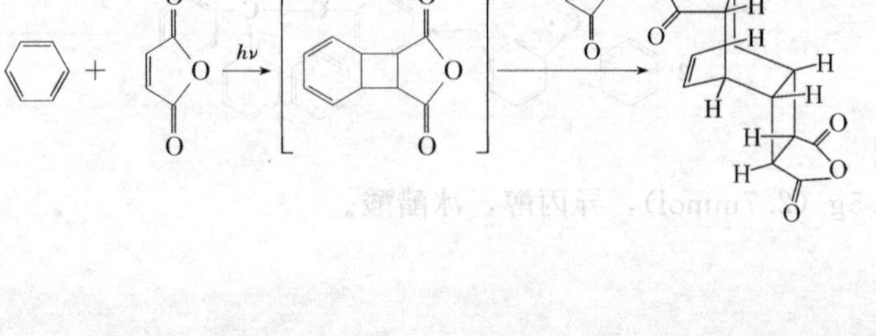

【主要试剂】
马来酸酐 5g（51 mmol），苯。
【实验步骤】
在通风橱中，将 5g 马来酸酐加入 100mL 磨口锥形瓶中，再加入 50mL 干燥的苯使其完全溶解。盖上瓶塞并将此锥形瓶置于朝南的窗台上，在阳光下放置数天（放置时间长短依赖于光的强度）。最终光照加成产物无色晶体析出。抽滤收集晶体。待彻底抽干后，将其置于空气中干燥 15min。记录产率和熔点（产物的熔点大于 300℃）。

思 考 题

1. 生成中间体 1∶1 加成产物和最终的 1∶2 加成产物的反应是什么类型反应？
2. 如果实验中使用的不是干燥的苯，则反应液在放置几小时后即有熔点为 135℃的固体析出。你认为这可能是什么化合物？

3.16 有机电化学反应

有机电解合成（Electroorganic Synthesis）是利用电解反应来合成有机化合物。有机电解合成是绿色化学中有机合成洁净技术的重要组成部分，电化学合成法与常规合成法相比有以下特点：①可自动控制。电化学过程中的两大参数电流与电压信号，易测定和自动控制。②反应条件温和，能量效率高。电化学反应可在较低温度下进行。由于不经过卡诺循环，能量利用率高。③环境相容性高。电化学过程中使用的主要试剂是电子，是最洁净的试剂，不会对环境产生不良影响。④经济合算。所需设备简单，操作费用低。设计合理的电解池结构，利用先进电极材料，可达到零排放的要求。目前，电化学方法在有机合成上的应用已引起人们的广泛注意，日益为化学、化工界所重视。

实验 47　碘仿的合成

碘仿又称黄碘，为亮黄色晶体，在医药和生物化学中作防腐剂和消毒剂。
【反应式】
主反应
　阴极：　　　　　　　　　　　$2H^+ +2e \longrightarrow H_2$
　阳极：　　　　　　　　　　　$2I^- -2e \longrightarrow I_2$
　　　　　　　　　　　　　　$I_2 +2OH^- \Longleftrightarrow IO^- +I^- +H_2O$
　　　　　　　　　$CH_3COCH_3 +3IO^- \longrightarrow CHI_3\downarrow +CH_3COO^- +2OH^-$
副反应　　　　　　　　　　　$3IO^- \longrightarrow IO_3^- +2I^-$
【主要试剂】
碘化钾 2.2g（13.3mmol），丙酮 0.5mL（0.4g，6.9mmol），乙醇。
【实验步骤】
用 50mL 的小烧杯作电解槽，两支 1#电池碳棒作电极，把它们垂直固定在硬纸板或有机玻璃板上（见图 3-20）。烧杯中加 40mL 水、2.2g KI 和 0.5mL 丙酮。将烧杯放置在电磁搅拌器上，开动电磁搅拌器使药品溶解，接通电源（接 6V 直流电），在室温下电解。随着反应的进行有黄色沉淀生成。反应 1h 左右转换电极，以加快反应速率。反应 2.5h 后，切断电源，停止反应，再继续搅拌 1~2min，然后抽滤，滤饼用少量水洗涤两次，空气中自然干燥后即得粗产品。粗产品用乙醇重结晶（装置图见图 2-45）后得到纯品，产品经晾干后，称量，测熔点，计算产率。

纯碘仿为亮黄色晶体，m. p. 119℃，升华，不溶于水。

【注意事项】

① 为了减少电流通过介质的损失，两电极应尽可能的靠近。

② 电极表面积越大，反应速率越快。要保证电极浸入反应液的面积。

③ 纯净的碘仿为亮黄色晶体，但用石墨作电极时，析出的晶体呈灰绿色，是因为混有石墨，需要精制。

思 考 题

1. 本电解实验过程中，为什么电解液的pH逐渐增大？
2. 除用重结晶法提纯碘仿外，还可用什么方法提纯？

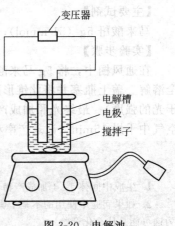

图 3-20　电解池

3.17　天然产物的提取

天然产物（Natural Products）指的是从天然动、植物体内衍生出来的有机化合物。天然产物种类繁多，广泛存在于自然界中。多数天然产物的提取物具有特殊的生理效能，有的可用作香料和染料，有的甚至具有神奇的药效。天然产物的分离提取和鉴定是有机化学中十分活跃的研究领域。在天然产物的研究过程中，首先要解决的问题是天然产物的提取、纯化和鉴定。随着现代色谱和波谱技术的发展，对天然产物的分离和鉴定变得更为有利和方便。

实验 48　从茶叶中提取咖啡因

茶叶中含有多种生物碱、丹宁、色素、维生素、蛋白质等物质。咖啡因（Caffeine）是其中一种生物碱，在茶叶含量约为1%～5%，此外还有少量的茶碱和可可豆碱，属于杂环化合物嘌呤的衍生物，它们的结构式如下。

嘌呤　　　咖啡因　　　可可豆碱　　　茶碱

咖啡因的化学名称为1,3,7-三甲基-2,6-二氧嘌呤，是具有绢丝光泽的无色针状结晶，含一个结晶水，在100℃时失去结晶水开始升华，在120℃时升华相当显著，至178℃时升华很快，升华为针状晶体。无水咖啡因的熔点为235℃，是弱碱性物质，味苦。易溶于热水（约80℃）、乙醇、乙醚、丙酮、二氯甲烷、氯仿，难溶于石油醚。

可可豆碱的化学名称为3,7-二甲基-2,6-二氧嘌呤，在茶叶中约含0.05%，是无色针状晶体，味苦，熔点为342～343℃，于290℃升华。能溶于热水，难溶于冷水、乙醇，不溶于醚。

茶碱的化学名称为1,3-二甲基-2,6-二氧嘌呤，是可可豆碱的同分异构体，白色微小粉末结晶，味苦，熔点为273℃。易溶于沸水，微溶于冷水、乙醇，易溶于氯仿。

茶叶中的生物碱对人体具有一定程度的药理功能。咖啡因可兴奋神经中枢，消除疲劳，有强心作用；茶碱功能与咖啡因相似，兴奋神经中枢较咖啡因弱，而强心作用则比咖啡因强；可可豆碱功能也与咖啡因类似，兴奋神经中枢较前两者弱，而强心作用则介于前两者之间。

咖啡因在医学上用作心脏、呼吸器官和神经系统的兴奋剂，也是治感冒药 APC（阿司匹林-非那西丁-咖啡因）组成成分之一。过度使用咖啡因会增加抗药性和产生轻度上瘾。

咖啡因不仅可以通过测定熔点和光谱法加以鉴别，还可以通过制备咖啡因水杨酸盐衍生物进一步得到确认。作为弱碱性化合物，咖啡因可与水杨酸作用生成熔点为137℃的水杨酸盐。

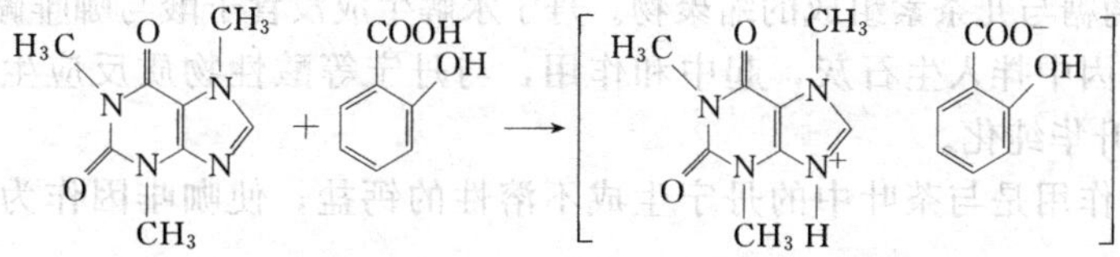

咖啡因　　水杨酸　　　　咖啡因水杨酸

【主要试剂】

方法 1　茶叶 3g，95%乙醇 60mL，生石灰粉。
方法 2　茶叶 10g，碳酸钙 10g，氯仿。
水杨酸 30mg，甲苯，石油醚（60～90℃）。

【实验步骤】

(1) 咖啡因提取

方法 1　称取 3g 茶叶放入 Soxhlet 提取器（见图 2-39）的滤纸筒中，加入 40mL 95%乙醇，在圆底烧瓶中再加入 20mL 乙醇，水浴加热回流提取，直到提取液颜色较浅时为止。待冷凝液刚刚虹吸下去时，立即停止加热。稍冷后改成蒸馏装置，回收抽取液中的大部分乙醇。趁热把瓶中残液倾倒入蒸发皿中，拌入 1～1.5g 生石灰粉，与萃取液拌和成茶砂，在蒸汽浴上蒸干成粉状（不断搅拌，压碎块状物），最后将蒸发皿移至石棉网上用煤气灯小火焙炒片刻，务必使水分全部除去。冷却后，擦去沾在蒸发皿壁上的粉末，以免升华时污染产物。

在上述蒸发皿上盖一张刺有许多小孔且孔刺向上的滤纸，再在滤纸上罩一个大小合适的玻璃漏斗，漏斗颈部塞一小团疏松的棉花［见图 2-48(a)］，用砂浴小心加热升华。当滤纸上出现白色毛状结晶时，暂停加热，冷却至 100℃左右，揭开漏斗和滤纸，仔细地把附在滤纸上及器皿周围的咖啡因用刮刀刮下。若残渣为绿色，可将残渣经拌和后用较大的火再加热片刻，使升华完全，直至残渣为棕色。合并两次升华收集到的咖啡因，称量所得的产物，测其熔点，计算咖啡因在茶叶中的含量。

方法 2　在 250mL 烧杯中加入 100mL 水和 10g 粉末状碳酸钙。称取 10g 茶叶，用纱布包好后放入烧杯中煮沸 30min，取出茶叶。压干，趁热抽滤，滤液冷却后用 15mL 氯仿分两次萃取，合并萃取液（萃取液若混浊，颜色较浅，则加入少量蒸馏水洗涤至澄清），水浴蒸去氯仿，残液留作升华实验用。提取液的升华实验同方法 1。

(2) 咖啡因水杨酸盐衍生物的制备

在试管中加入 40mg 咖啡因、30mg 水杨酸和 2.5mL 甲苯。在水浴上加热振摇使固体溶解，然后加入 1.5mL 石油醚（60～90℃）。在冰浴中冷却结晶。如无结晶析出，可用玻璃棒或刮刀摩擦管壁。抽滤，收集产物，干燥后测定熔点。纯盐的熔点为 137℃。

【注意事项】

① 滤纸套筒大小要适中，既要紧贴器壁，又要方便取放，并且其高度不得超过虹吸管。

② 萃取液和生石灰焙炒时，务必将溶剂全部除去。若不除净，在下一步加热升华时，在漏斗内会出现水珠。若遇此情况，则用滤纸迅速擦干漏斗内的水珠并继续升华。

③ 在升华过程中，要始终严格控制温度，温度太高会使被烘物冒烟炭化，导致产品不纯和损失。

④ 丹宁不是单一化合物，是相对分子质量在 500～3000 之间的一种酚类化合物。茶叶中丹宁通常分为两类：一类能水解；另一类不能水解。能水解的丹宁是由二没食子酰基与葡萄糖分子中的某一位中的羟基所形成的酯的混合物，能在热水中水解生成没食子酸和葡萄糖。不能水解的丹宁是葡萄糖与儿茶素组成的缩聚物。丹宁水解生成没食子酸与咖啡碱生成难溶盐。

⑤ 在粗咖啡因中拌入生石灰，起中和作用，与丹宁等酸性物质反应生成钙盐，游离的咖啡因就可通过升华纯化。

⑥ 碳酸钙的作用是与茶叶中的丹宁生成不溶性的钙盐，使咖啡因作为可溶性的生物碱留在水溶液中。

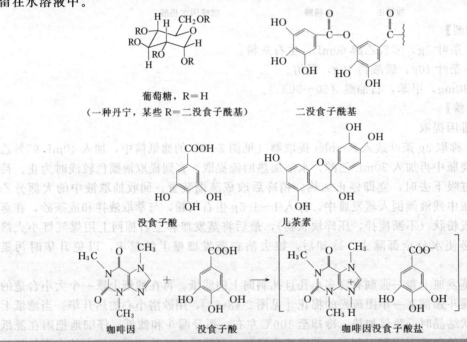

葡萄糖，R＝H
（一种丹宁，某些 R＝二没食子酰基）　　二没食子酰基

没食子酸　　儿茶素

咖啡因　　没食子酸　　咖啡因没食子酸盐

思 考 题

1. 提取咖啡因时，加入氧化钙和碳酸钙的目的是什么？
2. 咖啡因、茶碱和可可豆碱在结构上有何区别？它们对人体有何影响？
3. 用升华法提纯固体有什么优点和局限性？

3.18 金属有机化合物的制备

实验 49　二茂铁的合成

【反应式】

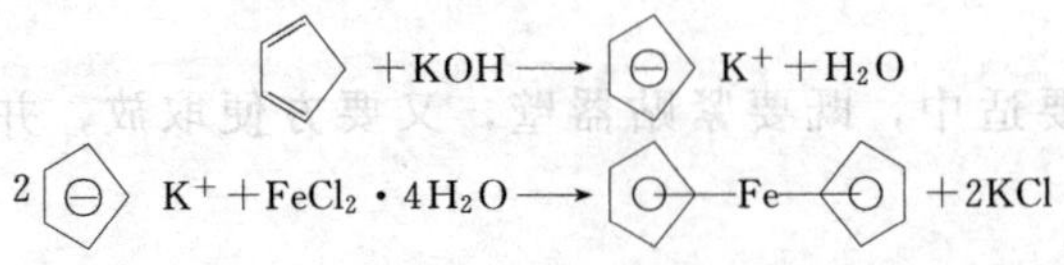

$$\text{C}_5\text{H}_6 + \text{KOH} \longrightarrow \text{C}_5\text{H}_5^- \text{K}^+ + \text{H}_2\text{O}$$

$$2\,\text{C}_5\text{H}_5^- \text{K}^+ + \text{FeCl}_2 \cdot 4\text{H}_2\text{O} \longrightarrow (\text{C}_5\text{H}_5)_2\text{Fe} + 2\text{KCl}$$

【主要试剂】

环戊二烯 2.6mL（2.1g，31.7mmol），KOH 1.3g（23.2mmol），$FeCl_2 \cdot 4H_2O$，二甲基亚砜（DMSO），氮气，2mol/L 盐酸。

【实验步骤】

在 100mL 三口瓶上配置电动搅拌器、y 形管和氮气导气管。y 形管的一口连上恒压滴液漏斗，另一口与鼓泡器相连接向反应瓶中加入 1.3g KOH、30mL DMSO 及 2.6mL 新解聚的环戊二烯，通入氮气，开动搅拌器。待形成环戊二烯钾黑色溶液后，滴加刚刚用 3.5g $FeCl_2 \cdot 4H_2O$ 和 25mL DMSO 配制的溶液，同时强搅拌，加完后再搅拌反应 20min。把反应液倾入 50g 冰-50g 水中，搅拌均匀，用 2mol/L 盐酸调反应液 pH 为 3~5，待黄色固体完全析出后，抽滤，分 4 次各用 10mL 水洗滤饼，抽干烘干，粗产品约 2.2g。

若要得到纯品，可采用升华法进行纯化。将粗产品放入干净且干燥的 400mL 烧杯中，盖上表面皿，用脱脂棉塞住烧杯嘴，缓缓加热烧杯，表面皿外边用湿布冷却，如此常压 100℃升华，可得到黄色片状光亮的晶体，m. p. 173~174℃。

【注意事项】

① 环戊二烯在室温下容易二聚，生成环戊二烯的二聚体。当二聚体在 170℃下沸腾时，环戊二烯单体与其二聚体之间能建立起一个平衡。

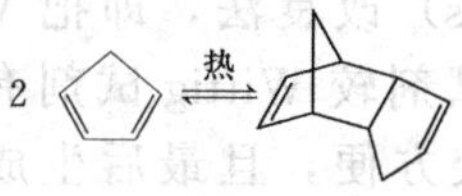

因此，纯净的环戊二烯可经二聚体的解聚、分馏而获得。二聚体解聚成环戊二烯单体的方法见实验 35 环戊二烯与马来酸酐的反应。新解聚的环戊二烯应立即使用或暂时存放在冰箱中低温保存。

② 在空气中，二茂铁能被氧化成蓝色的正离子 $Fe^{3+}(C_5H_5)_2$。$FeCl_2 \cdot 4H_2O$ 在 DMSO 中也会使 Fe^{2+} 氧化成 Fe^{3+}，因此需用氮气保护以隔绝空气。

③ $FeCl_2 \cdot 4H_2O$ 如果变成棕色可用乙醇或乙醚洗成淡绿色再用，用前应研细溶解。如药品中含有较多的棕色三价铁，则会降低实验效果。

④ KOH 应研细加入，动作要快，以防吸水。

思 考 题

1. 通过查阅文献谈一谈二茂铁有何作用。
2. 二茂铁比苯更易发生亲电取代反应，但用混酸（$HNO_3 + H_2SO_4$）使二茂铁发生硝化时，实验没有成功。为什么？

3.19 维悌希（Wittig）反应

磷内鎓盐（磷叶立德）与羰基化合物进行亲核加成生成烯烃的反应，称为维悌希（Wittig）反应。利用它由羰基化合物合成烯烃，其双键位置确定，一般不发生重排、转位等副反应。由 α,β-不饱和的醛或酮制备共轭烯烃十分有利，具有反应条件温和、产率高等优点。因此，Wittig 反应作为合成烯烃的一般方法在有机合成中得到了广泛的应用。

以具有亲核性的三苯基膦及卤代烃为原料得到季𬭸盐，再用强碱处理脱去烷基上的 α-氢原子可得到 Wittig 试剂。

$$(C_6H_5)_3P + BrCHR^3R^4 \longrightarrow (C_6H_5)_3\overset{\oplus}{P}CHR^3R^4 Br^- \xrightarrow[-LiBr]{C_6H_5Li} \atop -C_6H_6$$

$$(C_6H_5)_3\overset{\oplus}{P}-\overset{\ominus}{C}R^3R^4 \longleftrightarrow (C_6H_5)_3P=CR^3R^4$$
$$\text{ylid}$$

ylid 是一种亲核试剂，可以进攻醛酮的羰基碳，最后消除掉三苯基氧膦而生成烯烃。

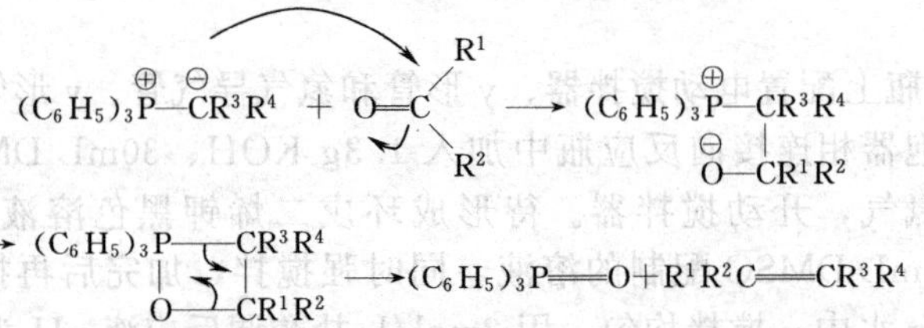

$$(C_6H_5)_3\overset{\oplus}{P}-\overset{\ominus}{C}R^3R^4 + \overset{R^1}{\underset{R^2}{O=C}} \longrightarrow (C_6H_5)_3\overset{\oplus}{P}-CR^3R^4 \atop \overset{\ominus}{O}-CR^1R^2$$

$$\longrightarrow \underset{O-CR^1R^2}{(C_6H_5)_3P-CR^3R^4} \longrightarrow (C_6H_5)_3P=O + R^1R^2C=CR^3R^4$$

Wittig 试剂在实际应用时也受到了一定的限制。首先，Wittig 试剂（氢或烷基与 α-碳原子相连）很活泼，与氧、水、氢卤酸、醇等反应，因此该反应需控制在没有这些物质存在的情况下进行。其次，当 α-碳原子连有—COR、—CN、—COOR、—CHO 等吸电子基团时，Wittig 试剂比较稳定，能与醛作用，而与酮作用较慢或不发生反应。此外，通常使用的强碱是不容易处理、较危险的有机锂化合物，需要在惰性气体保护下操作；使用的膦试剂的毒性较大，价格昂贵。

为了增强 α-碳原子的亲核性，人们做了许多改进 Wittig 反应的研究，其中最有效的是洪恩纳-埃孟斯（Horner-Emmons）改良法，即把 Wittig 试剂中的磷原子上的苯基用氧或乙氧基取代。洪恩纳-埃孟斯试剂较 Wittig 试剂有许多优点：它对碱性介质不很敏感，与氧和水反应迟缓，操作比较方便，且最后生成的磷酸二乙酯溶于水，容易与制得的烯烃分离。

实验 50 （E）-1,2-二苯乙烯的合成

【反应式】

$$PCl_3 + 3C_2H_5OH + 3C_6H_5N(CH_3)_2 \longrightarrow (C_2H_5O)_3P + 3C_6H_5N(CH_3)_2 \cdot HCl$$

$$(C_2H_5O)_3P + C_6H_5CH_2Cl \longrightarrow (C_2H_5O)_2\overset{O}{\overset{\|}{P}}CH_2C_6H_5 + C_2H_5Cl$$

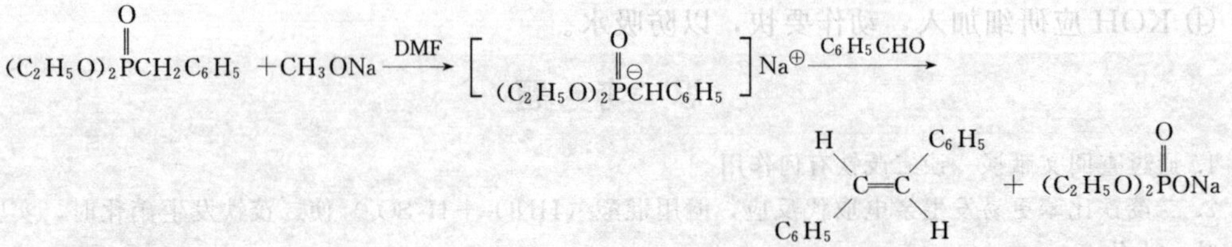

$$(C_2H_5O)_2\overset{O}{\overset{\|}{P}}CH_2C_6H_5 + CH_3ONa \xrightarrow{DMF} \left[(C_2H_5O)_2\overset{O}{\overset{\|}{P}}\overset{\ominus}{C}HC_6H_5\right] Na^{\oplus} \xrightarrow{C_6H_5CHO}$$

$$\underset{C_6H_5}{\overset{H}{}}C=C\underset{H}{\overset{C_6H_5}{}} + (C_2H_5O)_2\overset{O}{\overset{\|}{P}}ONa$$

【主要试剂】

绝对乙醇 4.6g（5.8mL，100mmol），新蒸馏过的 N,N-二甲基苯胺 12.1g（12.7mL，100mmol），新蒸馏过的三氯化磷 4.5g（2.9mL，33mmol），无水石油醚（30～60℃）。

氯化苄 6mL（6.60g，52mmol），亚磷酸三乙酯 9mL（8.70g，52mmol）。

甲醇钠 3.00g（55mmol），新蒸的苯甲醛 5.2mL（5.40g，51mmol），N,N-二甲基甲酰胺，甲醇，异丙醇。

【实验步骤】

(1) 亚磷酸三乙酯的制备 在 100mL 三口瓶上分别安装电动搅拌器、恒压滴液漏斗和顶端连有氯化钙干燥管的球形冷凝管，瓶中放置 4.6g 绝对乙醇、12.1g 新蒸馏过的 N,N-二甲基苯胺和 30mL 无水石油醚（30～60℃）。在滴液漏斗中放置 4.5g 新蒸馏过的三氯化磷和 15mL 无水石油醚（30～60℃）。在搅拌下，将三氯化磷石油醚溶液逐滴加入三口瓶中。控制滴加速度使反应液维持微沸，由于反应放热，必要时可用冷水浴冷却。随着三氯化磷溶液

的加入，三口瓶内出现白色沉淀（这是什么化合物?）。滴加完毕后，继续搅拌 0.5h，再在搅拌下用水浴加热回流 1h。待反应物冷至室温后，减压抽滤，沉淀用无水石油醚洗涤两次，抽干。滤液转移至蒸馏瓶内，在水浴上蒸去石油醚，然后进行减压蒸馏，收集 57~58℃/2.13kPa（16mmHg）的馏分。产品约 4.5g，产率 82%。

纯亚磷酸三乙酯沸点为 156.5℃。

(2) 苄基膦酸二乙酯的制备　在 50mL 干燥两口烧瓶中加入 6mL 氯化苄、9mL 亚磷酸三乙酯和几粒沸石，烧瓶侧口装回流冷凝管，其上端连氯化钙干燥管，中间口插入 300℃温度计。在石棉网上小火加热，在 130~140℃可发现氯乙烷释放出来，约在 165℃开始沸腾，继续加热 1.5~2h，温度最终达到 200℃以上。停止加热，使反应液温度降至室温待用。

(3) (E)-1,2-二苯乙烯的制备　在装有滴液漏斗、温度计和氯化钙干燥管的 100mL 干燥三口瓶中加入 3.00g 甲醇钠和 25mL 干燥的 N,N-二甲基甲酰胺（DMF）。将三口瓶置于冰水浴中冷却，然后加入（2）中得到的苄基膦酸二乙酯（生成 ylid 的反应是放热的）。在滴液漏斗中放置 5.2mL 新蒸的苯甲醛。当三口瓶中的反应混合物的温度低于 20℃时，滴加苯甲醛，加入速度以维持反应温度在 30~40℃为宜。加完后，移去冰水浴，使反应瓶在室温下放置 10min，使反应完全。在搅拌下加入 10mL 水，此时反式二苯乙烯晶体析出。再将三口瓶置于冰水浴中，搅拌片刻，使晶体析出完全。抽滤，收集产物得白色晶体。用少量等体积冷的甲醇-水溶液洗涤粗产物。干燥后的粗产物，用异丙醇（约 30~50mL）进行重结晶。熔点为 124~126℃，产量 5~6g。

纯 (E)-1,2-二苯乙烯的熔点为 125℃。

【注意事项】

① 三氯化磷有毒并具有刺激性，实验须在通风橱内进行，防止吸入其蒸气。若溅到皮肤上，立即用肥皂和水洗涤。N,N-二甲基苯胺有毒，防止吸入其蒸气。如果溅到皮肤上，应立即用 5%乙酸清洗后再用肥皂水清洗。

② 三氯化磷遇水分解，所以实验中所用仪器、溶剂、药品均应干燥无水。三氯化磷和 N,N-二甲基苯胺均应经蒸馏后再使用。市售无水乙醇在使用前用 5A 分子筛浸泡。

③ 制备亚磷酸三乙酯过程中，减压抽滤白色沉淀物时，应用干燥布氏漏斗和吸滤瓶，并尽量抽干。

④ 氯化苄是一种催泪剂，操作时须戴防护镜。亚磷酸三乙酯具有令人不愉快的气味。

⑤ 制苄基膦酸二乙酯时，加热时间不足 1h 和反应温度低于 220℃都会降低产率。副产物氯乙烷对人体有害，应导出室外。

⑥ 尽可能不让吸湿性强的甲醇钠暴露在空气中，应迅速称量并及时盖好瓶塞。

⑦ 异丙醇的用量取决于粗产物的干燥程度。

甲醇钠的制备

在 100mL 两口烧瓶上安装连有氯化钙干燥管的球形冷凝管，导气管导入通风管道。向两口瓶中加入 15mL 无水甲醇，用磨口塞塞紧。取金属钠一小块，用镊子夹住，用刀子切去外皮，迅速称量 6g，并用刀子切成薄片，直接投入盛有石油醚的 100mL 烧杯内。用镊子夹取一片钠投入甲醇中，立即有氢气逸出，待此片完全溶解后再加另一片。此操作应在通风橱内进行，近处不能有明火，更不能一次投入多片钠。金属钠全部反应后，烧瓶上装蒸馏装置，在热水浴中蒸出过量的甲醇。然后，在油浴温度 150℃及水泵减压下蒸出甲醇，得到较干燥的白色甲醇钠固体。将所得甲醇钠放入干燥的广口瓶内，用橡皮塞塞紧，放在真空干燥器内备用。

思 考 题

1. 用 Wittig 反应制备烯烃有哪些特点？写出反应机理？
2. 由醛酮制备烯烃还可通过哪些途径？
3. 久置的苯甲醛中含有什么杂质？用它来进行 Wittig 反应会有什么影响？

3.20 霍夫曼酰胺降解反应

实验 51 邻氨基苯甲酸的合成

【反应式】

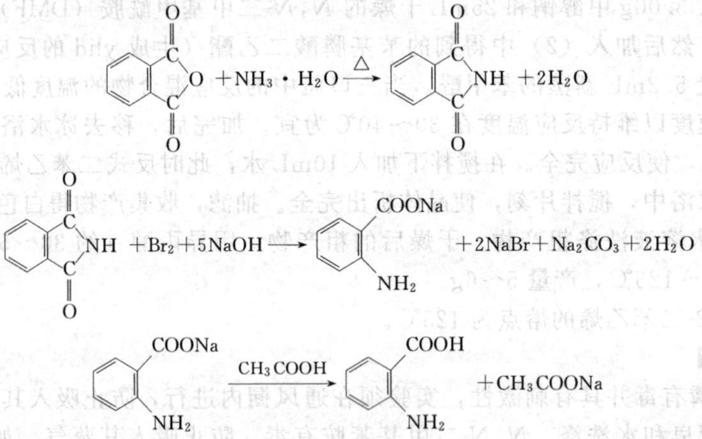

邻苯二甲酸酐 + $NH_3 \cdot H_2O$ $\xrightarrow{\triangle}$ 邻苯二甲酰亚胺 + $2H_2O$

邻苯二甲酰亚胺 + Br_2 + $5NaOH$ → 邻氨基苯甲酸钠 + $2NaBr + Na_2CO_3 + 2H_2O$

邻氨基苯甲酸钠 $\xrightarrow{CH_3COOH}$ 邻氨基苯甲酸 + CH_3COONa

反应机理

邻苯二甲酰亚胺 + Br_2 + $5NaOH$ → 邻氨基苯甲酸钠 + $2NaBr + Na_2CO_3 + 2H_2O$

经过中间体 [COONa, CONH₂] $\xrightarrow{NaOH}$ [COONa, CONHBr] $\xrightarrow{NaOBr/NaOH}$ [COONa, CON:] $\xrightarrow{-HBr}$ [COONa, N=C=O] $\xrightarrow{H_2O}$ 产物

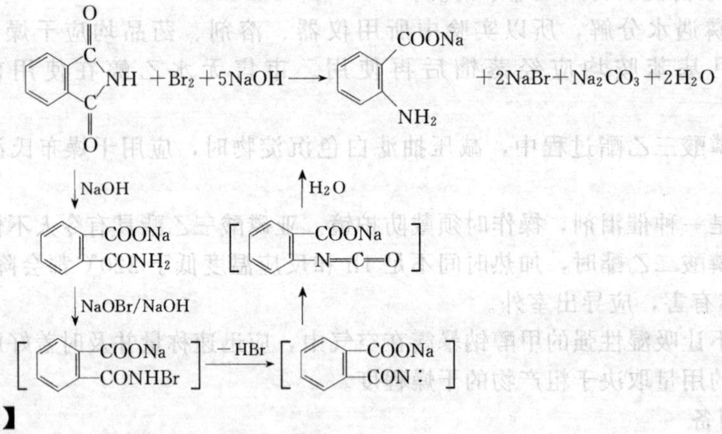

【主要试剂】

邻苯二甲酸酐 10g（67.5mmol），浓氨水 10mL，溴 2.1mL（6.5g，41mmol），氢氧化钠，浓盐酸，饱和亚硫酸氢钠溶液，冰醋酸。

【实验步骤】

（1）邻苯二甲酰亚胺的制备　在 100mL 两口烧瓶中，放入 10g 邻苯二甲酸酐和 10mL 浓氨水，装上空气冷凝管及一支 360℃ 温度计。先在石棉网上加热，然后用小火直接加热，温度逐渐升到 300℃。间歇摇动烧瓶。用玻璃棒将升华进入冷凝管的固体物质推入烧瓶里。趁热把反应物倒入搪瓷盘中。冷却后凝成的固体放在研钵中研成粉末。产量约 8g，m. p. 232～234℃。

(2) 邻氨基苯甲酸的制备　在 125mL 锥形瓶中，加入 7.5g 氢氧化钠和 30mL 水配制成碱液。将此锥形瓶放入冰盐浴中，冷却至 $-5\sim0℃$。往碱液中一次加入 2.1mL 溴，振荡锥形瓶，使溴全部反应。此时温度略有升高。将制成的次溴酸钠冷却到 0℃ 以下，放置备用。

在另一个小锥形瓶中，加入 5.5g 氢氧化钠和 20mL 水配制另一碱液。

取 6g 研细的邻苯二甲酰亚胺，加入少量水调成糊状物，一次全部加到冷的次溴酸钠溶液中，剧烈振荡锥形瓶。反应混合物应保持在 0℃ 左右。从冰盐浴中取出锥形瓶，再剧烈摇动直到反应物转为黄色清液。把配制好的氢氧化钠溶液全部迅速加入，反应温度自行升高。把反应混合物加热到 80℃ 约 2min。加入 2mL 饱和亚硫酸氢钠溶液，以还原剩余的次溴酸。冷却，减压过滤。把滤液倒入 250mL 烧杯中，放在冰水浴中冷却。在不断搅拌下小心地滴加浓盐酸，使溶液恰好呈中性（pH=7，用石蕊试纸检验，约需 15mL 盐酸），然后再缓慢地滴加 5~7mL 冰醋酸，使邻氨基苯甲酸完全析出。减压过滤，用少量冷水洗涤，晾干，产量约 4g。灰白色粗产物用水进行重结晶，可得白色片状晶体。

纯邻氨基苯甲酸为白色片状晶体，m.p.145℃。

【注意事项】

① 溴具有强腐蚀性和刺激性，必须在通风橱中量取，取溴时应戴防护镜和橡皮手套，并且注意不要吸入溴蒸气。

② 邻氨基苯甲酸既能溶于碱，又能溶于酸，故过量的盐酸会使产物溶解。若加了过量的盐酸，需加氢氧化钠中和。

③ 邻氨基苯甲酸的等电点约为 3~4。为使邻氨基苯甲酸完全析出，必须加入适量的醋酸。

思　考　题

1. 邻氨基苯甲酸在合成和分析上有哪些应用？
2. 假若溴和氢氧化钠的用量不足或有较大的过量，对反应各有何影响？
3. 邻氨基苯甲酸的碱性溶液，加盐酸使之恰好呈中性后，为什么不再加盐酸而是加适量醋酸使邻氨基苯甲酸完全析出？

3.21　多步合成

3.21.1　以苯酚为原料的多步合成

【多步合成路线 1】　4-碘苯氧乙酸（4-Iodophenoxyacetic acid）

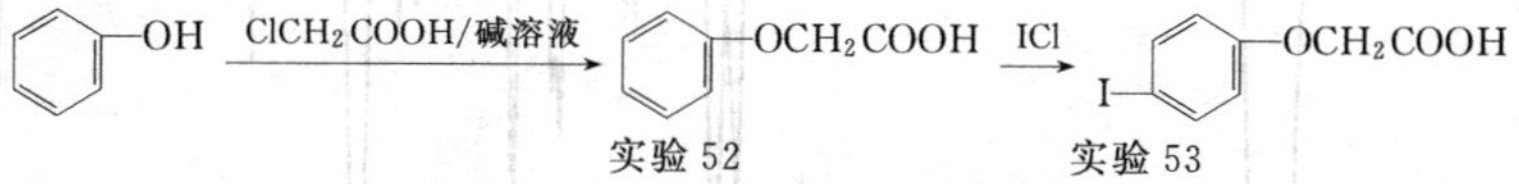

$$\text{PhOH} \xrightarrow{\text{ClCH}_2\text{COOH/碱溶液}} \text{PhOCH}_2\text{COOH} \xrightarrow{\text{ICl}} \text{I-C}_6\text{H}_4\text{-OCH}_2\text{COOH}$$

实验 52　　　　　　　　　　　实验 53

实验 52　苯氧乙酸的合成

苯氧乙酸是一种白色片状或针状晶体，可用于合成染料、药物、杀虫剂，还可直接用作植物生长调节剂，且对人畜无害，因而应用较为广泛。苯氧乙酸可用苯酚、一氯乙酸在碱性溶液中进行威廉姆逊反应而制得。

【总反应式】

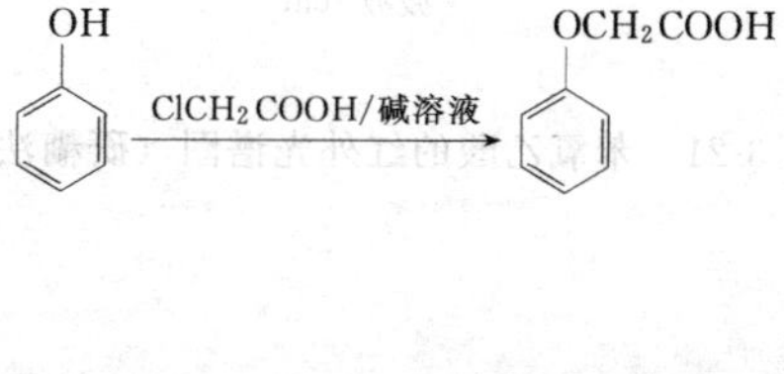

$$\text{PhOH} \xrightarrow{\text{ClCH}_2\text{COOH/碱溶液}} \text{PhOCH}_2\text{COOH}$$

【分步反应式】

$$\text{C}_6\text{H}_5\text{OH} + \text{NaOH} \longrightarrow \text{C}_6\text{H}_5\text{ONa} + \text{H}_2\text{O}$$

$$2\text{ClCH}_2\text{COOH} + \text{Na}_2\text{CO}_3 \longrightarrow 2\text{ClCH}_2\text{COONa} + \text{H}_2\text{O} + \text{CO}_2$$

$$\text{C}_6\text{H}_5\text{ONa} + \text{ClCH}_2\text{COONa} \longrightarrow \text{C}_6\text{H}_5\text{OCH}_2\text{COONa} + \text{NaCl}$$

$$\text{C}_6\text{H}_5\text{OCH}_2\text{COONa} + \text{HCl} \longrightarrow \text{C}_6\text{H}_5\text{OCH}_2\text{COOH} + \text{NaCl}$$

【主要试剂】

苯酚 2.8g (30mmol),氢氧化钠 1.3g (32.5mmol),一氯乙酸 3.1g (32.8mmol),碳酸钠 2g (18.9mmol),20%盐酸,乙醚,15%氯化钠溶液。

【实验步骤】

(1) 配制氯乙酸钠溶液 依次将 3.1g 一氯乙酸和 10mL 15%氯化钠溶液加入到 100mL 的烧杯中,搅拌下少量多次慢慢加入 2g 碳酸钠,加入速度以反应混合物温度不超过 40℃为宜。此时溶液 pH 为 7~8,如不足此值,再改用饱和碳酸钠水溶液将反应混合液 pH 调至 7~8。

(2) 配制苯酚钠溶液 在 100mL 的三口烧瓶上配置搅拌器、回流冷凝管和温度计。向三口烧瓶中加入 1.3g 氢氧化钠、7.5mL 水和 2.8g 苯酚,开动搅拌器搅拌使固体溶解,冷却后待用。

(3) 苯氧乙酸的合成 将配好的一氯乙酸钠溶液加入到上述苯酚钠溶液的三口烧瓶中,开动搅拌器,在石棉网上小火加热,使反应温度保持在 100~110℃之间,回流 2h。

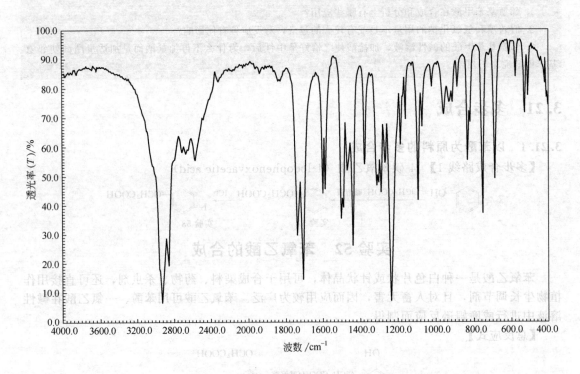

图 3-21 苯氧乙酸的红外光谱图(研糊法)

反应结束后，趁热将反应混合物倒入 250mL 烧杯中，加入 30mL 水，搅拌均匀，用浓盐酸调节溶液 pH 为 1~2，冷却，析出白色晶体。抽滤，用 5mL 冷水洗涤粗产品，抽干后，将苯氧乙酸粗产品倒入 250mL 烧杯中，加入 30mL 水，加入固体碳酸钠使苯氧乙酸固体溶解。将溶液转入到分液漏斗中，加入 10mL 乙醚，振荡、静置分层，除去乙醚层。水层再用 20%盐酸酸化至 pH 为 1~2，静置、冷却结晶，得到白色晶体，抽滤后用冷水洗涤滤饼两次，经干燥后即得精制产物。称量，测熔点，并计算产率。

纯苯氧乙酸为白色针状结晶，m.p. 98~99℃。

纯苯氧乙酸的红外光谱见图 3-21。

【注意事项】

① 一氯乙酸和苯酚具有腐蚀性，避免触及皮肤。

② 安装好搅拌器，避免损坏玻璃仪器。

③ 配制一氯乙酸钠溶液时，采用食盐水有利于抑制一氯乙酸钠的水解。中和反应温度超过 40℃时，一氯乙酸易发生水解。

④ 合成苯氧乙酸反应刚开始时，反应混合物 pH 为 12，随着反应的进行，其 pH 逐渐变小，直至 pH 为 7~8，反应即告结束。

⑤ 乙醚是用来萃取未反应而游离出来的少量酚。

思 考 题

1. 以酚钠和一氯乙酸作原料制醚时，为什么要先使一氯乙酸成盐？可否用苯酚和一氯乙酸直接反应制备醚？
2. 用碳酸钠中和一氯乙酸时为何要加食盐水？
3. 在苯氧乙酸合成过程中，为何 pH 会发生变化，以 pH 7~8 作为反应终点的依据是什么？
4. 苯氧乙酸的红外光谱是采用在样品中加入液体石蜡（Nujor），以研糊法所测得，试对其红外光谱加以解析。
5. 通过查阅文献，找出苯氧乙酸的其他合成方法。

实验 53 4-碘苯氧乙酸的合成

4-碘苯氧乙酸商品名为增产灵，是一种植物生长素，它能有效地提高植物的结果率，防止过早落果，使作物增产。

【反应式】

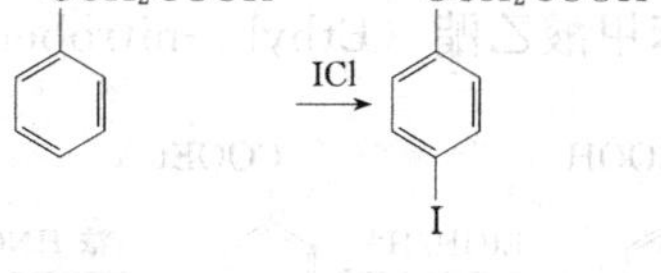

$$\underset{}{\text{C}_6\text{H}_5\text{OCH}_2\text{COOH}} \xrightarrow{\text{ICl}} \underset{}{4\text{-I-C}_6\text{H}_4\text{OCH}_2\text{COOH}}$$

【主要试剂】

苯氧乙酸 0.5g（3.3mmol），氯化碘 0.58g（3.6mmol），浓盐酸 2mL。

【实验步骤】

在 100mL 三口烧瓶上配置搅拌器、恒压滴液漏斗和温度计。依次将 0.5g 苯氧乙酸和 38mL 水加入三口烧瓶中，水浴加热并搅拌，使苯氧乙酸溶解。然后继续搅拌，在 40℃浴温条件下，自滴液漏斗向反应瓶滴加 0.58g 氯化碘与 2mL 浓盐酸配成的溶液。滴加完毕，提高浴温至 80℃，继续加热搅拌 1h。将反应液冷却至室温，有浅粉色针状晶体析出，抽滤，粗产物用少量水洗涤两次后，将粗产物用水或 50%乙醇重结晶，得到白色针状晶体。产物经干燥后称量，测熔点并计算产率。

4-碘苯氧乙酸为白色针状晶体，m.p. 154~156℃。

4-碘苯氧乙酸的红外光谱见图 3-22。

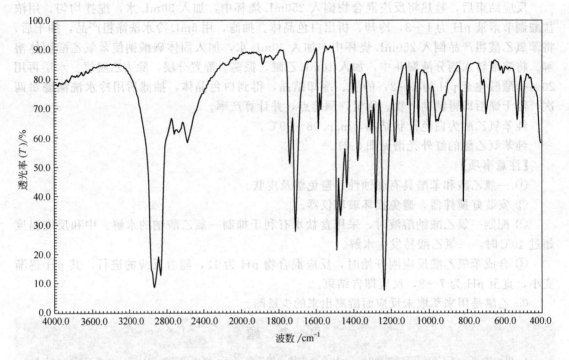

图 3-22　4-碘苯氧乙酸的红外光谱图（研糊法）

【注意事项】
① 氯化碘和浓盐酸均具有腐蚀性并且易挥发，操作时要小心，应在通风橱内量取。
② 氯化碘见光易分解并且对橡胶有强侵蚀性，应贮于棕色具塞玻璃瓶中。化学试剂商店里有氯化碘出售。

思 考 题

在苯氧乙酸分子中，—OCH_2COOH 基团对苯环具有活化作用。在苯氧乙酸的碘化反应中理应生成 2-碘苯氧乙酸和 4-碘苯氧乙酸两种异构体。但是，实验表明，4-碘苯氧乙酸是主要产物，为什么？

3.21.2　以甲苯为原料的多步合成

【多步合成路线 2】间硝基苯甲酸乙酯（Ethyl 3-nitrobenzoate）

实验 54　间硝基苯甲酸乙酯的合成

利用硝酸和硫酸将硝基引入芳环是用硝𬭩离子 NO_2^+ 进行芳环亲电取代的一个例子。乙酸乙酯中的乙氧羰基钝化苯环，使硝基进入苯甲酸乙酯的间位。

【反应式】

【主要试剂】

苯甲酸乙酯 1.8mL（1.88g，12.5mmol），浓硫酸 5.5mL（10.13g，101.2mmol），浓硝酸 1.5mL（2.12g，23.5mmol）。

【实验步骤】

向 25mL 锥形瓶中加入 1.8mL 苯甲酸乙酯，在振摇下加入 4mL 浓硫酸。然后在冰浴中冷却该混合物。在另一锥形瓶中加入 1.5mL 浓硝酸和 1.5mL 浓硫酸，混合均匀并置于冰浴中冷却。在振摇下将此硝酸溶液滴入苯甲酸乙酯溶液中。同时用冰浴保持反应温度在 0～10℃之间。此过程约需 30min。滴加完毕后，将溶液在室温放置 10min，然后在搅拌下将其倒入盛有碎冰的烧杯中，继续搅拌直至沉淀变为小颗粒，抽滤，用冷水洗涤。粗产物用乙醇重结晶后即得到无色固体。记录产率和熔点数据。

纯的间硝基苯甲酸乙酯是无色固体。m.p. 40～43℃，b.p. 297～298℃。

思 考 题

写出用 HNO_3/H_2SO_4 硝化甲苯的反应机理。

3.21.3 以苯胺为原料的多步合成

【多步合成路线 3】对溴苯胺（p-Bromoaniline）

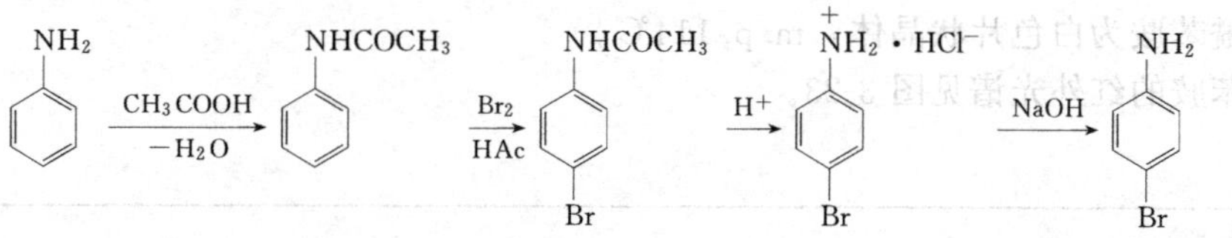

实验 55　　实验 56　　　　　　实验 57

在此实验中，苯胺首先被转化为乙酰苯胺，然后再在其对位进行溴代。溴代反应之后，水解除去乙酰基得到所需产物对溴苯胺。

可以利用乙酰氯、乙酸酐或冰醋酸（利用除去反应中生成的水的方法）使苯胺乙酰化。用乙酰氯反应比较剧烈，乙酸酐水解较慢，所以通常采用乙酸酐进行乙酰化。冰醋酸比较便宜，但需要较长的反应时间。

乙酰化常用于保护伯胺和仲胺的氨基。氨基被乙酰化后不易被氧化。对芳环的活化能力降低，又因其碱性较弱，使之不易发生许多氨基特有的反应。在多步合成之后，可在酸或碱性条件下水解恢复氨基。

【多步合成路线 4】对硝基苯胺（p-Nitroaniline）

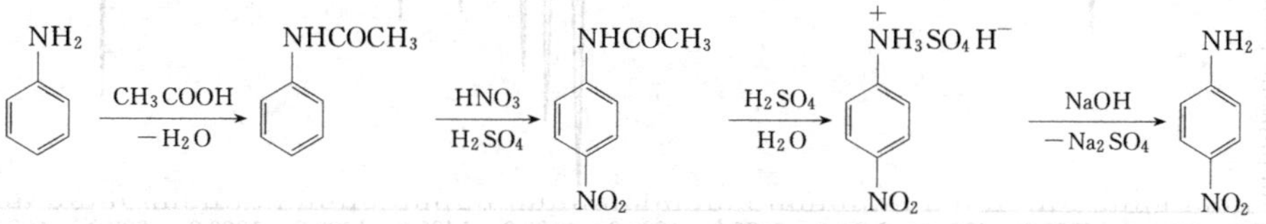

实验 55　　实验 58　　　　　　实验 59

第一步，苯胺转变为乙酰苯胺在实验 55 中已完成。

氨基对苯环的亲电取代致活能力很强，以至于苯胺的直接硝化难于控制。因此关于对硝基乙酰苯胺的制备是采用氨基乙酰化来降低氨基对苯环的致活能力，然后再在酸性条件下水解除去乙酰基。

对硝基苯胺被用于制备对苯二胺和特殊染料，如对位红。也可被用作实验室合成一些难以得到的化合物的中间体。

实验 55　乙酰苯胺的合成

【反应式】

$$\underset{}{\underset{}{\text{C}_6\text{H}_5\text{NH}_2}} \xrightarrow[-\text{H}_2\text{O}]{\text{CH}_3\text{COOH}} \underset{}{\text{C}_6\text{H}_5\text{NHCOCH}_3}$$

【主要试剂】

苯胺 2.5mL（2.55g，27.5mmol），冰醋酸 3.8mL（3.98g，66mmol），锌粉 0.03g。

【实验步骤】

在 10mL 干燥的圆底烧瓶中加入 2.5mL 新蒸馏的苯胺，3.8mL 冰醋酸和少许锌粉（约 0.03g），摇匀。装上分馏柱，柱的侧管处连接接引管，用一个 5mL 圆底烧瓶接收蒸出的水和乙酸，柱顶插上温度计。在石棉网（或加热套）上小火加热，保持反应液微沸约 10min，逐渐升温，使反应温度维持在 100~105℃约 1h，反应生成的水和剩余的乙酸被蒸出，当温度不断下降时（有时反应瓶中会出现白雾），可认为反应已经结束。

在搅拌下，趁热将反应物倒入盛有 20mL 冷水的 50mL 烧杯中，即有白色固体析出，搅拌，冷却后抽滤，并压碎晶体，用少量冷水洗涤晶体，以除去残留的酸液，抽干。粗产品可用水重结晶，得到白色片状晶体，抽滤、烘干后称量，计算产率，测熔点。

纯乙酰苯胺为白色片状晶体，m.p. 114℃。

乙酰苯胺的红外光谱见图 3-23。

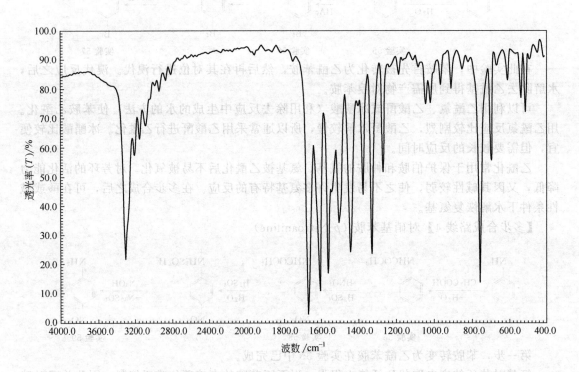

图 3-23　乙酰苯胺的红外光谱图

【注意事项】

① 苯胺易被氧化，久置后含有杂质，这将影响乙酰苯胺的质量，所以需使用新蒸馏的无色或淡黄色的苯胺。

② 加锌粉的目的是防止苯胺在反应过程中被氧化。
③ 若使反应液直接冷却，产物易黏附在烧瓶上，不易处理，故趁热倒入冷水中。
④ 不同温度下，乙酰苯胺在100mL水中的溶解度为（g/℃）：0.46/20；0.56/25；0.84/50；3.45/80；5.5/100。
⑤ 若粗产品有颜色，重结晶时应加入活性炭脱色。

思 考 题

1. 本实验采取什么措施来提高产率？为什么要用分馏装置？
2. 为什么要控制分馏柱上端的温度在100～105℃？
3. 常用的乙酰化试剂有哪些？比较它们的乙酰化能力。
4. 苯胺的乙酰化反应有什么用途？

实验56　对溴乙酰苯胺的合成

【反应式】

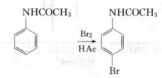

【主要试剂】

乙酰苯胺2.70g（20mmol），溴3.2g（1.0mL，20mmol），冰醋酸，亚硫酸氢钠，乙醇。

【实验步骤】

在50mL三口烧瓶上，配置电动搅拌器、温度计和恒压滴液漏斗，恒压滴液漏斗上口连接气体吸收装置以吸收反应中产生的溴化氢。向三口烧瓶中加入2.70g乙酰苯胺和6mL冰醋酸，用温水浴稍稍加热，使乙酰苯胺溶解。然后，在45℃浴温条件下，边搅拌边滴加1.0mL溴和1.2mL冰醋酸配成的溶液。滴加速度以棕红色溴能较快褪去为宜。滴加完毕，在45℃浴温下继续搅拌反应1h，然后将浴温升高至60℃，再搅拌一段时间，直到反应混合物液面不再有红棕色蒸气逸出为止。

在搅拌下，向反应混合物中加入40mL冷水，此时有固体析出。若有黄色，可加入固体亚硫酸氢钠，使黄色恰好褪去。待反应混合物冷却至室温后，减压抽滤收集产物，并用冷水充分洗涤，干燥，得到粗产品，用95％乙醇重结晶后得到无色针状晶体。产物晾干后，称量，测熔点，并计算产率。

纯对溴乙酰苯胺为针状晶体，m.p.167℃。

对溴乙酰苯胺的红外光谱见图3-24。

【注意事项】

① 溴-冰醋酸试剂必须在通风橱内配制。溴具有强烈的刺激性和腐蚀性，溴引起的灼伤很严重。使用时若不慎触及皮肤，应立即用大量水冲洗，然后用酒精擦洗至无溴液，涂以鱼肝油软膏。使用溴时应当特别小心，操作时应戴上防护手套与眼镜，以防与皮肤接触和对眼睛产生刺激。必须在通风橱中量取。

② 在反应条件下，能得到95％对溴乙酰苯胺和5％邻溴乙酰苯胺。通常两种异构体可用甲醇重结晶分开，也可以用乙醇-水重结晶分开，利用对位异构体比邻位异构体在同一溶剂中溶解度小的规律，经分级结晶将其分离。对溴乙酰苯胺的m.p.167℃，邻溴乙酰苯胺的m.p.99℃。

③ 滴加溴-冰醋酸溶液时，滴速不宜过快，否则反应太剧烈会导致一部分溴来不及参与

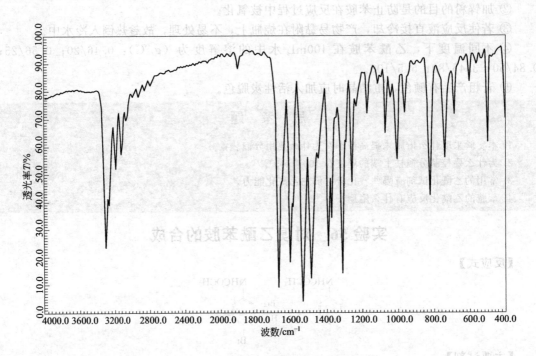

图 3-24 对溴乙酰苯胺的红外光谱图

反应就与溴化氢一起逸出，同时也可能会产生二溴代产物。

思 考 题

1. 乙酰苯胺的一溴代产物为什么以对位异构体为主？
2. 在溴代反应中，反应温度的高低对反应结果有何影响？
3. 在对反应混合物的后处理过程中，使用亚硫酸氢钠的目的是什么？
4. 产物中可能存在哪些杂质，如何除去？
5. 试解析对溴乙酰苯胺的红外光谱图，并指出 N—H 和 C═O 的吸收峰位置。

实验 57 对溴苯胺的合成

【反应式】

$$\underset{Br}{\underset{|}{C_6H_4}}\text{—NHCOCH}_3 \xrightarrow{HCl} \underset{Br}{\underset{|}{C_6H_4}}\text{—}\overset{+}{N}H_3Cl^- \xrightarrow{NaOH} \underset{Br}{\underset{|}{C_6H_4}}\text{—NH}_2$$

【主要试剂】

对溴乙酰苯胺 2.6g（12mmol），95％乙醇，浓盐酸，20％氢氧化钠水溶液。

【实验步骤】

在 50mL 三口烧瓶上，配置回流冷凝管和恒压滴液漏斗。向三口烧瓶中加入 2.6g 对溴乙酰苯胺、6mL 95％乙醇和几粒沸石。加热至沸腾使对溴乙酰苯胺溶解，然后自滴液漏斗慢慢滴加 3.4mL 浓盐酸。

加毕，回流 30min。加入 10mL 水使反应混合物稀释。将回流装置改为蒸馏装置，蒸出约 8mL 馏出液。然后将残余物对溴苯胺盐酸盐倒入盛有 20mL 冰水的烧杯中，将此烧杯置于冰水浴中，在搅拌下，滴加20%氢氧化钠水溶液使之刚好呈碱性，即见白色固体（对溴苯胺）析出。彻底冷却后，抽滤，用少量冰水洗涤，抽干后置于表面皿上晾干。产品晾干后称量，测熔点，计算产率。

对溴苯胺为无色晶体，m.p. 65~66℃。

【注意事项】
① 滴加浓盐酸不宜太快，以免反应太剧烈。
② 通过共沸蒸馏除去水解过程中产生的乙酸乙酯、乙醇和水。
③ 粗产品可用乙醇-水重结晶。

实验58 对硝基乙酰苯胺的合成

【反应式】

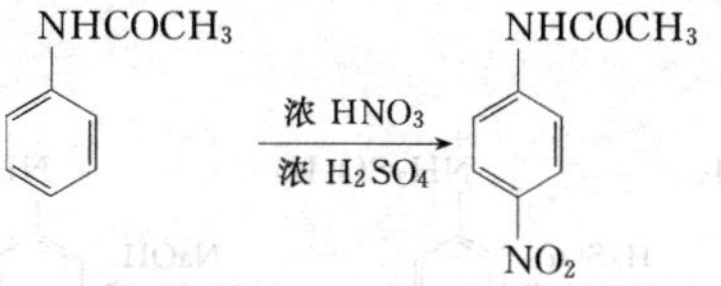

$$\underset{}{C_6H_5NHCOCH_3} \xrightarrow[\text{浓 } H_2SO_4]{\text{浓 } HNO_3} \underset{}{p\text{-}O_2N\text{-}C_6H_4\text{-}NHCOCH_3}$$

【主要试剂】

乙酰苯胺 6.75g（50mmol），浓硝酸 4mL（62.7mmol），浓硫酸 17.5mL（32.22g，322mmol），无水碳酸钠。

【实验步骤】

向一个盛有17.5mL浓硫酸的100mL烧杯中，室温搅拌下分批加入6.75g干燥的乙酰苯胺，充分搅拌使之完全溶解。把烧杯置于冰盐浴中，使乙酰苯胺的硫酸溶液冷却至0℃。然后在电动搅拌下用滴管慢慢地滴加4mL浓硝酸，使硝化温度不超过5℃。加完硝酸后，原温度下再继续搅拌30min。然后在搅拌下，把反应混合液以细流慢慢地倒入盛有25mL H_2O 和25g碎冰的250mL烧杯中，粗对硝基乙酰苯胺就立刻成固体析出。放置约10min，冷却后，减压抽滤，尽量挤压掉粗产品中的酸液，用60mL冰水分两次洗涤，以便除去酸。

将粗产物倒入一个盛50mL水的500mL烧杯中，在不断搅拌下，分几次加入碳酸钠粉末，直到混合物呈碱性（pH约为10），再加约50mL水。将反应混合物加热至沸腾。这时，对硝基乙酰苯胺不水解，而邻硝基乙酰苯胺则水解为邻硝基苯胺。混合物稍冷却后（不低于50℃），迅速进行减压过滤，尽量挤压掉溶于碱性溶液的邻硝基苯胺，再用热水（60~70℃）洗涤滤饼，尽量抽干。取出对硝基乙酰苯胺，放在表面皿上晾干。产品为白色粉末状固体，粗产品的产量约8g。

纯对硝基乙酰苯胺为无色晶体，m.p. 215.6℃。

【注意事项】
① 本实验可能存在下列副反应。

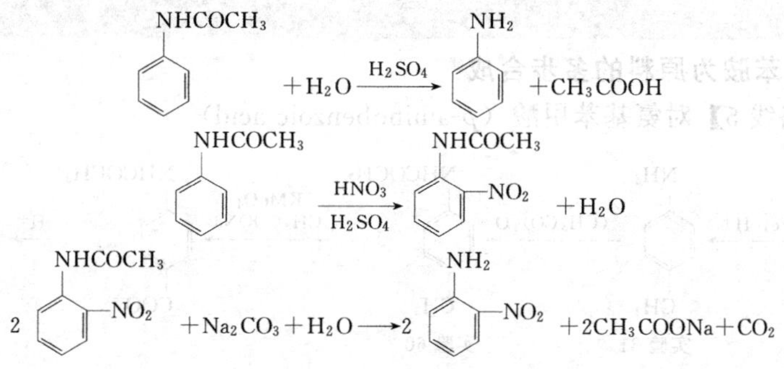

$$C_6H_5NHCOCH_3 + H_2O \xrightarrow{H_2SO_4} C_6H_5NH_2 + CH_3COOH$$

$$C_6H_5NHCOCH_3 \xrightarrow[H_2SO_4]{HNO_3} o\text{-}O_2N\text{-}C_6H_4\text{-}NHCOCH_3 + H_2O$$

$$2\,o\text{-}O_2N\text{-}C_6H_4\text{-}NHCOCH_3 + Na_2CO_3 + H_2O \longrightarrow 2\,o\text{-}O_2N\text{-}C_6H_4\text{-}NH_2 + 2CH_3COONa + CO_2$$

② 为了避免乙酰苯胺的水解，溶解乙酰苯胺时的温度不宜高于 25℃，在此条件下，完全溶解约需 20min。

③ 乙酰苯胺与混酸在 5℃以下作用，主要产物是对硝基乙酰苯胺；在 40℃作用，则生成约 25％的邻硝基乙酰苯胺。因此，要控制硝化时的温度不高于 5℃。

④ 当 pH 为 10 时，硝基乙酰苯胺与碱液共热，邻位异构体易水解成邻硝基苯胺，而后者在 50℃时又溶于碱液，故可用减压过滤方法除去邻硝基苯胺。

⑤ 根据邻硝基乙酰苯胺和对硝基乙酰苯胺在乙醇中的溶解度的不同，所得粗产品可用乙醇进行重结晶，从而除去溶解度较大的邻硝基乙酰苯胺。

⑥ 所得的对硝基乙酰苯胺不必干燥，直接进行下一步的水解反应制备对硝基苯胺。

实验 59　对硝基苯胺的合成

【反应式】

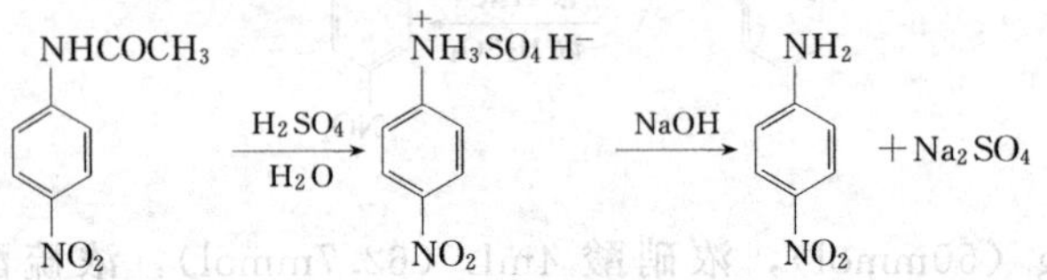

$$\underset{NO_2}{\underset{|}{C_6H_4}}-NHCOCH_3 \xrightarrow[H_2O]{H_2SO_4} \underset{NO_2}{\underset{|}{C_6H_4}}-NH_3SO_4H^- \xrightarrow{NaOH} \underset{NO_2}{\underset{|}{C_6H_4}}-NH_2 + Na_2SO_4$$

【主要试剂】

70％硫酸，20％氢氧化钠。

【实验步骤】

将实验 58 所得的对硝基乙酰苯胺粗品转入 100mL 的圆底烧瓶中，加入 37.5mL 70％硫酸及两粒沸石，装上回流冷凝管，在石棉网上加热回流 25min。反应混合物变为透明的溶液。冷却后，倒入 100mL 的冰水中。若有沉淀析出（可能为较难溶于稀硫酸的邻硝基苯胺）减压过滤除去。滤液为对硝基苯胺的硫酸盐溶液。待溶液彻底冷却后，加入 20％氢氧化钠溶液，使对硝基苯胺沉淀下来（pH 约为 8）。冷却至室温后，减压抽滤，滤饼用冷水洗涤，除去碱液后，取出晾干。如欲制得较纯的产品，可用水进行重结晶。产品产量约 4g。

纯对硝基苯胺为黄色针状晶体，m.p. 147.5℃。

【注意事项】

对硝基苯胺在 100mL 水中的溶解度：18.5℃，0.08g；100℃，2.2g。

思　考　题

1. 对硝基苯胺是否可从苯胺直接硝化来制备？为什么？
2. 本实验如何除去对硝基乙酰苯胺粗产品中的邻硝基乙酰苯胺？
3. 在对硝基乙酰苯胺的酸性水解过程中，加热回流后的溶液为什么是透明的？根据什么原理使产品析出？

3.21.4　以对甲苯胺为原料的多步合成

【多步合成路线 5】对氨基苯甲酸（p-aminobenzoic acid）

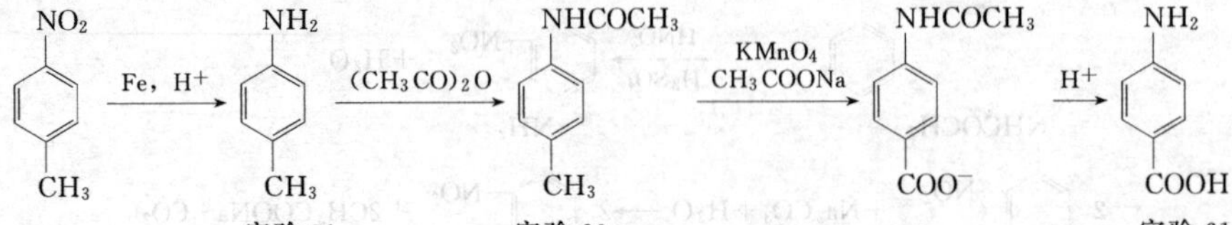

$$\underset{CH_3}{\underset{|}{C_6H_4}}-NO_2 \xrightarrow{Fe, H^+} \underset{CH_3}{\underset{|}{C_6H_4}}-NH_2 \xrightarrow{(CH_3CO)_2O} \underset{CH_3}{\underset{|}{C_6H_4}}-NHCOCH_3 \xrightarrow[CH_3COONa]{KMnO_4} \underset{COO^-}{\underset{|}{C_6H_4}}-NHCOCH_3 \xrightarrow{H^+} \underset{COOH}{\underset{|}{C_6H_4}}-NH_2$$

　　　　　　实验 31　　　　　　　　　　　实验 60　　　　　　　　　　实验 61

对氨基苯甲酸（PABA）是与维生素B络合物有关的物质之一。作为叶酸（维生素B_{10}）的中心单元存在。维生素B_{10}是由蝶啶单元，对氨基苯甲酸单元和谷氨酸单元组成的。利用细菌合成叶酸时需要对氨基苯甲酸，磺胺类药物可以阻碍这一合成。没有叶酸，细菌就不能合成增殖所需的核酸，其结果抑制了细菌的增殖，而使人体的免疫系统恢复正常并杀死细菌。

由于PABA能够吸收日光的紫外线，所以也可用作防晒剂。

目前，对氨基苯甲酸的合成方法是比较简单的且适合于小规模的实验室制备，它是以对甲苯胺为原料，经乙酰化制得对甲基-N-乙酰苯胺，再用高锰酸钾氧化对甲基-N-乙酰苯胺制得对乙酰氨基苯甲酸，然后再经酸性水解就可得到对氨基苯甲酸。

实验60 对甲基-N-乙酰苯胺的合成

【反应式】

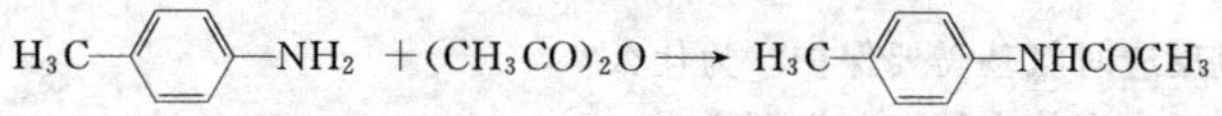

$$H_3C-C_6H_4-NH_2 + (CH_3CO)_2O \longrightarrow H_3C-C_6H_4-NHCOCH_3$$

【主要试剂】

对甲苯胺2.00g（18.7mmol），乙酸酐2.4mL（25.4mmol）。

【实验步骤】

在10mL的圆底烧瓶中，加入2.00g对甲苯胺和2.4mL乙酸酐，装上回流冷凝管，反应立即发生并放热，使固体完全溶解。加热回流10min，趁热将反应混合物倒入50mL的冷水中，边加边搅拌，立即析出淡黄色固体。彻底冷却后，抽滤，用少量冷水洗涤晶体三次，干燥后得粗品2.60g，产率93%，m.p.147～149℃。粗产品可用乙醇-水进行重结晶。

【注意事项】

① 本实验所用仪器均是干燥的。

② 所用乙酸酐为新蒸馏的，收集138～139℃的馏分。

实验61 对氨基苯甲酸的合成

【反应式】

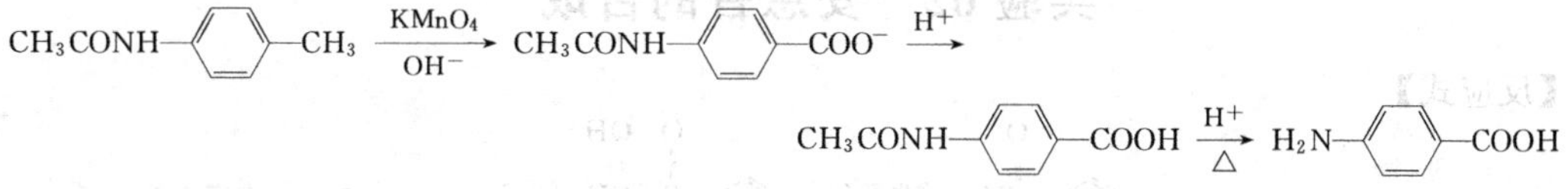

$$CH_3CONH-C_6H_4-CH_3 \xrightarrow[OH^-]{KMnO_4} CH_3CONH-C_6H_4-COO^- \xrightarrow{H^+} CH_3CONH-C_6H_4-COOH \xrightarrow[\triangle]{H^+} H_2N-C_6H_4-COOH$$

【主要试剂】

对甲基-N-乙酰苯胺（自制）2.60g（17.4mmol），高锰酸钾8.00g（50.6mmol），乙酸钠2.00g（24.4mmol），盐酸，氢氧化钠。

【实验步骤】

在250mL烧杯中加入2.60g自制的对甲基-N-乙酰苯胺、2.00g乙酸钠、8.00g高锰酸钾和60mL水，室温下电动搅拌均匀。在不断搅拌下，小火加热约30min，反应液呈深褐色并有大量沉淀物生成。在减压下趁热过滤，并用少量热水洗涤沉淀，合并滤液，冷却至室温。用20%稀硫酸酸化，使pH为1～2。此时，析出大量白色固体，抽滤，尽量抽干，收集产物。纯产品m.p.250～252℃。

将所得粗产品置于50mL圆底烧瓶中，再加入30mL 1∶1的盐酸（水与盐酸体积比）和2粒沸石，安装球形冷凝管，并在冷凝管上端连上尾气吸收装置，然后在石棉网上加热回流 0.5h。彻底冷却后将瓶内反应混合物转移至 200mL 烧杯中，加入 20mL 水，用冷的 20% 氢氧化钠中和此溶液至石蕊试纸刚刚变蓝。根据溶液的体积，按每 30mL 加入 1mL 冰醋酸的比例酸化，在冰浴中析出结晶，收集产品，晾干。产品为淡黄色针状结晶，产量 1.50g。纯产品 m.p. 186~187℃。

【注意事项】

① $KMnO_4$ 氧化时要小火加热，否则反应太剧烈（此反应放热），难于控制。搅拌速度要适中。

② 如滤液呈紫红色，可能有过量高锰酸钾存在，可加入少量亚硫酸氢钠使红色褪去，使滤液呈近无色的清澈液体。

③ 注意各步酸碱度的调节。红色石蕊试纸刚变蓝时 pH 为 6.4，中和时注意溶液 pH 的变化，并尽量使溶液量不能过大，否则产物不易结晶出来。

思 考 题

1. 在进行哪些有机反应时氨基需要保护？为什么？
2. 用于保护基团的反应有什么特点？举例说明。

3.21.5 以苯甲醛为原料的多步合成

【多步合成路线 6】 5,5-二苯基乙内酰脲与赤-1,2-二苯基-1,2-乙二醇

(5,5-diphenylhydantion and *erythro*-1,2-diphenyl-1,2-ethandiol)

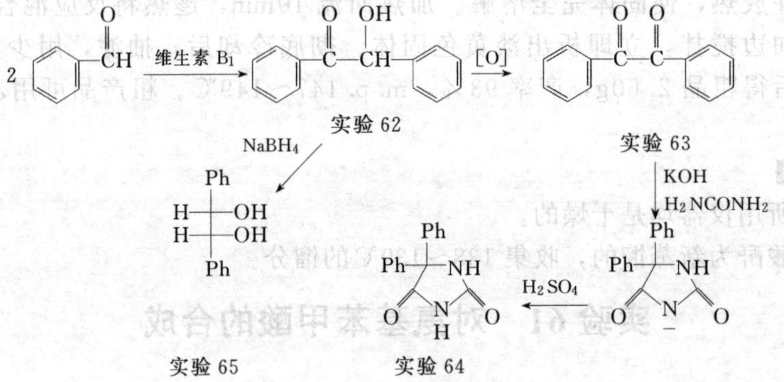

实验 62 安息香的合成

【反应式】

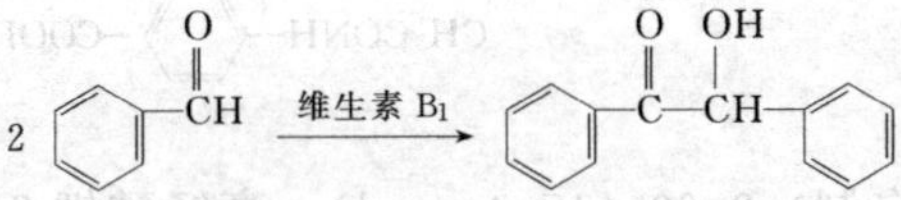

【主要试剂】

苯甲醛 1.5mL (1.56g，14.7mmol)，维生素 B_1 0.30g，95%乙醇。

【实验步骤】

在 30mL 锥形瓶中，加入 0.30g 维生素 B_1 和 1.0mL 蒸馏水，待固体维生素 B_1 溶解后，再加入 3.0mL 95%乙醇。塞上瓶塞，在冰盐浴中充分冷却。将已冷却的 1.0mL 2.5mol/L 氢氧化钠溶液逐滴加入到上述冰盐浴的锥形瓶中，使溶液的 pH 为 10~11。然后迅速加入 1.5mL 新蒸馏的苯甲醛，充分混匀，室温放置一天，有白色针状晶体析出，待结晶完全后，抽滤，用少量冷水洗涤晶体，晾干后得到粗产品。粗产品可用 95%乙醇重结晶。纯产品为白色针状晶体，m.p. 134~136℃。产量 0.6g，产率 38.5%。

安息香的红外光谱见图 3-25。

第 3 章 半微量有机化合物的合成实验

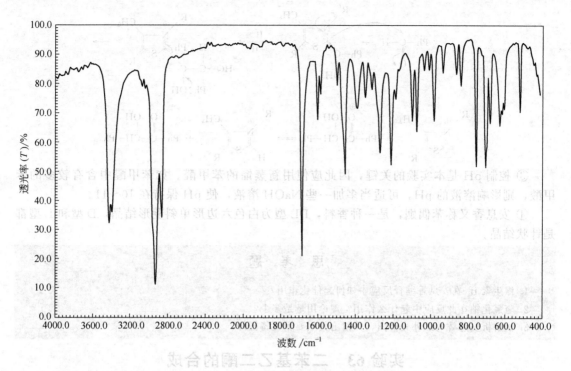

图 3-25 安息香的红外光谱图（研糊法）

【注意事项】

① 维生素 B_1（VB_1）市售品以其盐酸盐的形式储存。维生素 B_1 在酸性条件下稳定，但易吸水，受热易变质，在水溶液中易被空气氧化失效。遇光和 Cu、Fe、Mn 等金属离子均可加速氧化，所以应置于冰箱内保存。在 NaOH 溶液中噻唑环易开环失效，因此 VB_1 溶液、NaOH 溶液在反应前必须用冰水充分冷却，否则 VB_1 在碱性条件下会分解，这是实验成功的关键。没用完的 VB_1 应尽快密封保存在阴凉处。

② 维生素 B_1 是一种辅酶，可代替剧毒的氰化钠作催化剂进行安息香缩合，VB_1 的结构式如下。

VB_1 分子中右边噻唑环上的氮原子和硫原子之间的氢有较大的酸性，在碱的作用下易被除去形成碳负离子，从而催化安息香的形成。本实验的反应机理如下。

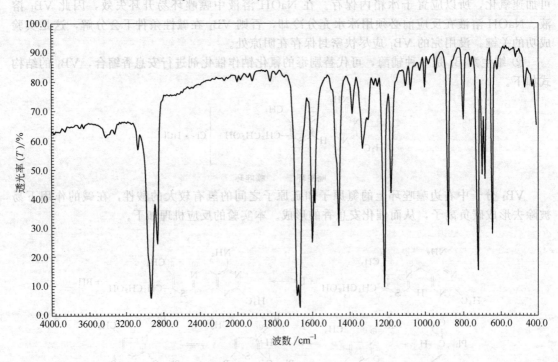

③ 控制 pH 是本实验的关键，因此应使用新蒸馏的苯甲醛。如苯甲醛中含有较多的苯甲酸，则影响溶液的 pH，可适当多加一些 NaOH 溶液，使 pH 保持在 10～11。

④ 安息香又称苯偶姻，是一种香料，DL 型为白色六边形单斜菱形结晶，D 型和 L 型都是针状结晶。

思 考 题

1. 维生素 B_1 在安息香缩合反应中如何起催化作用？
2. 氢氧化钠在此反应中起什么作用？理论用量是多少？
3. 试解析安息香的红外光谱图，并指出其主要吸收特征峰？

实验 63　二苯基乙二酮的合成

【反应式】

图 3-26　二苯基乙二酮的红外光谱图（研糊法）

【主要试剂】
安息香 0.4g (1.9mmol)，$FeCl_3 \cdot 6H_2O$ 1.7g (6.3mmol)。

【实验步骤】
在 30mL 圆底烧瓶中，加入 2mL 冰醋酸、1mL 水及 1.7g $FeCl_3 \cdot 6H_2O$，装上回流冷凝管，小火加热至沸，且不时地加以振荡。停止加热，待沸腾平息后，加入 0.4g 安息香，继续加热回流 1h。加入 10mL 水再沸腾后，冷却反应液，有黄色固体析出。抽滤出此固体粗产品，并用少量冷水洗涤固体 3 次。粗产品用 75% 乙醇重结晶，得到黄色针状晶体 0.3g，产率 75.2%，m.p. 94~95℃。

纯二苯基乙二酮 m.p. 95℃。

二苯基乙二酮的红外光谱见图 3-26。

实验 64　5,5-二苯基乙内酰脲的合成

【反应式】

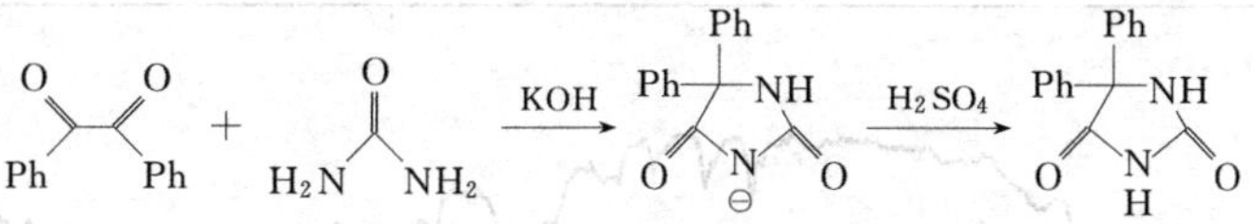

【主要试剂】
二苯基乙二酮 0.3g (1.43mmol)，尿素 0.18g (3.0mmol)，15.2 mol/L 氢氧化钾溶液。

【实验步骤】
在 25mL 三口烧瓶中，加入 0.3g 二苯基乙二酮和 0.18g 尿素，再加入 3.0mL 95% 乙醇及 1.0mL 15.2mol/L 氢氧化钾溶液，装上回流冷凝管，在电动搅拌下，水浴加热回流 2.5h，反应过程中有少许不溶固体产生。反应液冷却后过滤除去不溶物。滤液在冰水浴中进一步冷却，并用 6mol/L 硫酸慢慢酸化至 pH≈3，得到粉末状白色固体，抽滤出此固体，并用水充分洗涤以除去无机盐，烘干即得粗产品。粗产品用 95% 乙醇重结晶，得到白色针状晶体。产量 0.2g，产率 55.6%，m.p. 286~295℃。

【注意事项】
5,5-二苯基乙内酰脲的钠盐（dilantin）是抗痉挛药物，通过静脉注射可控制严重的癫痫病患者的病情。5,5-二苯基乙内酰脲的合成反应过程如下。

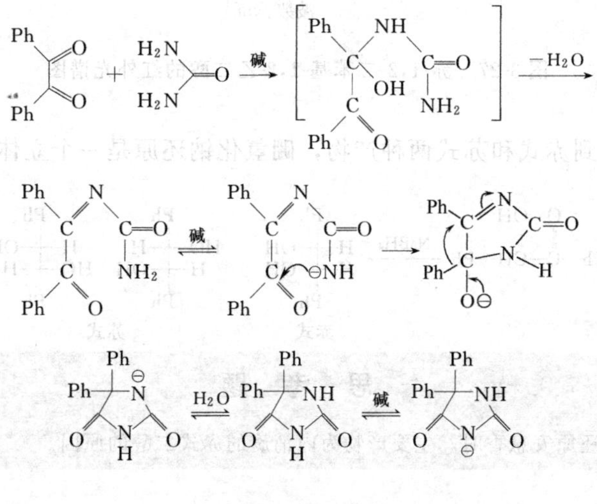

实验 65　赤-1,2-二苯基-1,2-乙二醇的合成

【反应式】

$$\text{Ph-CO-CH(OH)-Ph} \xrightarrow{\text{NaBH}_4} \begin{array}{c} \text{Ph} \\ \text{H}{-}\text{OH} \\ \text{H}{-}\text{OH} \\ \text{Ph} \end{array}$$

【主要试剂】

安息香 0.5g（2.36mmol），硼氢化钠 0.05g（1.3mmol），95%乙醇，18%盐酸。

【实验步骤】

在 30mL 锥形瓶中，加入 0.5g 安息香和 5mL 95%乙醇，在电磁搅拌下，加入 0.05g 硼氢化钠。室温下继续搅拌 15min 后，将反应瓶置于冰水浴中，边搅拌边加入 2.0mL H_2O，分解过量的硼氢化钠，再滴入 0.25mL 18%盐酸，溶液中逐渐有白色固体生成，停止搅拌，继续冷却至晶体全部析出。抽滤，得到白色晶体，用少量冷水洗涤粗产品。得到粗产品 0.4g，产率 79.1%。粗产品可用丙酮-石油醚或丙酮-水重结晶。纯品 m.p. 138℃。

赤-1,2-二苯基-1,2-乙二醇的红外光谱见图 3-27。

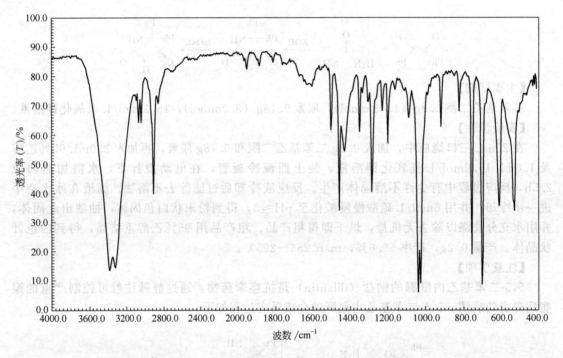

图 3-27　赤-1,2-二苯基-1,2-乙二醇的红外光谱图

【注意事项】

安息香还原可得到赤式和苏式两种产物，硼氢化钠还原是一个立体选择性反应，主要生成内消旋的赤式二醇。

$$\text{Ph-CO-CH(OH)-Ph} \xrightarrow{\text{NaBH}_4} \text{赤式} + \text{苏式}$$

思　考　题

试解释用硼氢化钠还原安息香时，主要产物为内消旋的赤式二醇的原因。

Chapter 3 Semimicro Synthetic Experiment of Typical Organic Compounds

3.1 Preparation of alkanes and alkenes

The aliphatic hydrocarbons are divided into three classes: ①the **alkanes**, ②the **alkenes**, and ③the **alkynes**. The alkanes have only single bonds in the molecule and are said to be *saturated* hydrocarbons. They are relatively inert. The alkenes have at least one double bond in the molecule, and the alkynes have at least one triple bond. Both the alkenes and the alkynes are said to be *unsaturated* hydrocarbons and both are relatively reactive. The three classes of hydrocarbons react differently with certain reagents, and the differences may be used to distinguish between the three classes in many instances. The following experiments illustrate the reactions of typical saturated and unsaturated hydrocarbons.

Alkenes are important organic chemical material. Alkenes are produced mainly by the pyrolysis in the petroleum industry. Sometimes alkenes are also obtained by the dehydration of alcohols at high temperature in the presence of metallic oxide catalysts (aluminum oxide *et al*). In the laboratory, alkenes are obtained from alcohols or alkyl halides by elimination reactions. Treatment of an alcohol with acid (sulfuric or phosphoric acid) effects dehydration, and treatment of an alkyl halide with a strong base (a solution of potassium hydroxide in ethanol) effects dehydrohalogenation.

Experiment 1 Preparation and properties of methane

In this experiment you will learn the method of preparing methane in the laboratory and verify the properties of alkanes.

Reaction equation

Main reaction

$$CH_3COONa + NaOH \xrightarrow{\triangle} CH_4\uparrow + Na_2CO_3$$

Secondary reaction

$$2CH_3COONa \xrightarrow{\triangle} CH_3COCH_3 + Na_2CO_3$$

Materials	
anhydrous sodium acetate (FW 82)	5.00 g (61 mmol)
soda lime	3.00 g
sodium hydroxide (FW 40)	2.00 g (50 mmol)
1 percent bromine in carbon tetrachloride	
potassium permanganate solution (0.1%)	
concentrated sulfuric acid (FW 98.08)	
sulfuric acid (10%)	
heptane (FW 100)	

Procedure

(1) *Preparation of methane*

Place 5 g of anhydrous sodium acetate, 3 g of soda lime and 2 g of sodium hydroxide in a mortar box, mix uniformly and mill them quickly into powder. The milled mixture is packed into a column by a piece of tin leaf and put into a hard big test tube. After checking the airtightness of the apparatus, the whole test tube is heated slowly and uniformly with a small flame, then the reactant near the mouth of the test tube is heated strongly, and then the flame is moved gradually to the bottom of the test tube. After the air in the system is discharged completely, collect methane in three small test tubes and a big sidearm test tube respectively with the draining method, and then do the following experiments.

(2) *Property tests*

① Substitution reaction Add 2~3 drops of a 1% solution of bromine in the carbon tetrachloride to the two small test tubes containing methane respectively. Stopper the tubes with rubber stoppers and shake the tubes well. Place one tube which has been wrapped with a piece of black cloth in your laboratory locker and the other in bright sunlight. Allow both tubes stand for 15~20 minutes. Compare whether the color of the solution in the two tubes are the same. What changes do they have? Why?

② Potassium permanganate test Add 2~3 drops of a dilute potassium permanganate solution (0.1%) and 2 mL of dilute sulfuric acid (10%) to the test tube containing methane. Stopper the tube with a rubber stopper, and shake the tube well. Observe the color of the solution in the tube. What change does it have? What does this indicate?

Take 0.5 mL of heptane and carry on the property tests of alkane according to the procedure① and ②. Observe the results.

③ Flammability test Invert the big sidearm test tube containing methane, install rapidly the water-filled dropping funnel to it, and then erect the tube. After the tube was fixed on the support stand, open the piston of the dropping funnel and the clamp and ignite methane gas near the sharp mouth of the sidearm test tube. Observe and record the character of the flame.

Notes

1. The used reagents should be dry and the operation should be quick in order to avoid the reagents absorbing moisture. The apparatus must be tight.
2. The tin leaf should be flat and should not be broken.
3. The mixture of anhydrous sodium acetate, soda lime and sodium hydroxide should be milled, put on the tin leaf flatly and packed loosely. The tin leaf packed the mixture is rolled into a column and stayed a gas outlet.
4. Move the gas lamp from the mouth to the bottom of the test tube to avoid cracking the test tube when it is heated.
5. After collecting methane, or before stopping the reaction anytime, take the gas outlet tube out the surface of the water, then disconnect the rubber tube between the sidearm test tube and the hard big test tube, and then put the fire out. The purpose is to avoid the sulfuric acid and water being sucked into the test tube.
6. Before igniting methane, the dropping funnel should be installed quickly. Drop water while opening the clamp, and immediately ignite methane near the sharp mouth of the

Chapter 3 Semimicro Synthetic Experiment of Typical Organic Compounds

outlet tube.
7. After the experiment, carefully pour the residue in the sidearm test tube into a waste liquid receptacle.

Questions

1. What are the key steps to preparing methane in the laboratory?
2. Why is the mouth of test tube lower than the bottom of it in the apparatus of preparing methane? What is the correct operation when stop heating?
3. What is the composition of soda lime? What is the action of it in preparing methane?
4. What are the advantages of packing samples by tin leaf? What are the factors of affecting the color of flame while burning methane?
5. Explain the reaction of methane with bromine by free radical mechanism.

Experiment 2 Preparation and properties of ethylene

In this experiment you will learn the method of preparing ethylene in the laboratory and verify the properties of unsaturated hydrocarbon.

Reaction equation

Main reaction

$$CH_3CH_2OH + H_2SO_4 \xrightarrow{170\ ^\circ C} C_2H_4 \uparrow + H_2O$$

Secondary reactions

$$CH_3CH_2OH + H_2SO_4 \begin{array}{c} \nearrow CO + CO_2 + SO_2 + H_2O \\ \searrow CH_3COOH + SO_2 + H_2O \end{array}$$

Materials	
ethanol (95%) (FW 46.07)	4 mL (68.5 mmol)
concentrated sulfuric acid (FW 98.08)	12 mL (225 mmol)
1 percent bromine in carbon tetrachloride	
potassium permanganate solution (0.1%)	
sulfuric acid (10%)	
sodium hydroxide solution (10%)	

Procedure

(1) *Preparation of ethylene*

Place 4 mL of ethanol (95%) in a dry 125 mL distilling flask. Add 12 mL of concentrated sulfuric acid slowly while shaking and cooling the flask. Introduce a few boiling stones into the flask. Stopper a rubber stopper provided with a 300 ℃ thermometer on the flask. After checking the airtightness of the apparatus, heat strongly the mixture in order to rise the temperature of the reaction mixture to 160~170 ℃ quickly. Adjust the flame and maintain the range of temperature to obtain ethylene current uniformly. After the air in the system is discharged completely, do the following experiments.

(2) *Property tests*

① Addition reaction Add 0.5 mL of a 1% solution of bromine in carbon tetrachloride

175

to a small dry test tube. Bubble ethylene through the bromine solution and observe what phenomenon occurs.

② Oxidation reaction   Add 0.1 mL of a dilute potassium permanganate solution (0.1%) and 2 mL of a dilute sulfuric acid (10%) to a small test tube. Bubble ethylene through the mixture solution and observe what phenomenon occurs.

③ Flammability test   Remove the test tube in the apparatus of preparing ethylene, ignite ethylene gas near the sharp mouth of the outlet tube, and observe the character of the flame. Compare it to the flame of methane and explain the reason.

Notes
1. Shake and cool the flask while adding the concentrated sulfuric acid to the flask containing ethanol.
2. The reaction temperature should be raised beyond 140 ℃ quickly, and kept at 170 ℃.
3. After finishing the experiment, disconnect the rubber tube between the flask and gas-washing bottle (the sidearm tube), then put out the fire to avoid the solution in the gas-washing bottle being sucked into the flask.
4. After the experiment, carefully pour the residue in the flask into a waste liquid receptacle.

Questions
1. Why should the temperature of the reaction for preparing ethylene be raised to 160~170 ℃ quickly? If not, what is the result?
2. What impurities will be produced in the preparation of ethylene? In which part of the apparatus can they be taken off?
3. What is the difference between the flame of ethylene and that of methane?

Experiment 3   Preparation of cyclohexene

Reaction equation

$$\text{C}_6\text{H}_{11}\text{OH} \xrightarrow{H_2SO_4} \text{C}_6\text{H}_{10} + H_2O$$

Equipment   Apparatus for: fractional distillation, extraction/separation, distillation.

Materials	
cyclohexanol (FW 100)	5 mL (4.81 g, 48 mmol)
concentrated sulfuric acid (FW 98)	0.5 mL
sodium carbonate solution (10%)	
sodium chloride solution (saturated)	
anhyarous calcium chloride	

Procedure

Place 4.81 g of cyclohexanol in a dry 25 mL round-bottomed flask and dropwise add 0.5 mL of concentrated sulfuric acid with swirling and cooling the flask. Mix the contents thoroughly and add two boiling stones. Set up a fractional distillation apparatus as illustrated in Figure 1-4. Immerse the receiving flask in an ice-water bath because the product is very volatile and flammable.

Heat the flask gently over an asbestos gauge with a small flame (or heating mantle).

The temperature at the top of the column should not be allowed to rise above 90 ℃. Collect the fraction boiling below 85 ℃. The distillate is an azeotropic mixture of cyclohexene and water. Stop the distillation when no further cyclohexene distills.

Transfer the distillate to a separatory funnel. Wash it with 2×5 mL portions of saturated sodium chloride solution, draw off the lower aqueous layer. Wash the organic layer with 3 mL of 10% sodium carbonate solution to neutralize traces of acid (test with pH indicator paper). Draw off the lower aqueous layer and then pour the upper layer (crude cyclohexene) through the mouth of the funnel (why?) into a small dry Erlenmeyer flask. Add anhydrous calcium chloride to the flask and swirl the mixture occasionally until the solution becomes clear (about 30 minutes). Decant the dried product into a dry 5 mL of round-bottomed flask, add two boiling stones, and distill from a water bath. The weighed receiving flask should again be placed in an ice-water bath. Collect the fraction boiling at 82~84 ℃ in the weighed receiving flask. The yield is 2.40 g. The percentage yield is 60.8%.

Pure cyclohexene is a colorless and transparent liquid, b. p. 83 ℃, d_4^{20} 0.8102, n_D^{20} 1.4465.

Notes
1. Cyclohexanol is a sticky liquid (m. p. 24 ℃). Weigh its mass to reduce the error produced from weighing the volume with graduated cylinder.
2. While mixing sulfuric acid and cyclohexanol, swirl and mix the mixture thoroughly to avoid the local carbonization and polymerization while heating. Phosphoric acid (85%) may be also used as the catalyst in this experiment and the effect is also very good, but the used amount of it must be above two times of sulfuric acid.
3. In this reaction, cyclohexene and water from an azetrope containing 10% water, the boiling point is 70.8 ℃, cyclohexanol and cyclohexene from an azetrope containing 30.5% cyclohexanol, the boiling point is 64.9 ℃, cyclohexanol and water from an azetrope containing 80% water, the boiling point is 97.8 ℃. So the temperature of the reaction can not be too high while heating and the speed of the distillation can not be excessively quickly to avoid the unreacted cyclohexanol being distilled off.
4. When cyclohexene is collected and transferred, take care to cool fully in order to minimize the loss by evaporation.
5. Before distilling the dried product, make sure that all the glassware you will use for distilling is clean and dry.

Questions

1. What measures have been taken to raise the yield in this experiment? Which step will decrease the yield due to improper operation? What is the key operation in this experiment?
2. What is the purpose of adding the saturated sodium chloride solution to the distillate?
3. What is the advantage of anhydrous calcium chloride as drying agent?
4. What is the controlled optimal temperature on the top of the fractional column during the reaction?
5. Write the reaction mechanism for the dehydration of cyclohexanol catalyzed by acid.

3.2 Preparation of alkyl halides

Halohydrocarbons, especially alkyl halides, are a kind of important intermediates in organic synthesis. Many kinds of useful compounds can be prepared by the nucleophilic substitution reaction of alkyl halides such as nitriles, amines, ethers and so on. Alkyl halides react with magnesium metal in anhydrous ether solvent to yield organomagnesium halides. That is Grignard reagents. Grignard reagents react with carbonyl compounds, such as aldehydes, ketones and carbon dioxide, to yield the corresponding alcohols and carboxylic acids. The important method for preparing alkyl halides is to make them from alcohols by treatment either with HX (for tertiary alcohol) or with $SOCl_2$ or PBr_3 (for primary and secondary alcohols).

Experiment 4 Preparation of ethyl bromide

In the laboratory ethyl bromide usually is prepared from the corresponding alcohol (ethanol) reacting sodium bromide and sulfuric acid.

Reaction equation

Main reactions

$$NaBr + H_2SO_4 \longrightarrow HBr + NaHSO_4$$
$$C_2H_5OH + HBr \rightleftharpoons C_2H_5Br + H_2O$$

Secondary reactions

$$2C_2H_5OH \xrightarrow{H_2SO_4} C_2H_5OC_2H_5 + H_2O$$
$$2C_2H_5OH \xrightarrow{H_2SO_4} C_2H_4 \uparrow + H_2O$$
$$2HBr + H_2SO_4 \longrightarrow Br_2 + SO_2 + 2H_2O$$

Equipment Apparatus for: extraction/separation, distillation.

Materials	
ethanol (95%) (FW 46.07)	3.3 mL (2.60 g, 53.6 mmol)
sodium bromide (FW 102.9)	5.00 g (48.6 mmol)
concentrated sulfuric acid (FW 98.08)	6.3 mL (11.60 g, 116 mmol)

Procedure

In a 30 mL round-bottomed flask place 3.3 mL of 95% ethanol and 3 mL of water. Cool the mixture in a cold water bath and slowly add 6.3 mL of concentrated sulfuric acid with thorough mixing (swirling) and cooling in the cold water bath. To the cold mixture add 5.00 g of finely pulverized sodium bromide while shaking slightly the flask to avoid formation of lumps. Introduce two boiling stones into the flask. Assemble the apparatus as shown in the Figure 1-3. Place a small amount of ice-water in the receiving flask immersed in an ice-water bath to avoid the loss of the product by evaporation. The side tube of the vacuum adapter is connected with a rubber tube led to the sewer (why?). Heat the flask gently over an asbestos gauze with a small flame (or heating mantle) until distillate does not contain the oily matter (about 40 minutes).

Transfer the distillate to a separatory funnel. Draw off the organic phase (which layer?) into a dry Erlenmeyer flask immersed in an ice-water bath. To remove the impurities in the crude product, such as diethyl ether, ethanol and water, add dropwise concentrated sulfuric acid to the flask while swirling the contents until the layers have separated sharply. Separate the sulfuric acid carefully and completely (which layer?) using a dry separatory funnel. Transfer the crude ethyl bromide to a dry distilling flask (how to transfer?), add a boiling stone and distill from a water bath using a dry apparatus. Collect the pure ethyl bromide in a weighed bottle, which has been cooled in an ice-water bath. Collect the fraction boiling at 37~40 ℃. The yield is 3.20 g; the percentage yield is 60.4%.

Pure ethyl bromide is a colorless liquid, b.p. 38.4 ℃, d_4^{20} 1.4604, n_D^{20} 1.4239.

Notes
1. Add a small amount of water to the reacting flask to prevent the formation of a large amount of foam during the reaction, decrease the formation of by-product diethyl ether and avoid the volatilization of hydrobromic acid.
2. Control strictly the reaction temperature in order that the reaction can carry on reposefully and avoid the reactants rushing out the distilling flask due to the formation of a large amount of the foam at the beginning of the reaction. The boiling point of ethyl bromide is lower; distill slowly to avoid the loss of the product.
3. Place a small amount of ice-water in the receiving flask immersed in an ice-water bath to avoid the volatilization of the product while receiving the crude product.
4. Separate the aqueous layer as completely as possible otherwise the product will lose due to the exothermicity while washing the organic layer with concentrated sulfuric acid.
5. The concentrated sulfuric acid is used to get rid of the impurities, such as diethyl ether, ethanol and water. When the organic layer is not washed completely, it will contain a small amount of water and ethanol. Both of them and ethyl bromide can form an azetrope respectively. Ethyl bromide and water can form an azetrope containing 1% water, the boiling point is 37 ℃, ethyl bromide and ethanol can form an azetrope containing 3% ethanol, the boiling point is 37 ℃.

Questions
1. What is the main cause of the yield to be lower in this experiment?
2. Which starting material is used to calculate the yield in this experiment?
3. Which impurities does the crude product have? How are they removed?
4. Which measures have been taken in this experiment in order to reduce the loss by the volatilization of ethyl bromide?
5. Write the reaction mechanism of forming ethyl bromide.

Experiment 5 Preparation of *n*-butyl bromide

n-Butyl bromide can be easily prepared by allowing *n*-butyl alcohol reacting with sodium bromide and sulfuric acid.

Reaction equation

Main reactions

$$NaBr + H_2SO_4 \longrightarrow HBr + NaHSO_4$$

$$n\text{-}C_4H_9OH + HBr \rightleftharpoons n\text{-}C_4H_9Br + H_2O$$

Secondary reactions

$$2C_4H_9OH \xrightarrow{\text{conc. } H_2SO_4} C_4H_9OC_4H_9 + H_2O$$

$$2C_4H_9OH \xrightarrow[\triangle]{\text{conc. } H_2SO_4} CH_3CH_2CH_2=CH_2 + CH_3CH=CHCH_3 + 2H_2O$$

Equipment Apparatus for: reflux with a gas absorption trap, extraction/separation, distillation.
Instruments Abbé refractometer, IR.

Materials	
n-butyl alcohol (FW 74.12)	2.5 mL (2.02 g, 27.3 mmol)
sodium bromide (FW 102.9)	3.33 g (32.4 mmol)
concentrated sulfuric acid (FW 98.08)	4.0 mL (7.36 g, 73.6 mmol)
sodium hydroxide solution (5%)	
sodium bicarbonate solution (saturated)	
anhydrous calcium chloride	

Procedure

In a 30 mL round-bottomed flask place 3.3 mL of water and slowly add 4.0 mL of concentrated sulfuric acid with thorough mixing and cooling the diluted acid to room temperature. Add 2.5 mL of n-butyl alcohol and 3.33 g of sodium bromide to the cold diluted acid while shaking the flask to avoid formation of lumps. Introduce two boiling stones into the flask, attach an upright condenser, and connect the top of the condenser to a gas absorption trap to absorb hydrogen bromide that is evolved during the reaction [see Figure 1-5(c). **caution**: the inverted funnel should just touch the surface of the liquid in the beaker, cannot be immersed in the liquid to avoid the backflow of the liquid]. The 5% sodium hydroxide solution is used as absorbing agent.

Heat the flask gently over an asbestos gauge with a small flame (or a heating mantle) until the mixture begins to reflux gently. Heat the mixture under reflux for 30 minutes and shake the flask occasionally during the reflux. When the reaction has finished, remove the heating source and allow the flask to cool slightly. Remove the condenser and reassemble the apparatus for distillation. Add a boiling stone to the flask. Distill the mixture until the distillate appears to be clear.

Transfer the distillate to a separatory funnel using a capillary pipet, add 3.3 mL of water to it, and wash the mixture. Draw off the organic layer (which layer?) into another dry separatory funnel. Wash the organic layer by shaking it thoroughly with 2.0 mL of concentrated sulfuric acid. Separate the sulfuric acid carefully and completely (which layer?). Save the organic layer. Wash the organic layer with 5 mL of water, then with 5 mL of saturated sodium bicarbonate solution, and finally with 5 mL of water. Separate the "wet" n-butyl bromide into a dry Erlenmeyer flask. Dry the crude n-butyl bromide using a small amount of anhydrous calcium chloride. Stopper the flask and swirl the contents occasionally until the liquid is clear. Decant carefully the dried product into a dry distilling flask of proper size, add a boiling stone and distill the crude product over an asbestos gauge in a dry apparatus. Collect in a weighed bottle the portion boiling at 99~103 ℃. The yield is 1.65 g. The percentage yield is 44.1%. Determine the refractive index and the infrared spectrum.

Pure n-butyl bromide is a colorless and transparent liquid, b.p. 101.6 ℃, d_4^{20} 1.2760, n_D^{20} 1.4399.

Notes

1. The follow three methods may be used to judge weather the crude n-butyl bromide has

Chapter 3 Semimicro Synthetic Experiment of Typical Organic Compounds

been distilled completely.

① The distillate has changed from cloudy to clear.

② The upper oily layer in the distillation flask has disappeared.

③ Collect a few drops of distillate in a small test tube containing some water and note whether the distillate contains droplets of oil or is composed only of water. If the distillate does not produce droplets of oil in the water, then the crude n-butyl bromide has been distilled completely.

2. In accordance with the density of the liquid while separating, decide whether the product would appear as the upper or lower layer in the separatory funnel. If it is determined difficultly for a time, save the layers.

3. After washing the product with water, if the product shows to be red, it indicates that the product contains bromine formed by concentrated sulfuric acid to oxidate sodium bromide. The bromine can be removed by adding a small amount of saturated sodium bisulfite solution to the organic phase.

$$2NaBr + 3H_2SO_4 \longrightarrow Br_2 + SO_2 + 2NaHSO_4 + 2H_2O$$
$$Br_2 + 3NaHSO_3 \longrightarrow 2NaBr + NaHSO_4 + 2SO_2 + H_2O$$

4. Because concentrated sulfuric acid may dissolve unreacted n-butyl alcohol and the by-products, such as 1-butene and dibutyl ether from the crude n-butyl bromide, use a dry separatory funnel while separating. To remove unreacted n-butyl alcohol and avoid the formation of constant boiling mixture which is formed by n-butyl alcohol and n-butyl bromide during distilling, wash the crude product by shaking it thoroughly with concentrated sulfuric acid and separate the sulfuric acid carefully and completely.

5. Observe the water surface in the separatory funnel whether the oily product floats on the surface of water while washing the crude product with water each time. If has, after drawing off the lower oily layer, swirl the separatory funnel gently to make it to sink, then draw off it and combine it into the organic layer in order to reduce the loss of product.

Questions

1. What role does concentrated sulfuric acid play in this experiment? What influences do its used amount and the concentration have to this experiment?
2. What secondary reactions does this experiment have possibly? How to reduce the occurrence of them?
3. If you didn't operated according to the order of adding the starting materials in the procedure, but mix sodium bromide and concentrated sulfuric acid at first in the flask, then add n-butyl alcohol and water to it, what phenomenon would present?
4. When separating the crude product from the reaction mixture, why does the distillation be used, rather than separate it directly with a separatory funnel?
5. What is the purpose of each washing step in the post-treatment procedure for n-butyl bromide? Why must the organic layer be washed with water before washing it with saturated sodium bicarbonate solution?
6. Why must the drying agent be removed before the final distillation?

Experiment 6 Preparation of 2-chlorobutane (sec-butyl chloride)

Reaction equation

$$\underset{\underset{OH}{|}}{C_2H_5CHCH_3} + HCl \xrightarrow{ZnCl_2} \underset{\underset{Cl}{|}}{C_2H_5CHCH_3} + H_2O$$

Equipment   Apparatus for: reflux with a gas absorption trap, extraction/separation, fractional distillation.

Materials	
anhydrous zinc chloride (FW 136.28)	16 g (117.4 mmol)
sec-butyl alcohol (FW 74.12)	5 mL (4.03 g, 54.4 mmol)
concentrated hydrochloric acid (FW 36.46)	7.5 mL (9 g, 90 mmol)
sodium hydroxide solution (5%)	
anhydrous calcium chloride	

Procedure

Set up a 50 mL round-bottomed flask with a reflux condenser connected to a gas absorption apparatus. Place 16 g of anhydrous zinc chloride and 7.5 mL of concentrated hydrochloric acid in the flask. Shake the mixture until the zinc chloride is dissolved completely. Cool the solution in the flask to room temperature, add 5 mL of sec-butyl alcohol and two boiling stones. Then boil the mixture gently over an asbestos gauge with a small flame for 40 minutes. Cool the reaction mixture slightly, remove the condenser, and reassemble the apparatus for distillation. Add a boiling stone to the flask, distill and collect the fraction boiling below 115 ℃. Transfer the distillate to a separatory funnel, separate the organic layer, and wash it with 6 mL of water, 2 mL of 5% aqueous sodium hydroxide solution and 6 mL of water respectively. Decant the organic layer into a small dry flask and dry it with anhydrous calcium chloride. Decant the dried liquid into a small dry distilling flask and add two boiling stones into the flask. Distill the product from a dry fractional distillation set using a hot water bath.

Collect the fraction boiling at 67~69 ℃ in a weighed receiver. Weigh the product and calculate the percentage yield.

Pure 2-chlorobutane is a colorless and transparent liquid. b.p. 68.25 ℃, d_4^{20} 0.8732, n_D^{20} 1.3971.

Notes

1. When Lucas reagent is prepared, the anhydrous zinc chloride must be dried thoroughly by fusion method. The operation of weighing the dried anhydrous zinc chloride must be quick because it is easily deliquescent.
2. 2-Chlorobutane does not dissolve in hydrochloric acid. When the oily matter presents in the upper layer of the solution in the flask, it indicates that the reaction has occurred. The by-products, alkenes, have been removed from the product during distilling.
3. Because the boiling point of 2-chlorobutane is lower, the operations must be quick in order to avoid the loss of the product.

Questions

1. Why should the reflux be gentle during the reflux reaction?
2. Why is the product collected by the apparatus of the fractional distillation instead of that of the distillation?
3. What factors will reduce the yield of the product in this experiment?

Chapter 3 Semimicro Synthetic Experiment of Typical Organic Compounds

Experiment 7 Preparation of 2-methyl-2-chloropropane (*t*-butyl chloride)

Reaction equation

$$\underset{\underset{\text{OH}}{|}}{\overset{\overset{\text{CH}_3}{|}}{\text{CH}_3-\text{C}-\text{CH}_3}} + \text{HCl} \longrightarrow \underset{\underset{\text{Cl}}{|}}{\overset{\overset{\text{CH}_3}{|}}{\text{CH}_3-\text{C}-\text{CH}_3}} + \text{H}_2\text{O}$$

This reaction is an interesting contrast with experiment 5 in mechanisms. The *n*-butyl bromide synthesis proceeds by an S_N2 mechanism, while 2-methyl-2-chloropropane (*t*-butyl chloride) is prepared by an S_N1 reaction.

Equipment Apparatus for: extraction/separation, distillation.

Materials	
t-butyl alcohol (FW 74.12)	3.93 mL (3.1 g, 42 mmol)
concentrated hydrochloric acid (FW 36.46)	10.5 mL (12.39 g, 126 mmol)
sodium bicarbonate solution (5%)	
anhydrous calcium chloride	

Procedure

In a 50 mL separatory funnel, place 3.93 mL of *t*-butyl alcohol and 10.5 mL of concentrated hydrochloric acid. Shake the mixture constantly for 10~15 minutes. Allow the mixture to stand in the funnel until the layers have separated sharply, then draw off the water layer through the stopcock and discard it. Wash the organic layer with 3 mL of water, then with 3 mL of 5% sodium bicarbonate solution and finally with 3 mL of water. Dry the crude product with anhydrous calcium chloride until it is clear. Decant the dried liquid into a dry distilling flask, add a boiling stone, and distill the crude 2-methyl-2-chloropropane in a dry apparatus, using a hot water bath (about 70~75 ℃) collect the fraction boiling at 50~51 ℃ in a weighed bottle, cooled in an ice-water bath. Weigh the product and calculate the percentage yield.

Pure 2-methyl-2-chloropropane is a colorless and transparent liquid, b.p. 52 ℃, d_4^{20} 0.847, n_D^{20} 1.3877.

Notes

1. The solidifying point of *t*-butyl alcohol is 25 ℃. It is often solidified in the bottle at lower temperature. It may be melted by placing the bottle in a warm water bath before using it.
2. After placing the concentrated hydrochloric acid in the separatory funnel containing *t*-butyl alcohol, stopper the funnel. Allow the mixture to stand until the layers have separated sharply, and then vigorously shake the mixture. From time to time, relieve the internal pressure in the funnel regularly during the shaking period.
3. The boiling point of the product is lower; the operations should be quick in order to avoid the loss of the product by volatilization.

Questions

1. Why the yield is lower in this experiment?
2. When washing the crude product, if the concentration of the sodium bicarbonate solution was higher or the time of washing the product was longer, what influence would it occur to the product?

3.3 Preparation of ethers

There are many methods for ether preparation, for example intermolecular dehydration of an alcohol catalyzed by concentrated sulfuric acid, the reaction of dialkyl sulfate with alkoxide or phenoxide and Williamson synthesis. The dehyration method is usually limited to the preparation of symmetrical ethers such as methyl ether and diethyl ether. This method is unsuited to the preparation of unsymmetrical ether from two different alcohols, except alkyl aryl ether, because varieties of ethers will produce and the separation of the product is more difficult. As the toxicity of alkylation agent such as dimethyl sulfate and diethyl sulfate is very big, the more convenient and general method for unsymmetrical ether is the Williamson synthesis. Either symmetrical or unsymmetrical ether can be prepared by the Williamson synthesis, which involves the reaction between a primary or secondary alkyl halide and an alkoxide or a phenoxide ion. The Williamson synthesis is mainly used to prepare unsymmetrical (mixed) ether, especially alkyl aryl ether. Mechanistically, the Williamson synthesis is a nucleophilic substitution reaction (S_N2 reation) that occurs by nucleophilic substitution of a halide ion from an alkyl halide by the alkoxide or phenoxide ion.

Ethers are generally colourless, volatile, flammable and explosive liquid. When the ethers are stored for long periods, they may form small amount of explosive peroxides because of the autoxidation. Examine the ether before using whether it contains the peroxide or not.

Ethers are often used as organic solvents in organic synthesis as they can dissolve most organic compounds and some oragnic reactions must be performed in ether, for example Grignard reaction.

Experiment 8   Preparation of diethyl ether

Reaction equations

Main reactions

$$CH_3CH_2OH + H_2SO_4 \xrightleftharpoons{100 \sim 130\ ℃} CH_3CH_2OSO_2OH + H_2O$$

$$CH_3CH_2OSO_2OH + CH_3CH_2OH \xrightleftharpoons{135 \sim 145\ ℃} CH_3CH_2OCH_2CH_3 + H_2SO_4$$

Total reaction

$$2CH_3CH_2OH \xrightleftharpoons[H_2SO_4]{140\ ℃} CH_3CH_2OCH_2CH_3 + H_2O$$

Secondary reactions

$$CH_3CH_2OH \xrightarrow[\ [O]\ ]{H_2SO_4} \begin{array}{l} \xrightarrow{170\ ℃} CH_2=CH_2 + H_2O \\ \rightarrow CH_3CHO + SO_2 + H_2O \end{array}$$

$$CH_3CHO \xrightarrow{H_2SO_4} CH_3COOH + SO_2 + H_2O$$

$$SO_2 + H_2O \longrightarrow H_2SO_3$$

Equipment   Heating mantle; apparatus for: distillation at atmospheric pressure, extraction/separation.

Chapter 3 Semimicro Synthetic Experiment of Typical Organic Compounds

Materials

ethanol (95%) (FW 46.07)	18.5 mL (317 mmol)
concentrated sulfuric acid (FW 98.08)	6 mL (108 mmol)
sodium hydroxide solution (5%)	
sodium chloride solution (saturated)	
calcium chloride solution (saturated)	
anhydrous calcium chloride	

Procedure

Place 6 mL of 95% ethanol in a dry 50 mL 3-neck flask. Cool the flask in a cold water bath and slowly add 6 mL of concentrated sulfuric acid to the flask with swirling. Introduce a few boiling stones into the flask. The end of the dropping funnel and the thermometer must be immersed below the surface of the liquid and the distance to the bottom of the flask is about 0.5~1 cm. The receiving flask should be immersed in an ice-water bath. The side tube of the vacuum adapter is connected with a rubber tube led to the sewer.

Heat the flask over an asbestos gauge (or with a heating mantle) and allow the temperature of the reaction solution rises rapidly to 140 ℃. Then add ethanol dropwise from the dropping funnel to the reaction mixture at such a rate that approximates to the speed of the distillate (one drop per second), maintaining the temperature of the reaction at 135~145 ℃, this process needs about 30~45 minutes. After addition of ethanol is complete, continue to heat the flask for 10 minutes. When the temperature of the contents rise to 160 ℃, remove the heat source and allow the reaction mixture to cool.

Transfer the distillate to a separatory funnel and wash it with 6 mL of 5% sodium hydroxide solution, then with 6 mL saturated sodium chloride solution and finally with 2 × 6 mL portions saturated calcium chloride solution.

Separate the diethyl ether layer and dry it with anhydrous calcium chloride until the ether layer is clear (caution: the container still needs to be cooled in an ice-water bath). Decant carefully the dried diethyl ether into a dry distilling flask and add two boiling stones. Assemble a distillation apparatus as shown in the Figure 1-3(b). Distill the crude diethyl ether with a hot water bath (about 50 ℃) and collect fraction boiling at 33~38 ℃ in a weighed receiver. The yield is 3.5~4.5 g. The percentage yield is about 35%.

Pure diethyl ether is a colorless and transparent liquid, b. p. 34.5 ℃, d_4^{20} 0.7137, n_D^{20} 1.3526.

Notes

1. The apparatus must be tight.
2. Ethyl ether is extremely volatile and flammable. When the mixture of its vapour and air in the range of certain proportionment meets flame, it extremely explodes. Take care that no flame is nearby. After the reaction has finished, stop heating first, cool the reaction mixture to room temperature, then remove the receiver.
3. Distilling diethyl ether, use the hot water bath which has been made to reach the required temperature at another place. And do not distil it while heating by open flame.
4. Control the temperature of the reaction and the rate of adding ethanol (one drop per second). If the rate of the addition exceeds over the rate of the distillate obviously, not on-

ly the unreacted ethanol is distilled out, but also the temperature of the reaction solution will fall down sharply and the yield of diethyl ether will be reduced.

5. Take care to the order of the washing and no flame in the laboratory.
6. After the ether layer is washed with sodium hydroxide solution, the basicity of the ether layer is stronger. If the ether layer were washed with calcium chloride solution directly, the precipitation of calcium hydroxide would be separated. So wash the ether layer with the saturated sodium chloride solution before using calcium chloride solution in order to reduce the solubility of diethyl ether in water. So that the residue of the alkali can be washed away and the ethanol can be also removed partly.
7. Separate the layers thoroughly after washing the ether layer each time. Dry the washed ether with anhydrous calcium chloride about 30 minutes. Because calcium chloride and ethanol can form a complex, $CaCl_2 \cdot 4CH_3CH_2OH$, the unreacted ethanol can be removed.
8. Because the diethyl ether and water can form an azeotropic mixture containing 1.26% water (the boiling point is 34.15 ℃) and the distillate contains a small amount of ethanol, the boiling range is longer.

Questions

1. What measures should be taken to remove the impurities in the crude product?
2. What is the effect on the reaction while the temperature of the reaction solution is higher or lower?
3. Why should the end of the thermometer and the dropping funnel be immersed in the reaction solution?
4. Why can not the ethanol be distilled out when the temperature of the mixture consisted ethanol and concentrated sulfuric acid is above the boiling point of the ethanol at the beginning of the reaction?
5. Why is the anhydrous calcium chloride used as drying agent? What is the effect on the product when the drying time is too short?
6. Why is the crude diethyl ether washed with the saturated sodium chloride solution, rather than water?

Experiment 9 Preparation of *n*-butyl ether

Reaction equation

Main reaction

$$2CH_3CH_2CH_2CH_2OH \xrightleftharpoons[134\sim135\,℃]{H_2SO_4} (CH_3CH_2CH_2CH_2)_2O + H_2O$$

Secondary reaction

$$CH_3CH_2CH_2CH_2OH \xrightarrow[>135\,℃]{H_2SO_4} CH_3CH_2CH=CH_2 + H_2O$$

Equipment Apparatus for: reflux with a water segregator, extraction/separation, distillation.
Instruments Abbé refractometer, IR.

Materials	
n-butanol (FW 74.12)	5.2 mL (4.21 g, 56.8 mmol)
concentrated sulfuric acid (FW 98.08)	0.8 mL (1.47 g, 15 mmol)
sulfuric acid (50%)	
anhydrous calcium chloride	

Chapter 3 Semimicro Synthetic Experiment of Typical Organic Compounds

Procedure

Place 5.2 mL of n-butanol in a dry 50 mL two neck flask, add 0.8 mL of concentrated sulfuric acid while shaking and cooling the flask. Add two boiling stones to the flask. A thermometer is placed in one neck and immersed below the liquid level while a water segregator in another neck. On the top of the water segregator, a condenser is fitted. Add $(V-0.7)$ mL of water into the water segregator whose volume is V. Heat the flask gently over an asbestos gauze with a small flame to make the solution boil slightly. Continue to heat the flask and maintain the solution to reflux. Along with the reaction advances, the liquid level in the water segregator rises incessantly and the temperature of the reaction solution also rises unceasingly. Because the water that is formed in the reaction and unreacted n-butanol collect in the water segregator after cooling through the condenser, the liquid level rises gradually. Water is in the lower layer, n-butanol whose density is lower than that of water forms the upper layer. n-Butanol overflows into the reaction flask when the water segregator is full of liquid. Continue to heat the flask until the temperature of the contents increases to 134~135 ℃. When the water segregator is filled with water, it shows that the reaction is nearly completed (this process needs about 1 h.), stop heating right now.

Allow the mixture to cool, and then pour the mixture and water from the water segregator into a separatory funnel contained 7 mL of water. Shake the funnel fully, and then let it stand for a moment to make sure the two layers are well separated. Separate the upper layer as the crude product, n-butyl ether. Wash it with 2×3 mL portions of 50% sulfuric acid, then with 5 mL of water, and finally dry it over anhydrous calcium chloride.

Decant the dried crude product into a distilling flask, distil and collect the fraction boiling at 139~142 ℃ in a weighed receiver. Weigh the product, calculate the yield, determine the refractive index and the infrared spectrum.

Pure n-butyl ether is a colorless and transparent liquid, b.p. 142 ℃, d_4^{20} 0.773, n_D^{20} 1.3992.

Notes

1. If n-butanol is not mixed uniformly with concentrated sulfuric acid during adding them to the flask, the reaction solution in the flask easily becomes black after it is heated.
2. The theoretical yield of water that is formed in the reaction is about 0.5 g according to the equation, but the real yield is more than the theoretical yield because of by-product from monomolecular dehydration. When the $(V-0.7)$ mL of water in the water segregator, which has been added to the water segregator before the experiment starts, together with the water that is formed in the reaction are full of the water segregator, n-butanol condensed through the condenser overflows into the flask. Just so the purpose of the automatic separation is achieved.
3. In this experiment, the water which is formed in the reaction is removed from the reaction system continuously by the water segregator via the azeotropic distillation. In the reaction solution, n-butyl ether and water form an azeotropic mixture containing 33.4% water, the boiling point is 94.1 ℃, n-butanol and water form an azeotropic mixture containing 45.5% water, the boiling point is 93 ℃, n-butyl ether and n-butanol form a constant boiling binary mixture containing 82.5% n-butanol, the boiling point is 117.6 ℃. In addition, n-butyl ether with n-butanol and water form a constant boiling ternary mix-

ture containing 34.6% n-butanol and 29.9% water, the boiling point is 90.6 ℃. These azeotrpic mixture containing water enter the water segregator after condensated. In the water segregator, the upper layer mainly is n-butanol and n-butyl ether, and the lower layer mainly is water. The organic compounds in the water segregator may flow to the flask.

4. Because of the presence of the azeotropic mixture, the temperature of the solution can not reach 135 ℃ immediately at the beginning of the reflux. However, as the water is separated from the flask gradually, the temperature of the reaction rises gradually, and reaches above 135 ℃ finally. Stop heating right now. If not, the reaction solution will become black and the by-product butylene will be produced.

5. Pour 20 mL of concentrated sulfuric acid into 34 mL of water slowly in order to obtain 50% sulfuric acid solution.

6. n-Butanol can dissolve in the 50% sulfuric acid, but n-butyl ether dissolves rarely. So the n-butanol in the crude product can be removed by 50% sulfuric acid.

Questions

1. Calculate the theoretical yield of water. If the real yield of the water which is formed in the reaction is more than the theoretical one, please analyze the reasons.
2. How do you know the reaction has been completed?
3. At the end of the reaction, why should the reaction mixture in the flask be poured into the 8 mL of water? What is the purpose of each washing step?

Experiment 10   Preparation of phenetole (ethyl phenolate)

Reaction equations

$$PhOH + NaOH \longrightarrow PhONa + H_2O$$

$$PhONa + CH_3CH_2Br \longrightarrow PhOCH_2CH_3 + NaBr$$

Equipment Mechanical stirrer; apparatus for: stir with dropping, extraction/separation, distillation.

Materials	
phenol (FW 94.11)	3.75 g (39.8 mmol)
sodium hydroxide (FW 40)	2.00 g (50 mmol)
ethyl bromide (FW 109.0)	4.25 mL (6.21 g, 57 mmol)
diethyl ether	
anhydrous calcium chloride	
sodium chloride solution (saturated)	

Procedure

Set up a 50 mL 3-neck flask with a pressure-equalizing dropping funnel, reflux condenser, and mechanical stirrer. Add 3.75 g of phenol, 2 g of sodium hydroxide and 2 mL of water to the flask. Stir the mixture until the solids dissolve completely. Heat the flask in an oil bath and control the temperature of the oil bath between 80~90 ℃. From the dropping funnel, add ethyl bromide dropwise (require about 40 min). After the addition is complete,

Chapter 3 Semimicro Synthetic Experiment of Typical Organic Compounds

maintain the temperature while stirring for 1 h. Cool the mixture to room temperature. Add right amount water (about 5~10 mL) to the flask to dissolve the produced solid completely and transfer the solution to a separatory funnel. Separate and save the organic layer, then extract the aqueous layer with 4 mL of diethyl ether. Combine the extract with the reserved organic layer, wash it twice with a same volume of saturated sodium chloride solution. Separate the organic layer, then extract the aqueous layer with 3 mL of diethyl ether each time. Combine the organic layer with the extracts and dry it with anhydrous calcium chloride. Decant the dried liquid into a dry round-bottomed flask of proper size and distill off the ether from a water bath. Continue to distill the residue at atmospheric pressure. Collect the product boiling at 168~171 ℃ in a weighed bottle.

Pure phenetole is a colorless and transparent liquid, b. p. 170 ℃, d_4^{15} 0.9666, n_D^{20} 1.5076.

Notes
1. Because of the low boiling point of ethyl bromide, the flow of the condensated water must be big during the reflux to insure the enough ethyl bromide to participate the reaction.
2. Distilling diethyl ether, do not use the open flame to heat. The tail gas should be led to the sewer to prevent the fire by the leak of the diethyl ether vapor.

Questions

1. Preparing phenetole, what is the purpose of washing the organic phase containing the product with saturated sodium chloride solution?
2. What is the reflux liquid in the process of the reaction? What is the forming solid? Why the reflux is not obvious in the late reaction?

Experiment 11 Preparation of β-naphthyl ethyl ether

β-Naphthyl ethyl ether is also called nerolin, it is one kind of synthetic perfume, and its dilute solution has the fragrance of similar orange blossom and the robinia flower, accompanying the sweet taste and the aroma of the strawberry or pineapple. It is a colorless and flaky crystal. If it is added to some spices which is volatile, the volatile speed of the spices will be slowed down (the compound which has this character is called the perfume fixative), thus it is widely used as the spices in many soap and the perfume fixative of other spices (for example rose spice, lavender spice, lemon spice and so on). β-Naphthyl ethyl ether is mixed ether, alkyl aryl ether. If it is prepared by the dehydration catalyzed by concentrated sulfuric acid, diethyl ether is one of the by-products. If the reaction temperature is higher, ethylene will produce. Because of the lower boiling point of the by-products, they are easily removed. Therefore, alkyl aryl ether such as β-naphthyl ethyl ether may be prepared by both Williamson reaction and the dehydration catalyzed by concentrated sulfuric acid.

Reaction equation

Method 1 The dehydration catalyzed by concentrated sulfuric acid

$$\text{naphthyl-OH} + C_2H_5OH \xrightarrow{H_2SO_4} \text{naphthyl-OC}_2H_5 + H_2O$$

Method 2 The Williamson synthesis

$$\text{naphthol-OH} + \text{NaOH} \longrightarrow \text{naphthol-ONa} + H_2O$$

$$\text{naphthol-ONa} + C_2H_5Br \longrightarrow \text{naphthol-}OC_2H_5 + NaBr$$

Equipment
  Method 1　Apparatus for: reflux, suction filtration, recrystallization.
  Method 2　Apparatus for: reflux, distillation, suction filtration, recrystallization.
Instruments　Ultrasonic wave cleaning instrument, IR.

Materials	
Method 1	
β-naphthol (FW 144.19)	2.9 g (20 mmol)
absolute ethanol (FW 46.07)	6 mL (4.74 g, 102.8 mmol)
concentrated sulfuric acid (FW 98.08)	1 mL (1.84 g, 18.4 mmol)
sodium hydroxide solution (5%)	20 mL
ethanol (95%)	10 mL
Method 2	
β-naphthol (FW 144.19)	2.9 g (20 mmol)
absolute ethanol (FW 46.07)	20 mL
sodium hydroxide (FW 40.0)	0.9 g (23 mmol)
ethyl bromide (FW 109.0)	2.1 g (1.44 mL 19.3 mmol)

Procedure
Method 1

　　Place 2.9 g of β-naphthol and 6 mL of absolute ethanol in a 25 mL round bottomed flask. Carefully add 1 mL of concentrated sulfuric acid with swirling the flask to obtain good mixing. Add two boiling stones, fit a reflux condenser to the flask, and heat the mixture under reflux for 2 h in an oil bath (about 120 ℃). After the reaction has finished, pour the reaction mixture into a 100 mL beaker containing 30 mL of ice-water. The crystals separate. Decant the aqueous layer, grind the solid, wash it with 7~8 mL of 15% sodium hydroxide solution, and then with 2×10 mL portions of water while stirring with a glass rod. Remove the washing liquid by decantation to obtain the crude β-naphthyl ethyl ether. Dry the crude product, weigh, determine the melting point and calculate the percentage yield. The crude product may be recrystallized from 95% ethanol.

Method 2

　　Place 2.9 g of β-naphthol, 20 mL of absolute ethanol and 0.9 g of sodium hydroxide in a dry 50 mL round bottomed flask. Ultrasonic wave radiates the mixture in the flask for about 5 minutes at room temperature to dissolve the solid. Add 1.5 mL of ethyl bromide, and swirl the flask to mix the reagent. Add two boiling stones, fit the flask with a reflux condenser. Heat the reaction mixture under reflux for 2 hours, using a heating mantle as a heat source. After the reaction has finished, reassemble the apparatus for distillation and distill off a mass of ethanol. Pour the hot residue into a beaker containing 30 mL of ice-water mixture with stirring. After the mixture in the beaker has been cooled, decant the aqueous layer, and then wash the crude product twice with water. Collect the crude product by suction filtration, and wash it with a little cold water, and then recrystallize it from 95% ethanol. The white sheet-shaped crystals are obtained. After drying, weigh the product, calculate the percentage yield, determine the melting point and the IR spectrum.

Pure β-naphthyl ethyl ether is a white sheet-shaped crystal, m. p. 37~38 ℃, b. p. 281~282 ℃.

Notes

1. β-Naphthol is toxic and can severely burn the skin and mucosa. Weigh it carefully. Should β-naphthol be contacted with your skin, wash it off with soap immediately.
2. Concentrated sulfuric acid and sodium hydroxide are corrosive. Should they be contacted with your skin, wash it off with water immediately.
3. Take care to use ethyl bromide in the Williamson synthesis because the vapor of it has narcotic and is irritating to eyes and the breath system. The boiling point of ethyl bromide is lower, and then the flow of the condensated water must be larger to ensure the participation reaction for enough ethyl bromide during reflux. The solid will be produced while reflux.

Questions

1. Which by-products can be produced when the β-naphthyl ethyl ether is prepared by the dehydration catalyzed by concentrated sulfuric acid? Whether these by-products affect the purification of the product? Why is the crude product washed with the dilute sodium hydroxide solution?
2. Why can not 2-bromonaphthalene and ethanol be used as the starting materials when β-naphthyl ethyl ether is prepared by Williamson synthesis?
3. When sodium β-naphtholate is prepared, why should the ethanol sodium hydroxide solution be used, rather than aqueous sodium hydroxide solution?
4. In the final treatment, if the reaction mixture is poured into the water directly, rather than distilled off the ethanol in it at first, what influence will it be to the experimental result?

3.4 The Friedel-Crafts reaction

The reaction of an aromatic substrate with a carboxylic acid chloride, an anhydride or an alkyl halide in the presence of anhydrous aluminum chloride catalyst is called the Friedel-Crafts reaction. An alkyl group or acyl group is substituted for the hydrogen atom on the aromatic ring respectively, the former is called the Friedel-Crafts alkylation reaction, and the latter is called the Friedel-Crafts acylation reaction.

$$\text{C}_6\text{H}_6 + \text{R}-\overset{\text{O}}{\text{C}}-\text{Cl} \xrightarrow{\text{AlCl}_3} \text{C}_6\text{H}_5-\overset{\text{O}}{\text{C}}-\text{R} + \text{HCl}$$

$$\text{C}_6\text{H}_6 + \text{RCl} \xrightarrow{\text{AlCl}_3} \text{C}_6\text{H}_5-\text{R} + \text{HCl}$$

Alkylation and acylation reaction of aromatic hydrocarbons are used to prepare alkyl aromatic hydrocarbons and aromatic ketones respectively.

In alkylation reaction, alkenes and alcohols may be used as alkylating agents except alkyl halides. Using alkyl halides as the alkylating agents, anhydrous aluminum chloride is the most effective "Friedel-Crafts catalyst", and it is used only the catalytic amount in the reaction. In addition, other catalysts have the similar activity such as anhydrous zinc chloride, boron trifluoride, etc. Using alkenes and alcohols as the alkylating agents, the protic acids are usually used as the catalysts, for example hydrofluoric acid, sulfuric acid and phosphoric acid.

In the alkylation reaction, because alkyl groups can activate the aromatic ring, the reaction can not stop after the monoalkylation step. The initial alkylation product is more reactive than the precursor and preferentially alkylates. Consequently, alkylation frequently produces complex mixtures of polyalkylated products. Moreover, the alkylation of aromatic hydrocarbons is occurred by the formation and attack of alkyl carbocation intermediates, thus the products of rearrangement will produce due to the rearrangements of the carbocation intermediates. It makes the alkylation reaction limited use for the preparation of straight chain above two carbon atoms alkylated aromatics.

In the acylation reaction, because acyl groups can deactivate the aromatic ring, the reaction may stop after the monoacylation step. It is favorable for the selective preparation of monosubstituted aromatics. Because the acylium ion is especially stable and do not undergoes rearrangement in this reaction, the acylation reaction is of special application value in the synthesis of straight chain alkylated aromatics and branched chain alkylated aromatics. Acyl chlorides and acid anhydrides are the most frequently employed acylation agents. The Lewis acids can be used as the catalysts, among them anhydrous aluminum chloride is the most effective catalyst.

The process of the acylation reaction is shown as follows.

First step

$$R-\overset{:O:}{\underset{}{C}}-\overset{..}{\underset{..}{Cl}}+AlCl_3 \longrightarrow R-\overset{:O:}{\underset{}{C}}-\overset{..}{\underset{..}{Cl}}-AlCl_3 \rightleftharpoons \overset{-}{AlCl_4} + [R-\overset{+}{C}=\overset{..}{\underset{..}{O}} \longleftrightarrow R-C\equiv\overset{+}{O:}]$$

Second step

[reaction scheme showing benzene attacking acylium ion to form arenium intermediate, then loss of proton with Cl—AlCl$_3$ to give aryl ketone·AlCl$_3$ complex, plus AlCl$_3$ + HCl]

In alkylation and acylation reactions, the used amount of anhydrous aluminum chloride is different. In alkylation reaction, the used anhydrous aluminum chloride is only catalytic amount, but in acylation reaction with acid chloride a slight excess over 1 mol of aluminum chloride is used (general excess 10%), since a 1 : 1 addition compound is formed by reaction of aluminum chloride with the ketone produced in the reaction. Acid anhydrides react in a similar way to produce ketones, usually in better yields than are obtained from acid chlorides, but it is necessary to use 2 moles of aluminum chloride because the acid chloride formed in the reaction reacts with aluminum chloride. In practice, the used amount of aluminum chloride is slight excess over 2 moles (excess 10%~20%).

$$R-\overset{O}{\underset{}{C}}-O-\overset{O}{\underset{}{C}}-R + AlCl_3 \longrightarrow R-\overset{O}{\underset{}{C}}-Cl + R-\overset{O}{\underset{}{C}}-OAlCl_2$$

Chapter 3 Semimicro Synthetic Experiment of Typical Organic Compounds

The Friedel-Crafts reaction is highly exothermic, so it will be carried out by slowly adding the acylation agent solution to the reaction flask containing aromatic compound solution. The common solvents include carbon disulfide, nitrobenzene, nitromethane, etc. If the starting material is a liquid aromatic hydrocarbon, such as benzene, toluene, etc., an excess of the aromatic hydrocarbon is generally used as both the reactant and the solvent. Because hydrogen chloride is evolved during the reaction, the reaction apparatus should be connected with an apparatus of absorbing hydrogen chloride.

Experiment 12 Preparation of acetophenone

The important method for the preparation of aromatic ketones is Friedel-Crafts acylation. Acetophenone is synthesized by the reaction of benzene with acetic anhydride using Lewis acid (anhydrous aluminum chloride) as the catalyst. Acetic anhydride is used as the acylating agent. Although its acylating ability is weaker, it is cheaper.

Reaction equation

$$C_6H_6 + (CH_3CO)_2O \xrightarrow{AlCl_3} C_6H_5COCH_3 + CH_3COOH$$

Equipment Mechanical stirrer; apparatus for: reflux, extraction/separation, distillation.
Instruments Abbé refractometer, IR.

Materials	
benzene (FW 78)	8 mL (7 g, 90 mmol)
acetic anhydride (FW 102.09)	2 mL (2.15 g, 21 mmol)
anhydrous aluminum chloride (FW 133.34)	6.5 g (48.7 mmol)
concentrated hydrochloric acid	
sodium hydroxide	
petroleum ether (b. p. 60~90 ℃)	
anhydrous magnesium sulfate	
pH indicator paper	

Procedure

Equip a 50 mL 3-neck flask with a mechanical stirrer, pressure-equalizing dropping funnel and reflux condenser protected with a calcium chloride drying tube. Connect the end of the drying tube to a gas absorption trap to absorb hydrogen chloride. Rapidly add 6.5 g of anhydrous aluminum chloride and 8 mL of anhydrous benzene to the flask. Add 2 mL of acetic anhydride dropwise through the dropping funnel to the flask with stirring. First add a few drops of acetic anhydride. After the reaction starts, continue to add the remaining acetic anhydride dropwise at such a rate that the flask is warm, this process needs about 10 minutes. This reaction is exothermic. If necessary, cool the flask with a cold water bath in order to avoid the reaction being too vigorous. After all of the acetic anhydride has been added and the reaction has commenced to slacken, heat to reflux the mixture with a boiling water bath while stirring until no hydrogen chloride evolves. This process needs about 50 minutes.

Remove the water bath, allow the flask to cool. Pour the reaction mixture into a beaker containing 10 mL of concentrated hydrochloric acid and 20 g of crushed ice with stirring (the operation should be carried on in the hood). If some solids still exist, add a little more hydrochloric acid to the solution to dissolve any basic aluminum salt that are precipitated. Transfer the mixture to a separatory funnel, separate the upper organic layer and save it. Extract the lower aqueous layer with 2×15 mL portions of petroleum ether, and then combine the extracts with the saved organic layer. Wash the combined solution with 5 mL of 10% sodium hydroxide solution, then with 5 mL of water until it is neutral, and finally dry it over anhydrous magnesium sulfate. Remove the petroleum ether and benzene by distillation in a water bath. Distill the residual oil from a small distilling flask at atmospheric pressure (or under reduced pressure), using an air-cooled condenser, collecting the fraction boiling in the range of 198~202 ℃. The product is a colorless, transparent and oily liquid. The percentage yield is about 65%.

Pure acetophenone is a colorless, transparent and oily liquid, m. p. 20.5 ℃, b. p. 202 ℃, d_4^{15} 1.0281, n_D^{20} 1.53718.

Notes

1. This experiment must be carried out in an anhydrous condition. All chemicals and apparatus used must be thoroughly dried before use. Anhydrous aluminum chloride reacts violently with moisture in the air, being decomposed. The operation of weighing and adding aluminum chloride must be quick. Fix the covers of the flask and reagent bottle timely. Benzene must be dried overnight with anhydrous calcium chloride before use. Acetic anhydride must be redistilled before use and the fraction boiling at 137~140 ℃ can be used.
2. Using anhydrous aluminum chloride, avoid contacting with your skin in order to prevent the burn.
3. Install rightly the hydrogen chloride absorbing apparatus to prevent the back flow.
4. When acetic anhydride is added to the reaction mixture, the reaction is exothermic. Add slowly acetic anhydride at the beginning, otherwise the bumping takes place. Control the reaction temperature below 60 ℃. The yield may be raised by the method of extending the reaction time.
5. Add hydrochloric acid to the reaction mixture in order to decompose the acetophenone-aluminum chloride complex and release the free acetophenone.

$$\text{C}_6\text{H}_5\text{-CO} \cdot \text{AlCl}_3 \xrightarrow{\text{H}^+, \text{H}_2\text{O}} \text{C}_6\text{H}_5\text{COCH}_3 + \text{AlCl}_3$$

6. The purification of acetophenone also can be performed by vacuum distillation. The correlations of boiling points with pressures are listed in the following table:

Pressure/Pa	533	667	800	933	1067	1120	1333	3333
b. p. /℃	60	64	68	71	73	76	78	98
Pressure/Pa	4000	5333	6666	7999	13332	19998	26664	
b. p. /℃	102	110	115.5	120	134	146	155	

Questions

1. What are the respective characteristics of Friedel-Crafts acylation and alkylation? What are the differences on the used amount of anhydrous aluminum chloride and aromatic hydrocarbon in the two reactions? Why?

2. Why are excessive benzene and anhydrous aluminum chloride used in this reaction?
3. Why should the product be decomposed by the mixture of hydrochloric acid and ice water after the reaction has finished?
4. Interpret the infrared spectrum of acetophenone.

Experiment 13 Preparation of benzophenone (Friedel-Crafts acylation method)

Benzophenone is a colorless and lustrous crystal with the sweet taste and rose fragrance, thus it can be used as the spice and endow with the sweet taste to the essence. It is applied to make perfumes, fancy soaps and essences. It can be used as a starting material for the manufacture of organic dye and insecticides, as well as the production of benztropine hydrobromide and diphenhydramine hydrochloride in pharmaceutical industry.

There are many methods for preparing benzophenone. It can be prepared by either benzyl chloride as starting material via the reaction of alkylation, oxidation etc., or by benzene as starting material via the reactions of alkylation, hydrolysis, etc. In this experiment, benzophenone is synthesized by the acylation reaction using benzoyl chloride and benzene as the starting materials.

Reaction equation

$$\text{C}_6\text{H}_5\text{COCl} + \text{C}_6\text{H}_6 \xrightarrow{\text{AlCl}_3} \text{C}_6\text{H}_5\text{COC}_6\text{H}_5 + \text{HCl}$$

Equipment Mechanical stirrer; apparatus for: reaction with simultaneous stirring and addition, reflux with a gas absorption trap, extraction/separation, distillation under reduced pressure.

Instrument IR.

Materials	
anhydrous aluminum chloride (FW 133.34)	3.75 g (28 mmol)
benzene (anhydrous) (FW 78.12)	15 mL (13.5 g, 0.17 mol)
benzoyl chloride (FW 140.5)	3 mL (3.65 g, 25 mmol)
aqueous sodium hydroxide solution (5%)	
concentrated hydrochloric acid	
anhydrous magnesium sulfate	

Procedure

Set up a dry 100 mL 3-neck flask with a mechanical stirrer, a reflux condenser and a Claisen adapter. Place a thermometer in the central opening of the adapter and a pressure-equalizing dropping funnel in the side arm. Attach a drying tube filled with calcium chloride at the top of the condenser and connect the open end of the drying tube to a gas absorption trap to absorb hydrogen chloride that evolves during the reaction. The trap is made up of an inverted funnel in the beaker. Place the 5% aqueous sodium hydroxide solution used as the absorbing agent in the beaker. The inverted funnel should just touch the surface of the liquid in the beaker (the distance between the lower edge of the funnel and the liquid surface is about 1~2 mm), not be immersed in the liquid to prevent the back flow of the liquid in the beaker.

Place 3.75 g of anhydrous aluminum chloride and 15 mL of dried benzene into the flask quickly, add dropwise 3 mL of benzoyl chloride from the pressure-equalizing dropping funnel to the flask while stirring at room temperature. Control the dropping rate to maintain the temperature of the reaction at 40 ℃. The mixture in the flask reacts intensely while producing hydrogen chloride and the reaction solution becomes brown gradually. After the addition, heat the solution in a hot water bath at 60 ℃ while stirring until no hydrogen chloride is evolved from the flask. This process requires about 1.5 h.

Allow the solution to cool, and then pour the contents in the flask into a beaker containing 25 mL of ice water in the hood. The precipitate separates from the solution at the same time. Add dropwise 1~2 mL of concentrated hydrochloride acid into the beaker while stirring until the precipitate is decomposed completely. Transfer the solution to a separatory funnel, separate the organic phase, and extract the aqueous phase with 2×10 mL portions of benzene. Combine the extracts with the organic phase. Wash the organic phase with 10 mL of water, then with 10 mL of 5% sodium hydroxide solution, and finally with water three times (10 mL of water each time) until the organic phase is neutral. Dry it over anhydrous magnesium sulfate. Remove the solvent by distillation to obtain the crude product. Distill the residue from a small distillation set under reduced pressure, collecting the fraction boiling at 187~189 ℃/2.00 Pa (15 mmHg). The product is a colorless and transparent liquid. After it is cooled in a refrigeratory, it solidifies and the white crystals are obtained. The crude product is also purified by recrystallization from petroleum ether (b.p. 60~90 ℃) instead of the vacuum distillation. After drying, weigh it, determine the melting point and the infrared spectrum, calculate the yield.

Pure benzophenone is a colorless crystal, m.p. 47~48 ℃, b.p. 305.4 ℃.

Notes
1. The notes of this experiment are similar to that of preparing acetophenone, see the experiment 12.
2. Benzophenone has many kinds of crystal forms. The melting points of them are different. It is 49 ℃ in α type, 26 ℃ in β type, 45~48 ℃ in γ type, 51 ℃ in δ type. Among them, the α type is more stable.

Questions
1. Whether the polyacylated aromatic hydrocarbon is formed easily in the acylation reaction?
2. Comparing with aliphatic ketone, which direction will the infrared absorption peak of carbonyl group in arone move toward (high wave number or low wave number)? Why?
3. Why can nitrobenzene be used as the solvent in Friedel-Crafts reaction? Why is it unfavourable to the Friedel-crafts reaction that a substituent group, such as OH, OR, or NH_2, etc., exists in the aromatic ring?
4. Why should the heating bath be used in vacuum distillation, rather than heat the flask over an asbestos gauge directly, even if the boiling point of the compound is high?

Experiment 14　Preparation of benzophenone (Friedel-Crafts alkylation method)

Reaction equation

Chapter 3 Semimicro Synthetic Experiment of Typical Organic Compounds

Equipment Mechanical stirrer; apparatus for: reaction with simultaneous stirring and addition, reflux with a gas absorption trap, distillation, extraction/separation, distillation under reduced pressure, recrystallization.

Materials

benzene (anhydrous) (FW 78.12)	4.5 mL (3.95 g, 51 mmol)
tetrachloromethane (FW 153.82)	11 mL (17.5 g, 114 mmol)
anhydrous aluminum chloride (FW 133.34)	3.35 g (25 mmol)
anhydrous magnesium sulfate	
sodium hydroxide solution (5%)	

Procedure

Set up a dry 250 mL 3-neck flask with a mechanical stirrer, a reflux condenser and a Claisen adapter. Place a thermometer in the central opening of the adapter and a pressure-equalizing dropping funnel in the side arm. Attach a drying tube filled with calcium chloride at the top of the condenser and connect the open end of the drying tube to a gas absorption trap to absorb hydrogen chloride that evolves during the reaction. The trap is made up of an inverted funnel in the beaker. Place the 5% sodium hydroxide solution used as the absorbing agent in the beaker. The inverted funnel should just touch the surface of the liquid in the beaker (the distance between the lower edge of the funnel and the liquid surface is about 1~2 mm), not be immersed in the liquid.

In the flask place 3.35 g of anhydrous aluminum chloride and 7.5 mL of tetrachloromethane quickly. Cool the flask in an ice-water bath so that the internal temperature is about 5~10 ℃, and add slowly a 2 mL solution prepared by 4.5 mL of anhydrous benzene and 3.5 mL of tetrachloromethane while stirring. After the reaction starts, hydrogen chloride forms, the temperature of the reaction mixture rises gradually. Control the reaction temperature with an ice-water bath between 5 ℃ and 10 ℃ during the reaction. As soon as the initial reaction moderates, add slowly the rest of the benzene solution to the flask at such a rate as to maintain the temperature in the flask between 5 ℃ and 10 ℃. After the addition (about 10 min), allow the mixture to stir for 1 h and maintain the reaction temperature at about 10 ℃. Cool the reaction mixture with an ice-water bath, and add slowly 50 mL of water with stirring from the funnel in order to hydrolyze the mixture. Arrange the apparatus for distillation, distil off the excessive tetrachloromethane with a water bath, and then heat the flask directly over an asbestos gauze for 0.5 h to remove the residual tetrachloromethane and promote the complete hydrolysis of dichloro-diphenylmethane.

Cool the reaction mixture to room temperature, transfer it to a separatory funnel, separate the organic layer, and extract the aqueous layer with 10 mL of tetrachloromethane once. Combine the extract with the organic layer and dry them over anhydrous magnesium sulfate. Filter off the drying agent, distill off the tetrachloromethane with a hot water bath at atmospheric pressure, then distil the residue from a small distillation set under reduced pressure, collecting the fraction boiling at 144~148 ℃/280 Pa (2.1 mmHg). The product is colorless, transparent and oily liquid. After the oily liquid is cooled in a refrigeratory, it solidifies to form the white crystals. Recrystallize it from petroleum ether (b.p. 60~90 ℃). Pure benzophenone melts in the range 45~46 ℃. The percentage

yield is about 70%.
Notes
1. See the Experiment 12.
2. When the reaction temperature is below 5 ℃, the reaction is slow. When the reaction temperature is above 10 ℃ the tarry resinous substance will form easily.

Questions

1. How many types does Friedel-Crafts reaction have? What is the respective character? Which type should this experiment belong to?
2. Which materials can be used as the catalysts of Friedel-Crafts reaction?
3. Why should this experiment be carried out under the anhydrous condition?
4. Why is carbon tetrachloride excessive, rather than benzene in this experiment? What is the result if benzene is excessive?

Experiment 15 Preparation of 2-t-butyl hydroquinone

There are many types of antioxidants, phenols are one of them. 2-t-Butyl-hydroquinone (TBHQ) is a kind of phenolic antioxidant. It has the properties of the antioxidation, polymerization inhibition, etc. It is lower toxic and inexpensive. Therefore, it is widely used as the antioxidants of the rubber and the plastics as well as food additives. It is prepared by the alkylation of the highly activated ring in hydroquinone with t-butyl alcohol and phosphoric acid. t-Butyl alcohol is used as alkylating agent. Phosphoric acid is the catalyst. Alkyl halides and alkenes are also used as the alkylating agents.

Reaction equation

$$\underset{\underset{OH}{|}}{\overset{\overset{OH}{|}}{C_6H_4}} + (CH_3)_3COH \xrightarrow{H_3PO_4} \underset{\underset{OH}{|}}{\overset{\overset{OH}{|}}{C_6H_3}}-C(CH_3)_3 + H_2O$$

Equipment Magnetic stirrer; apparatus for: reaction with addition, extraction/separation, steam distillation, suction filtration, recrystallization.

Materials	
hydroquinone (FW 110.11)	2.2 g (20 mmol)
t-butanol (FW 74.12)	1.58 g (2 mL, 21 mmol)
concentrated phosphoric acid (FW 98)	8 mL
toluene	10 mL

Procedure

Set up a 100 mL 3-neck flask with a pressure-equalizing dropping funnel, thermometer, reflux condenser and magnetic stirrer bar. Add 2.2 g of hydroquinone, 8 mL of concentrated phosphoric acid and 10 mL of toluene to the flask. Heat the mixture with a water bath while stirring and allow the temperature of mixture to rise to *ca.* 90 ℃. From the dropping funnel, slowly add 2 mL of t-butanol and control the reaction temperature in the range 90~95 ℃. After the addition is complete, continue to stir the mixture for 0.5 h at 90 ℃ until the solid in the mixture dissolves completely.

After the reaction has finished, pour the hot reaction mixture into a separatory funnel. Draw off and discard the lower, phosphoric acid. Transfer the organic layer to the 3-neck flask, add 20 mL of water to the flask, and distill the mixture until the solvent has been removed by steam distillation (see the Figure 2-34). After the distillation, filter the hot residual aqueous solution in the flask by suction filtration, discard the insolubles and transfer the filtrate to a beaker. If the volume of the residual liquid is less than 20 mL, supply hot water in order that the product dissolves completely in hot water. Allow the filtrate to stand at room temperature. The white crystals separate in this period. Collect the product by suction filtration, wash it with cold water twice, and dry it. Weigh the product, determine the melting point and calculate the yield. The crude product may be recrystallized from hot water. Pure 2-t-butyl hydroquinone is a white needle-shape crystal, m. p. $127 \sim 129$ ℃.

Notes
1. Phosphoric acid is very corrosive. Great care must be taken when measuring it.
2. Toluene is used as the reaction solvent. Comparing hydroxyl group with methyl group, hydroxyl group can activate more effectively the benzene ring. So long as the alkylating agent is not excessive, the toluene may be served as the inert solvent.
3. Hydroquinone can not dissolve in toluene, but 2-t-butyl-hydroquinone can. So when hydroquinone has dissolved completely, it indicates that the reaction is finished.
4. The Büchner funnel should be preheated before using to prevent it being plugged.
5. 2-t-Butyl hydroquinone dissolves in hot water, slightly in cold water, but the disubstituted compound, 2,5-t-butyl hydroquinone does not dissolve in hot water.

Questions
1. Whether the product of toluene alkylation can be produced in this experiment?
2. Which secondary reactions may occur in this experiment?
   What measures have been taken in this experiment to reduce the secondary reactions?
3. How do you judge the end of steam distillation? Why should the residual liquid be filtered when it is hot after steam distillation?

3.5 Esterification reaction

In human's daily life, most of esters have the widespread use. Some esters may be used as the solvents of the edible oil, fat, plastics and paint. Many esters have the pleasant-smelling fragrance, and are inexpensive spices.

An carboxylic acid and an alcohol or a phenol react in the presence of an inorganic or a organic acid catalyst to produce ester and water, this process is called esterification reaction. The commonly used catalysts are concentrated sulfuric acid, dry hydrogen chloride, organic acid or cation exchange resin. The esterification reaction can also achieve in the absence of a acid catalyst if the ability of the carboxylic acid participating reaction is stronger, such as formic acid, oxalic acid.

$$RCOOH + R'OH \xrightleftharpoons{H^+} RCOOR' + H_2O$$

The esterification reaction is reversible. If equimolar amounts of acid and alcohol are used and the equilibrium has been established, the yield is only 67%. To increase the amount of ester formed, the use of an excess of the alcohol or organic acid will make the equilibrium shift to the right. The choice of reactant to be used in excess will depend upon factors such as availability, cost, and ease of removal of excess reactant from the product. In addition, driving an esterification to completion by removal of the water or ester formed in the reaction is a common practice. Water formed in the reaction can be removed by zeotropic distillation using, for example, benzene, toluene or chloroform, etc. As the solvent, which forms an azeotrope with water. Ester may be also prepared by the alcoholysis of acid chlorides or acid anhydrides.

Experiment 16 Preparation of ethyl acetate

Reaction equation

$$CH_3COOH + CH_3CH_2OH \xrightleftharpoons{H_2SO_4} CH_3COOCH_2CH_3 + H_2O$$

Equipment Heating mantle; apparatus for: reflux, extraction/separation, distillation.

Materials

glacial acetic acid (FW 60.05)	6 mL (6.0 g, 0.10 mol)
anhydrous ethanol (FW 46.07)	9.5 mL (9.2 g, 0.20 mol)
concentrated sulfuric acid (FW 98)	2.5 mL
sodium carbonate solution (saturated)	
sodium chloride solution (saturated)	
calcium chloride solution (saturated)	
anhydrous magnesium sulfate	

Procedure

Place 9.5 mL of anhydrous ethanol and 6 mL of glacial acetic acid in a dry 50 mL round-bottomed flask. Carefully add 2.5 mL of concentrated sulfuric acid dropwise to the mixture, swirl the flask to ensure complete mixing. Introduce two boiling stones to the flask and fit the flask with a reflux condenser to the flask.

Heat the mixture gently with a heating mantle and gently reflux it for 0.5 h. Cool the contents in the flask slightly and rearrange the apparatus for distillation in which the receiving flask is immersed in a cold water bath. Distill off the ethyl acetate formed in the reaction until the volume of the distillate is about half of total volume for the reaction liquid.

Slowly add saturated sodium carbonate solution to the distillate. Shake the mixture gently until carbon dioxide gas is no longer evolved. Transfer the solution to a separatory funnel and draw off the lower aqueous layer as completely as possible. Wash the organic layer contained in the separatory funnel with 5 mL of saturated sodium chloride solution, then with 5 mL of saturated calcium chloride solution, and finally with some water. Draw off the aqueous layer, pour the crude ester from the top of the separatory funnel into a dry Erlenmeyer flask and dry it with a small amount of anhydrous magnesium sulfate. Transfer the dried ester to a dry distillating flask, add two boiling stones into the flask, distill it with a heating mantle, and collect in a weighed bottle the fraction boiling at 73~78 ℃. Weigh the product and calculate the percentage yield.

Pure ethyl ester is a colorless, transparent and scented liquid, b. p. 77.06 ℃, d_4^{20} 0.9003, n_D^{20} 1.3723.

Notes

1. In the distillate, there are small amounts of the unreacted ethanol and acetic acid as well as the by-product diethyl ether except for ethyl acetate and water. The acid in the distillate can be removed by neutralization with saturated sodium carbonate solution. The unreacted ethanol can be removed by saturated calcium chloride solution. Otherwise it will affect the yield of the product.
2. If the organic layer which has been washed by saturated sodium carbonate solution further was washed by saturated calcium chloride solution, the floccule of calcium carbonate will produce. It makes the further separation difficult, so wash the organic layer with water between the two steps to separate easily. Because ethyl acetate has certain solubility in water, wash the organic layer with saturated sodium chloride solution instead of water to reduce the loss of the product.
3. Ethyl acetate and water or ethanol can form respectively constant boiling binary mixture. Ethyl acetate with water and ethanol can form a constant boiling ternary mixture. So ethanol and water in the organic layer must be removed thoroughly.

b. p. /℃	Composition(mass percent)/%		
	ester	ethanol	water
70.2	82.6	8.4	9.0
70.4	91.6	—	8.1
71.8	69.0	31.0	

Questions

1. What are the characters of the esterification reaction? What measures have been taken to drive the esterification toward completion in this experiment?
2. Is it right that excessive acetic acid is used in this experiment? Why?
3. Which impurities are contained in the crude ester? how to remove them?

Experiment 17 Preparation of *n*-butyl acetate

Reaction equation

$$CH_3COOH + n\text{-}C_4H_9OH \xrightleftharpoons{H_2SO_4} CH_3COOC_4H_9 + H_2O$$

Equipment Apparatus for: reflux with a water segregator, extraction/separation, distillation.

Instrument Abbé refractometer.

Materials	
n-butyl alcohol (FW 74.12)	5 mL (4.05 g, 54.5 mmol)
glacial acetic acid (FW 60.05)	3.5 mL (3.67 g, 61 mmol)
concentrated sulfuric acid	
sodium carbonate solution (10%)	
anhydrous magnesium sulfate	

Procedure

Set up a 50 mL 2-neck flask with a water segregator, a reflux condenser and thermometer. The water separator which volume is V has been added $(V-1.2)$ mL of water.

Add 5 mL of n-butyl alcohol, 3.5 mL of glacial acetic acid and 2~3 drops concentrated sulfuric acid to the flask, shake and mix them thoroughly, and then add two boiling stones into the flask. Heat the mixture with a heating mantle for 15 minutes at approximately 80 ℃, then reflux the mixture for 15 minutes, when the water segregator is filled with water, it shows that the reaction has finished.

Cool the reaction mixture in the flask to room temperature, remove the condenser, pour the reaction mixture and the liquid in the water segregator into a separatory funnel and remove the water layer. Wash the organic layer with 5 mL of 10% aqueous sodium carbonate, keeping the pH at 7, then with 5 mL of water and remove the water layer. Pour the organic layer into a dry Erlenmeyer flask and dry it with a small amount of anhydrous magnesium sulfate. Pour the crude dried product into a dry 10 mL round-bottomed flask and add two boiling stones into the flask. Distil and collect the fraction boiling at 124~126 ℃. Weigh the product and calculate the percentage yield. Determine the refractive index.

Pure n-butyl acetate is a colorless, transparent liquid, b.p. 126.3 ℃, d_4^{18} 0.8824, n_D^{20} 1.3947.

Notes

1. The water produced in the reaction is removed by forming constant boiling mixture in this experiment. n-Butyl alcohol, n-butyl acetate and water can form the following several azeotropes. After the azeotropes containing water have been condensed, the condensed liquids fall back into the water segregator. In the water segregator, the condensed liquids have divided into two layers. The upper layer is the ester and alcohol, and the lower layer mainly is water.

	Azeotrope	Boiling point/℃	Composition(mass percent)/%		
			n-butyl acetate	n-butyl alcohol	water
binary mixture	n-butyl acetate-water	90.7	72.9		27.1
	n-butyl alcohol-water	93.0		55.5	44.5
	n-butyl acetate-n-butyl alcohol	117.6	32.8	67.2	
ternary mixture	n-butyl acetate-n-butyl alcohol-water	90.7	63.0	8.0	29.0

2. Dripping concentrated sulfuric acid, shake and cool the flask to avoid the reactant being carbonized partially. The concentrated sulfuric acid is a catalyst in this reaction.

Questions

1. Calculate the amount of the removed water when the reaction is complete in the experiment.
2. What principles does this experiment adopt to raise the yield of n-butyl acetate? What is the function of concentrated sulfuric acid in this experiment?

Experiment 18   Preparation of n-pentyl acetate (n-amyl acetate)

Reaction equation

Main reaction

$$CH_3COOH + CH_3(CH_2)_3CH_2OH \xrightleftharpoons{H^+} CH_3COOCH_2(CH_2)_3CH_3 + H_2O$$

Secondary reactions

$$CH_3(CH_2)_3CH_2OH \xrightarrow[\Delta]{H^+} CH_3(CH_2)_2CH=CH_2 + H_2O$$

Chapter 3 Semimicro Synthetic Experiment of Typical Organic Compounds

$$2CH_3(CH_2)_3CH_2OH \xrightarrow[\triangle]{H^+} CH_3(CH_2)_3CH_2OCH_2(CH_2)_3CH_3 + H_2O$$

Equipment Apparatus for: reflux with a water segregator, extraction/separation, distillation.
Instrument Abbé refractometer.

Materials	
n-pentanol (FW 88)	5.5 g (6.75 mL, 62.5 mmol)
glacial acetic acid (FW 60.05)	4.5 g (4.3 mL, 75 mmol)
concentrated sulfuric acid (FW 98.08)	0.3 mL
sodium carbonate solution (5%)	
saturated aqueous sodium chloride	
anhydrous magnesium sulfate	
benzene	

Procedure

Place 6.75 mL of n-pentanol, 4.3 mL of glacial acetic acid and 7 mL benzene in a dry 30 mL round-bottomed flask. Add 0.3 mL of concentrated sulfuric acid slowly while shaking. Introduce two boiling stones into the flask, attach a water segregator to the flask, fill the water segregator with benzene, and connect the top of the water segregator to a condenser. Heat the mixture under reflux until no water is separated out (about two hours). During this process, the water that is formed in the reaction is separated from the flask continuously and flows into the water segregator.

After the reaction is complete, cool the reaction mixture to room temperature, pour the mixture into a separatory funnel. Rinse the flask with 7 mL of cold water, and then the rinsing solution is also poured into the funnel. Shake the funnel, and allow it to stand until the layers are separated. Discard the aqueous layer, wash the organic layer with 6 mL of 5% aqueous sodium carbonate solution, and then wash it with saturated sodium chloride solution to neutral. Separate out the aqueous layer, transfer the ester layer into a dry Erlenmeyer flask and dry it over anhydrous magnesium sulfate. Filter the dried crude product into a dry distilling flask. Add a boiling stone to the flask. Arrange an apparatus for fractional distillation. Collect the fraction boiling at 144~150 ℃ in a weighed receiver. Weigh the product, calculate the yield and determine the refractive index.

Pure n-pentyl acetate is a colorless and transparent liquid, b.p. 148.4 ℃, d_4^{20} 0.8970, n_D^{20} 1.4000.

Questions

1. What is the purpose of using the water segregator in this experiment?
2. Which step is benzene removed during the final treatment?

Experiment 19   preparation of isoamyl acetate (banana oil)

Reaction equation

$$CH_3COOH + HOCH_2CH_2\underset{\underset{CH_3}{|}}{C}HCH_3 \xrightleftharpoons{H^+} CH_3COOCH_2CH_2\underset{\underset{CH_3}{|}}{C}HCH_3 + H_2O$$

Equipment Apparatus for: reflux, extraction/separation, distillation.

Instrument   Abbé refractometer.

Materials

isoamyl alcohol (FW 88.2)	5 mL (4.05 g, 46.5 mmol)
glacial acetic acid (FW 60.1)	7 mL (7.35 g, 122.5 mmol)
concentrated sulfuric acid	1.0 mL
sodium carbonate solution (5%)	
saturated aqueous sodium chloride	
anhydrous magnesium sulfate	

Procedure

Pour 5 mL of isoamyl alcohol and 7 mL of glacial acetic acid into a dry 25 mL round-bottomed flask. Slowly add 1.0 mL of concentrated sulfuric acid to the contents of the flask with swirling. Add two boiling stones to the mixture. Fit a reflux condenser to the flask, and heat the mixture under reflux for 1.0 h. Then, remove the heating source and allow the mixture to cool to room temperature. Pour the cooled mixture into a separatory funnel. Rinse the reaction flask with 10mL of cold water and pour the rinsing into the separatory funnel. Stopper the funnel and shake it several times. Separate the lower aqueous layer, and wash the organic layer with 5 mL of 5% aqueous sodium carbonate, then with 5 mL of saturated aqueous sodium chloride until the organic layer is neutral. Separate the layers and discard the aqueous layer. Transfer the crude ester to a clean, dry 25 mL Erlenmeyer flask and dry it over anhydrous magnesium sulfate. Filter the dried product into a drying distilling flask, distill and collect the fraction boiling at 138~142 ℃ in a dry flask which has been immersed in an ice bath to ensure condensation and to reduce odors. Weigh the product, calculate the percentage yield and determine the refractive index.

Pure isoamyl acetate is a colorless, transparent liquid, b. p. 138~142 ℃, d_4^{20} 0.876, n_D^{20} 1.4000.

Notes

1. The reaction liquid is exothermal while adding concentrated sulfuric acid to the flask. Carefully shake the flask to rapidly diffuse the caloric.
2. When the crude product is washed by the sodium carbonate solution, carbon dioxide gas will evolve. Do not shake the funnel acutely, and release the inner pressure of the funnel.
3. Wash the organic layer with saturated aqueous sodium chloride until it is neutral. Otherwise the product is easily decomposed during distilling.

Questions

1. One method for favoring the formation of the ester is to add an excess of acetic acid. Suggest another method, involving the right-hand side of the equation, which will favor the formation of the ester.
2. Why is excess acetic acid easier to remove from the products than excess isoamyl alcohol?

Experiment 20   Preparation of ethyl benzoate

Reaction equation

$$\text{C}_6\text{H}_5\text{COOH} + \text{C}_2\text{H}_5\text{OH} \underset{}{\overset{\text{H}^+}{\rightleftharpoons}} \text{C}_6\text{H}_5\text{COOC}_2\text{H}_5 + \text{H}_2\text{O}$$

Chapter 3 Semimicro Synthetic Experiment of Typical Organic Compounds

Equipment Apparatus for: reflux, distillation, extraction/separation, vacuum distillation.
Instruments Abbé refractometer, IR.

Materials	
benzoic acid (FW 122.13)	1.53 g (12.5 mmol)
anhydrous ethanol (FW 46.07)	6.5 mL (5.13 g, 0.111 mol)
concentrated sulfuric acid (FW 98.08)	0.5 mL (0.92 g, 9.2 mmol)
diethyl ether	
sodium carbonate solution (10%)	
anhydrous magnesium sulfate	

Procedure

In a dry 30 mL round-bottomed flask place 1.53 g of benzoic acid, 6.5 mL of anhydrous ethanol, 0.5 mL of concentrated sulfuric acid and two boiling stones. Fit a condenser to the flask, and heat the flask gently over an asbestos gauge with a small flame. Reflux the mixture for 1.5 h. Cool the flask slightly, remove the condenser and arrange the apparatus for distillation. Add a boiling stone to the flask and distill off the excessive ethanol in a boiling water bath.

Transfer the cooled residue in the flask to a separatory funnel containing 10 mL of cool water and extract the aqueous layer with 2×5 mL portions of diethyl ether. Combine the organic layer, wash it respectively with 10 mL of 10% sodium carbonate solution and 2×5 mL portions of water. Separate the organic layer and dry it over anhydrous magnesium sulfate. Filter the transparent dried liquid into a dry flask, and distill off the diethyl ether from a hot water bath. The crude product is purified by vacuum distillation. Collect the fraction boiling at 73~74 ℃/533 Pa (4 mmHg) in a dry weighed bottle. Weigh the product and calculate the percentage yield. Determine its refractive index and the infrared spectrum.

Pure ethyl benzoate is a colorless and transparent liquid, b.p. 213 ℃, d_4^{20} 1.0468, n_D^{20} 1.5007.

Notes
1. Distilling diethyl ether from a water bath, the distillation should be carried on with no open flame and the temperature of water bath should be controlled.
2. The operation of vacuum distillation should be done according to the operating rules.

Questions
1. Why is excessive ethanol added in this reaction?
2. What principles and measures can be also adopted to raise the yield?

Experiment 21 Preparation of acetylsalicylic acid (aspirin)

Acetylsalicylic acid (aspirin) is one of the most generally useful drugs ever discovered. It was first synthesized in 1899 by Dreser and is wildly used as an analgesic, an antipyretic, an anti-inflammatory and an anti-rheumatic agent. Meanwhile, it also possesses the function of softening blood vessels. In recent years, the researches show that it can reduce the rate of bowel cancer about 30%~50%.

Salicylic acid (o-hydroxybenzoic acid) is a bifunctional compound. It is a phenol and a carboxylic acid. Hence it can undergo two different types of esterification reaction. In the presence of acetic anhydride, salicylic acid will react to form acetylsalicylic acid (aspirin), whereas in the presence of methyl alcohol it will yield methyl salicylate (oil of wintergreen). In this experiment we shall use the former reaction to prepare aspirin (The synthesis of oil of wintergreen is described in Experiment 22).

Reaction equation

Main reaction

$$\text{salicylic acid} + (CH_3CO)_2O \xrightarrow{H_2SO_4} \text{acetylsalicylic acid} + CH_3COOH$$

Secondary reaction

(structure showing salicylic acid trimer formation) $+ H_2O$

Equipment Apparatus for: suction filtuation, recrystallization.

Materials	
salicylic acid (FW 138.12)	0.5 g (3.6 mmol)
acetic anhydride (FW 102.09)	1.25 mL (13.25 mmol)
concentrated sulfuric acid (98%)	
ethanol (95%)	
ferric chloride solution (1%)	

Procedure

Weigh 0.5 g of salicylic acid crystals and place them in a dry 30 mL Erlenmeyer flask. Add 1.25 mL of redistilled acetic anhydride slowly, followed by 2 drops of concentrated sulfuric acid from a dropper, and then swirl the flask gently until the salicylic acid dissolves. Heat the flask for 10 minutes in a beaker of water heated to 90 ℃. Allow the flask to cool to room temperature and then place it in an ice-water bath, during which time the acetylsalicylic acid should begin to crystallize from the reaction mixture. If it does not, scratch the walls of the flask with a glass rod and cool the mixture slightly in an ice-water bath until crystallization has occurred. After the crystals separated, add 13 mL of water and continue to cool the mixture in an ice-water bath to crystallize completely. Collect the crude product by suction filtration on a Büchner funnel and rinse the crystals several times with small portions of cold water. Remove the crystals for air-drying.

Whether there is the unreacted salicylic acid which contains phenolic hydroxyl group in the crude product, it may be examined by 1% ferric chloride solution. Dissolve a few crystals of aspirin in several drops of ethanol in a test tube and add one drop of 1% ferric chloride solution. If the color reaction appears, the crude product should be recrystallized by a mixed solvent of ethanol and water.

Purify the crude product in the following way. In an Erlenmeyer flask dissolve the crude product in minimal volume of boiling ethanol. To the solution add hot water until the solution becomes turbidly, then heat the solution again until it becomes clear. Allow the mixture to stand and cool slowly. Collect the crystals by suction filtration on a Büchner funnel and allow it to dry in air. Weigh the dry aspirin, determine its melting point and calculate the percentage yield.

Pure acetylsalicylic acid is a white needle-shape crystal, m. p. 132~135 ℃.

Notes
1. Because of the molecular inner hydrogen bond in salicylic acid, the reaction temperature of preparing acetylsalicylic acid by the reaction of salicylic acid with acetic anhydride must be in the range 150~160 ℃. The purpose of adding concentrated sulfuric acid is to break the hydrogen bond in order that the reaction can carry on at lower temperature (90 ℃) and reduce the by-product effectively. So control the temperature in this experiment.
2. All apparatus and reagents should be dried thoroughly before the experiment is carried on. Acetic anhydride should be redistilled before use, collect the fraction boiling at 139~140 ℃.
3. When acetylsalicylic acid is heated, it easily decomposes. The temperature of the decomposition is at 126~135 ℃. Hence when drying or recrystallizing it, or determining the melting point, cannot heat for a long time.
4. If the crude product contains unreacted salicylic acid, it will be violet for the test of 1% ferric chloride solution.
5. The crude product may be also recrystallized from dilute hydrochloric acid (1 : 1) or a mixed solvent of benzene and petroleum (b. p. 30~60 ℃).

Questions

1. What is the function of concentrated sulfuric acid in this experiment?
2. What product will be obtained by the reaction of salicylic acid with ethanol in the presence of the concentrated sulfuric acid? Write the reaction equation.
3. Interpret the infrared spectrum of acetylsalicylic acid.

Experiment 22 Preparation of methyl salicylate (oil of wintergreen)

Methyl salicylate was first isolated in 1843 by extraction from the wintergreen plant (*Gaultheria*). Therefore it is called oil of wintergreen. It is a natural ester, exists in ylang-ylang oil, sesame oil and clove oil. It has a pleasant odor from the leaves of the wintergreen plant, and it is usually used as a flavoring. It is frequently used for food, toothpaste and cosmetics and so on. It also has analgesic and antipyretic character. Methyl salicylate may be prepared by esterifying salicylic acid with methyl alcohol, a reaction in which the catalyst is concentrated sulfuric acid.

Reaction equation

salicylic acid + CH_3OH $\xrightarrow{H_2SO_4}$ methyl salicylate + H_2O

Equipment Apparatus for: reflux, distillation, extraction/separation, vacuum distillation.

Instrument Abbé refractometer.

Materials

salicylic acid (FW 138.12)	3.45 g (25 mmol)
methanol (FW 32.04)	12 g (15 mL, 370.5 mmol)
concentrated sulfuric acid	1 mL
sodium bicarbonate solution (10%)	
diethyl ether	

Procedure

Set up a reflux apparatus like that shown in Figure 1-5(a), using a dry 30 mL round-bottomed flask. Add 3.45 g of salicylic acid and 15 mL of methanol to the flask. Swirl the flask to dissolve the salicylic acid. Slowly add 1 mL of concentrated sulfuric acid to the mixture while cooling and swirling to thoroughly mix the reactants. Add two boiling stones to the flask, and assemble the apparatus. Heat the mixture under reflux for 1.5 h. Remove the condenser and reassemble the apparatus for distillation.

Distill off the excess methanol in water bath. Cool the residual mixture in the reaction flask, and then add 10 mL of water. After swirling, transfer the mixture to a separatory funnel. Separate the organic layer. Extract the aqueous layer with 10 mL of diethyl ether. Combine the extract with the organic layer. Wash the ether solution with 10 mL of water, then with 10 mL of 10% sodium bicarbonate solution, and finally with water until the organic layer shows neutral. Dry the organic layer containing the product with anhydrous magnesium sulfate. Filter off the drying agent and distill off diethyl ether in water bath. Distill the crude product by vacuum distillation. Collect the fraction boiling at 115~117 ℃/2.7 kPa (20 mmHg) in a weighed bottle. Weigh the product, calculate the percentage yield, and determine the refractive index.

Pure methyl salicylate is a colorless and transparent liquid, m.p. $-8 \sim -7$ ℃, b.p. 222 ℃, d_4^{25} 1.1787, n_D^{20} 1.5360.

Notes

1. All apparatus must be dried thoroughly before use. Otherwise the yield of the esterification will be reduced.
2. If the flask does not be shaked in time while dropping concentrated sulfuric acid, sometimes the partial reactants may be carbonized.
3. Distill off methanol thoroughly, or the yield will be reduced because of increasing the solubility of the product in water.

Questions

1. Salicylic acid is esterified with methanol at the carboxyl group to yield methyl salicylate (oil of wintergreen). The esterification is an acid-catalyzed equilibrium reaction, what measures has been taken in this experiment in order that the equilibrium moves toward the right to favor the formation of the ester in high yield? What methods can be also adopted?
2. If the methanol was not distilled off, rather than the crude product was washed directly with water after the esterification reaction had been finished, what influence would it have on the experimental result?

Chapter 3 Semimicro Synthetic Experiment of Typical Organic Compounds

3. Why was 10% NaHCO₃ used to wash the crude product in the final treatment? What would have happened if the aqueous sodium hydroxide solution had been used?
4. What is the function of the sulfuric acid in this reaction? Write a mechanism for the acid-catalyzed esterification of salicylic acid with methanol.
5. Interpret the infrared spectrum of methyl salicylate.

Experiment 23 Preparation of glucose pentaacetate

Reaction equations

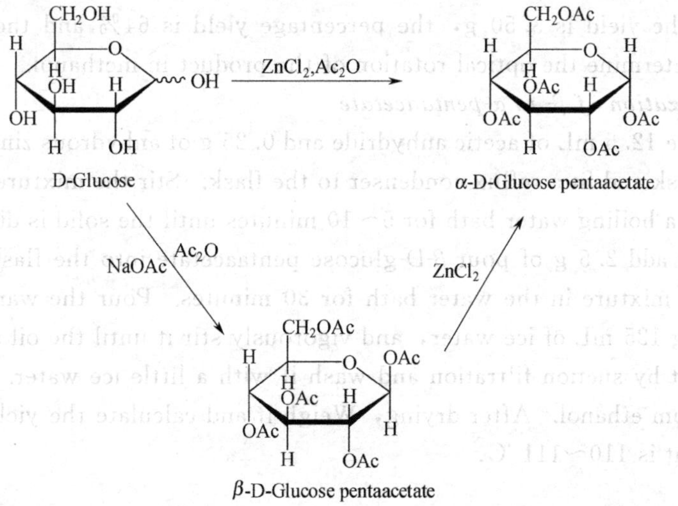

D-Glucose $\xrightarrow{ZnCl_2, Ac_2O}$ α-D-Glucose pentaacetate

D-Glucose $\xrightarrow[Ac_2O]{NaOAc}$ / $\xrightarrow{ZnCl_2}$ β-D-Glucose pentaacetate

Equipment Magnetic stirrer; apparatus for: reflux with stirring, suction filtration, recrystallization.

Instrument polarimeter.

Materials	
D-glucose (FW 180.16)	5.00 g (27.8 mmol)
acetic anhydride (FW 102.09)	37.5 mL
anhydrous zinc chloride (FW 136.28)	0.95 g
anhydrous sodium acetate (FW 82)	2.00 g
ethanol	

Procedure

(1) *Synthesis of α-D-glucose pentaacetate*

Set up a dry 50 mL 3-neck flask with a thermometer, a reflux condenser and a magnetic stirrer bar. Rapidly add 0.7 g of anhydrous zinc chloride and 12.5 mL of acetic anhydride into the flask. Stir the mixture magnetically and heat the flask in a boiling water bath for 5~10 minutes. Then add 2.5 g of powdered D-glucose in small portions to the flask. After all the glucose has been added, continuously heat the mixture in the boiling water bath for 1 h. Pour the cooled reaction mixture into a beaker containing 125 mL of ice water and stir it until the oil solidifies. Collect the white powder product by suction filtration and wash it with a little cold water. Recrystallize the crude product from ethanol or methanol. The yield is 3.50 g, the percentage yield is 64.6% and the melting point is in the range 110~111 ℃. Determine the optical rotation of the product in methanol.

(2) *Synthesis of β-D-glucose pentaacetate*

Place 2.00 g of anhydrous sodium acetate, 2.5 g of dry glucose and 12.5 mL of acetic anhydride in a dry 50 mL round bottomed flask, add a magnetic stirrer bar, and fit a reflux condenser to the flask. Stir the mixture magnetically and heat the flask in a water bath until the solution is clear, and continue to heat the reaction solution for 1 h. Slowly pour the solution into a beaker containing 150 mL of ice water with stirring. Allow the mixture to stand about 10 minutes until it solidifies. Collect the white powder product by suction filtration and wash it several times with cold water. Recrystallize the crude product from ethanol twice (25 mL×2). The yield is 3.50 g, the percentage yield is 64% and the melting point is 131~132 ℃. Determine the optical rotation of the product in methanol.

(3) *Isomerization of β- to α-pentaacetate*

Rapidly place 12.5 mL of acetic anhydride and 0.25 g of anhydrous zinc chloride in a dry 50 mL 3-neck flask and fit a reflux condenser to the flask. Stir the mixture magnetically and heat the flask in a boiling water bath for 5~10 minutes until the solid is dissolved completely. Then rapidly add 2.5 g of pour β-D-glucose pentaacetate into the flask and continue to heat the reaction mixture in the water bath for 30 minutes. Pour the warm solution into a beaker containing 125 mL of ice water, and vigorously stir it until the oil solidifies. Collect the crude product by suction filtration and wash it with a little ice water. Recrystallize the crude product from ethanol. After drying, Weigh it and calculate the yield of the product. The melting point is 110~111 ℃.

Notes

1. D-Glucose exists in two stereoisomeric cyclic forms of hemi-acetal type in nature (α-and β-anomers). When D-glucose reacts with excessive acetic acid or acetic anhydride in the presence of the catalyst, all five of the hydroxyl groups of D-glucose are acetylated, and α-and β-D-glucose pentaacetate are obtained. But using different catalysts, the main product is also different. When anhydrous zinc chloride is used as the catalyst, the α-D-glucose pentaacetate is main product. With anhydrous sodium acetate, the β-D-glucose pentaacetate predominates. The β-isomer is more stable than the α-isomer from the stereoisomeric configurations. The β-isomer can been converted to the α-isomer with anhydrous zinc chloride catalysis.

2. Remove the adsorbent water in the commercial anhydrous zinc chloride before using. Use the redistilled acetic anhydride. Anhydrous zinc chloride is deliquescent, when weigh and triturate it, the operation must be quick.

3. After the reaction mixture is poured into ice water, vigorously stir it to disintegrate the lumps of the crystalline precipitate into the powder. This will avoid the product being hydrolyzed partially during recrystallizing due to packing the solvent in the lumps of the product.

3.6 The Cannizzaro reaction

In the presence of strong alkalis, aldehyde which does not contain α-H undergoes disproportionation to form the corresponding primary alcohol and a salt of the carboxylic acid, this pross is called cannizzaro reaction.

In essence, Cannizzaro reaction is a nucleophilic addition to the carbonyl group. The

Chapter 3 Semimicro Synthetic Experiment of Typical Organic Compounds

process involves addition of hydroxyl ion to the carbonyl group of one molecule and transfer of hydride anion from the adduct to a second molecule of the aldehyde, accompanied by proton interchange to form the alcohol and the carboxylate. The mechanism is shown as follow.

$$Ar-\underset{}{\overset{O}{C}}-H + OH^- \rightleftharpoons Ar-\underset{OH}{\overset{O^-}{C}}-H$$

$$Ar-\underset{OH}{\overset{O^-}{C}}-H + Ar-\overset{O}{C}-H \rightleftharpoons Ar-\overset{O}{C}-OH + Ar-\underset{H}{\overset{O^-}{C}}-H \longrightarrow Ar-\overset{O}{C}-O^- + ArCH_2OH$$

In the cannizzaro reaction, 50% concentrated alkali is used generally and the used amount of alkali is usual over two times more than the aldehyde. Otherwise, the reaction is not complete and the separation of the unreacted aldehyde and the alcohol which is produced in the reaction is very difficult by distillation.

Experiment 24 Preparation of 2-furoic acid and 2-furancarbinol

Reaction equations

$$2 \langle O \rangle-CHO \xrightarrow{OH^-} \langle O \rangle-CH_2OH + \langle O \rangle-COO^-$$

$$\langle O \rangle-COO^- \xrightarrow{H^+} \langle O \rangle-COOH$$

Equipment Apparatus for: extraction/separation, distillation, suction filtration, recrystallization.

Materials	
furfuraldehyde (FW 96.08)	10 mL (11.6 g, 120.7 mmol)
sodium hydroxide solution (43%)	9 mL
diethyl ether	
hydrochloride acid (25%)	
anhydrous magnesium sulfate	
benzene	

Procedure

Place 9 mL of 43% aqueous sodium hydroxide in a 100 mL beaker. Put the beaker in an ice-salt bath and cool the solution to 5 ℃. Then add slowly 10 mL of redistilled furfuraldehyde from a Pasteur pipet, drop by drop, with ceaseless stirring using a glass rod. Maintain the reaction temperature in the range 8~12 ℃. After adding, continuously stir for about 25 minutes at room temperature, then a light yellow paste-like mixture is obtained.

To the mixture add right amount of water until the precipitate dissolves completely with stirring (require about 15 mL of water). The solution turns into a dark red. Transfer the solution to a separatory funnel and extract it with 4×10 mL portions of diethyl ether. Combine the ethereal extracts and dry it with 2 g of anhydrous magnesium sulfate. Filter off the drying agent under gravity and carefully distill off the diethyl ether in a water bath (**caution**: no open flame is nearby when the diethyl ether is distilled). Continuously distill the 2-furancarbinol at atmospheric pressure and collect the fraction boiling at 169~172 ℃ (or collect the fraction boiling at 88 ℃/4666 Pa by vacuum distillation). The yield is about 4~5 g, the percentage yield is 68%~84%.

Pure 2-furancarbinol is a colorless liquid, b.p. 17 ℃/100 kPa (750 mmHg), d_4^{20} 1.1296, n_D^{20} 1.4868.

The aqueous solution (from which the 2-furancarbinol has been extracted) mainly contains sodium 2-furoate. Acidify it with 25% hydrochloric acid until the pH is 3. Then the white precipitate is obtained. Cool it thoroughly in an ice-water bath to separate completely. Add 30 mL diethyl ether to the beaker to make the white precipitate dissolve in diethyl ether. Transfer the mixture into a separatory funnel, rinse the beaker with a little diethyl ether, and transfer these washings into the funnel. Separate the two layers and extract the aqueous layer with diethyl ether once again. Combine the ethereal extract with the organic layer and carefully distill off the diethyl ether in a water bath. Recrystallize the crude product from benzene to get the white crystals. Collect the crystals with suction filtration wash them with a little benzene and allow the product to dry in air. The yield is 5.5 g, the percentage yield is 81.5%.

The melting point of pure 2-Furoic acid is 133~134 ℃.

Notes
1. If stored for too long time, the furfuraldehyde will turn into chocolate brown or even black, and possibly contains water. So distill to purify it before use. Collect the fraction boiling at 160~162 ℃. Vacuum distillation is recommended. Collect the fraction boiling at 54~55℃/2.27 kPa (17 mmHg). The freshly distilled furfuraldehyde is colorless or light yellow liquid. Preserve it to evade the light and preferably use it after the distillation.
2. As the reaction takes place between two phases, stir thoroughly to react completely.
3. It is very important to control the temperature of the reaction in this experiment. If the temperature is over 12 ℃, it rises easily, the reaction is controlled difficultly and the reactant will become dark red. While below 8 ℃, the reaction carries on too slowly and the sodium hydroxide may be accumulated. Once the reaction occurs, it will be vigorous to make the temperature of the reaction rise quickly and the yield be reduced because of the side reactions. The method that the sodium hydroxide solution is added to the furfuraldehyde can be also used in this experiment. The yields of them are nearly the same.
4. Many sodium 2-furoate can be separated out during the reaction. Add some water to dissolve it and turn the cream-like paste mixture into claret-red transparent liquid. The water should not be more than required amount in order to avoid partial loss of the product.
5. The amount of acid must be sufficient to ensure the pH value of the solution is about 3 after the acidification in order to separate the 2-furoic acid thoroughly, which is the key to get high yield.
6. The crude product of 2-furoic acid is also recrystallized from water. If this method is used, do not heat the solution for a long time to prevent the 2-furoic acid being decomposed. The 2-furoic acid obtained from water is a leaf-like crystal. It will sublimate par-

tially at 100 ℃. It is better to dry it in air.

> **Questions**
>
> 1. What are the principles for separating and purifying the 2-furoic acid and 2-furancarbinol in this experiment?
> 2. What is the yellow paste material separated during the reaction?

Experiment 25 Preparation of benzoic acid and benzyl alcohol

Reaction equations

$$2C_6H_5CHO + NaOH \longrightarrow C_6H_5COONa + C_6H_5CH_2OH$$

$$C_6H_5COONa + HCl \longrightarrow C_6H_5COOH + NaCl$$

Equipment Apparatus for: extraction/separation, distillation, suction filtration, recrystallization.

Materials	
benzaldehyde (FW 106.13)	6.3 mL (6.6 g, 62.2 mmol)
sodium hydroxide (FW 40)	5.5 g (137.5 mmol)
diethyl ether	
concentrated hydrochloric acid	
saturated aqueous sodium bisulfite solution	
sodium carbonate solution (10%)	
anhydrous magnesium sulfate	

Procedure

In a 100 mL Erlenmeyer flask dissolve 5.5 g of solid sodium hydroxide in 5.5 mL of water with swirling. Cool the solution to room temperature. Add 6.3 mL of redistilled benzaldehyde to the sodium hydroxide solution in batches, and add about 1.5 mL of benzaldehyde each time. After adding each time, cork the flask firmly and shake the mixture thoroughly until a white emulsion is formed. During the process, if the reaction temperature is higher, timely cool the flask in a cold water bath. Allow the stoppered flask to stand for 24 h or longer.

Gradually add just enough water to the mixture (about 20~25 mL of water), tepefy and stir the solution to dissolve the precipitate of sodium benzoate. Cool the solution and pour it into a separatory funnel. Extract the alkaline solution with 3×5 mL portions of diethyl ether. Combine the ethereal extracts for isolation of benzyl alcohol and reserve the aqueous solution to obtain the benzoic acid. Wash the ethereal solution with 3 mL of saturated aqueous sodium bisulfite, 5 mL of 10% aqueous sodium carbonate and 5 mL of cold water, respectively. Separate the ethereal solution and dry it with anhydrous magnesium sulfate.

Decant the dried ethereal solution into a dry 25 mL round-bottomed flask and distill off the diethyl ether in a warm water bath. Attach a short air-cooled condenser, distill the benzyl alcohol and collect the fraction boiling at 198~204 ℃. The yield is about 2.2 g.

Pure benzyl alcohol is a colorless liquid, b. p. 205.4 ℃, d_4^{20} 1.045, n_D^{20} 1.5396.

Slowly pour the aqueous solution of sodium benzoate (from which benzyl alcohol has been extracted) into a vigorously stirred mixture of 20 mL of concentrated hydrochloric acid, 20 mL of water, and 12.5 g of chipped ice. Cool the product in an ice bath in order that benzoic acid separates completely. Collect the crystals of benzoic acid by suction filtration and wash it with a little cold water. Recrystallize the crude benzoic acid from hot water, collect the crystals and allow them to dry thoroughly. The yield is 3.5 g.

Pure benzoic acid is a colorless needle-shape crystal, m. p. 122.4 ℃.

Notes

1. Thoroughly shake the reaction mixture is the key to the success of the reaction. Do as this, the mixture in the flask will solidify after it is allowed to stand for 24 h and the odor of benzaldehyde will also disappear.
2. This is an exothermic reaction, so cool timely the flask in a cold water bath to avoid the reaction temperature to be higher and produce excessive benzoic acid.

Questions

1. Why employ the redistilled benzaldehyde? What impurity is contained in benzaldehyde which has been stored for a long time? If it has not been removed, what is the effect on this experiment?
2. What principle is employed to separate and purify benzyl alcohol and benzoic acid in this experiment? What is the purpose to wash the ethereal extract with saturated aqueous sodium bisulfite and 10% aqueous sodium carbonate?

3.7 Condensation reaction

Experiment 26 Preparation of cinnamic acid

Cinnamic acid is a significant intermediate of producing the drug of treating coronary heart disease, its ester derivatives are important raw materials of preparing essence or food flavor. Cinnamic acid also possesses extensive application in the production of fine chemical such as pesticides, plastics, photosensitive resin and so on.

When an aromatic aldehyde reacts with an acid anhydride in the presence of the potassium salt of the acid corresponding to the anhydride, an α,β-unsaturated acid can be prepared by the Perkin reaction.

In this experiment, potassium carbonate has substituted potassium acetate in the Perkin reaction according to the Kalnin method because it may shorten the reaction time and improve the product yield.

Reaction equation

$$PhCHO + (CH_3CO)_2O \xrightarrow[140\sim180\ ℃]{CH_3COOK} PhCH=CHCOOH + CH_3COOH$$

Equipment Apparatus for: reflux, steam distillation, suction filtration, recrystallization.

Materials	
benzaldehyde (FW 106.13)	1.5 mL (1.56 g, 15 mmol)
acetic anhydride (FW 102.1)	4.0 mL (4.33 g, 42.4 mmol)
anhydrous potassium carbonate (FW 138)	2.20 g (16 mmol)
concentrated hydrochloric acid	
sodium hydroxide solution (10%)	
ethanol	
decolorizing charcoal	

Procedure

Set up a 100 mL 2-neck flask with a thermometer and an air condenser. Place 1.5 mL of freshly redistilled benzaldehyde, 4 mL of acetic anhydride, 2.2 g of anhydrous potassium carbonate and two boiling stones in the flask. Heat the reaction mixtures cautiously to reflux over an asbestos gauze for 40 minutes. Control the temperature of the reaction in the range 150~180 ℃. Because carbon dioxide will be evolved in the process, the foaming will be produced at the initial stage of the reaction.

Cool the flask slightly, and then add 35 mL of warm water to the flask. Remove the air condenser and assemble an apparatus for steam distillation [See Figure 2-36 (f)]. Add a boiling stone to the flask and distill the mixture over an asbestos gauze until the unreacted benzaldehyde has been removed by steam distillation. Cool the flask again and add 10 mL of 10% sodium hydroxide solution. Make sure all the cinnamic acid is converted into the sodium salt and dissolve completely. Remove the insoluble material by suction filtration and wash the residue twice with a small amount of cold water (with 10 mL of water each time). Pour the filter liquor into a 250 mL beaker, introduce a small amount of decoloring charcoal, and boil the solution gently for 10 minutes. Cool the solution, filter it by suction filtration. Pour the filter liquor into a 250 mL beaker and cool it to room temperature. Acidify it with concentrated hydrochloric acid while stirring and check it with Congo-red test paper until it becomes blue. Cool the beaker in an ice-water bath. After the cinnamic acid has crystallized completely, collect the crystals with suction filtration and wash them with an appropriate amount of cold water. The crude product may be recrystallized from a mixed solvent of water and ethanol (the ratio is 5 : 1, V/V). The percentage yield is 68%.

Pure cinnamic acid is a colorless crystal, m. p. 135~136 ℃, b. p. 300 ℃, d_4^{20} 1.2475.

Notes

1. Benzoic acid may be produced by the automatic oxidation of benzaldehyde when benzaldehyde is stored for a long time. Benzaldehyde should be redistilled before use. Otherwise, the existence of trace benzoic acid not only influences the progress of reaction, but also is difficult to be removed from the product. So it will influence the purity of cinnamic acid.
2. Acetic anhydride may convert into acetic acid when it is exposed to moisture for a long time through hydrolysis. So the acetic anhydride should be redistilled before use.
3. Cinnamic acid has *cis* and *trans* isomers. The product prepared usually is the *trans* isomer, m. p. 135.6 ℃.
4. Do not heat vigorously at first to avoid the decomposition of acetic anhydride.

Questions

1. What product will be obtained by the reaction of benzaldehyde with propionic anhydride in the presence of anhydrous potassium propionate?
2. What is the purpose of steam distillation in this experiment? How to judge the end of the distillation?
3. Why can potassium carbonate substitute potassium acetate in the reaction?
4. What structural aldehyde can carry on the Perkin reaction? Write the mechanism of Perkin reaction.
5. Interpret the main absorption peaks in the infrared spectrum of cinnamic acid.

Experiment 27 Preparation of ethyl acetoacetate

An ester with a α hydrogen is treated with a base, a reversible condensation reaction occurs to yield a β-keto ester product. This reaction is called the Claisen ester condensation. If the used base is sodium ethoxide, the mechanism of the Claisen reaction is shown as follow.

$$RCH_2COOEt + C_2H_5O^- \rightleftharpoons EtOH + R\bar{C}HCOOEt \longleftrightarrow RCH=C\underset{\underset{O^-}{|}}{}-OC_2H_5 \quad (1)$$

$$RCH_2\overset{O}{\underset{\|}{C}}-OC_2H_5 + RCH=\overset{O^-}{\underset{|}{C}}-OC_2H_5 \rightleftharpoons RCH_2\overset{O}{\underset{\|}{C}}-\overset{O^-}{\underset{|}{C}}HCOC_2H_5 \rightleftharpoons RCH_2\overset{O}{\underset{\|}{C}}-\underset{\underset{R}{|}}{CH}COC_2H_5 + C_2H_5O^- \quad (2)$$

$$RCH_2\overset{O}{\underset{\|}{C}}-\underset{\underset{R}{|}}{CH}COC_2H_5 + C_2H_5O^- \rightleftharpoons RCH_2\overset{O}{\underset{\|}{C}}-\underset{\underset{R}{|}}{\bar{C}}COC_2H_5 \longleftrightarrow RCH_2\overset{O^-}{\underset{\|}{C}}=\underset{\underset{R}{|}}{C}COC_2H_5 \quad (3)$$

The equilibrium (1) inclines toward the left because the acidity of α hydrogen on the ester ($pK_a = 25$) is weaker than ethanol ($pK_a = 17$). The step (2) is also a equilibrium reaction. The ester condensation reaction can occur because the equilibrium (3) inclines toward the right. This is because the acidity of the hydrogen between two carbonyl groups in the formed β-keto ester ($pK_a = 11$) is stronger than the ethanol.

The sodium salt of β-keto ester may be neutralized by acetic acid to obtain β-keto ester. Ethyl acetoacetate is prepared by this reaction. The catalyst is sodium ethoxide, may also be the metallic sodium. Ethyl acetate and metallic sodium are served as the starting materials in this experiment.

The ethyl acetate is excess and used as the solvent. This is because a small amount of ethanol exists in ethyl acetate (less than 2%). Sodium ethoxide which is produced by the reaction of sodium with ethanol catalyzes this reaction. Ethanol is also produced while the condensation reaction carries on, then it continues to react with sodium. Thus, the reaction proceeds unceasingly until metalic sodium is consumed completely. But if a lot of ethanol and water exist in the ethyl acetate, the yield of ethyl acetoacetate will reduce obviously.

Reaction equation

$$2CH_3COOEt \xrightarrow{C_2H_5ONa} [CH_3COCHCOOEt]^- Na^+ \xrightarrow{H^+} CH_3COCH_2COOEt$$

Equipment   Apparatus for: reflux, extraction/separation, distillation at atmospheric pressure, distillation under reduced pressure.

Materials	
ethyl acetate (FW 88.12)	10 mL (9 g, 102.2 mmol)
soduium (FW 23.0)	0.9 g (39.1 mmol)
toluene (anhydrous)	
50% acetic acid	
saturated aqueous sodium chloride	
anhydrous magnesium sulfate	
benzene	
pH indicator paper	

Chapter 3 Semimicro Synthetic Experiment of Typical Organic Compounds

Procedure

(1) *Preparation of ethyl acetoacetate*

In a dry 50 mL round bottomed flask, place 10mL of anhydrous toluene and 0.9 g of sodium which has been removed any oxide at the surface. Fit the flask with a reflux condenser carrying a anhydrous calcium chloride drying tube, and heat the mixture under reflux until the sodium is melted completely (this process may need 10~15 minutes). Remove the heating source. After stop the reflux, remove the condenser, stopper tightly the flask with a rubber stopper, press the stopper and shake vigorously the flask several times in order that the sodium in the flask is dispersed into uniformly small beads (as the size of millet). The sodium beads solidify rapidly while the toluene in the flask is cooled gradually. After the toluene is cooled to room temperature, pour it into a recovery bottle (do not pour it into the trough to avoid fire hazard). Then add 10 mL of ethyl acetate to the flask immediately and fit a reflux condenser protected with calcium chloride drying tube to the flask quickly. The reaction starts at once, and the reaction solution boils gently. If the reaction does not start at once, heat the flask gently over an asbestos gauze with a small flame to promote the reaction. After the reaction has started, remove the heat source. If the reaction gets too violent, cool the flask with a water bath slightly to stem the reaction.

After the reaction has commenced to slacken, heat the mixture and maintain boil gently until the sodium disappears completely (about 2 hours). If the sodium does not disappear for a long time, add 2 mL of ethyl acetate again. At the end of the reaction the sodium salt of the ethyl acetoacetate formed in the flask will be a brown-red clear solution (sometimes a little yellow-white precipitate may separate in the solution).

Allow the reaction solution to cool, then remove the condenser and add 50% acetic acid (which is prepared by equal volume of acetic acid and water) until the solution is just slightly acid (pH=5~6) with shaking the flask (need about 8 mL). The solid in the solution should be dissolved thoroughly at the moment (if necessary add a little water to dissolve any solid). Transfer the reaction solution to a separatory funnel, add an equal volume of saturated aqueous sodium chloride solution, shake it thoroughly and let it stand for a few minutes. Separate the ester layer, and extract the aqueous layer with 8 mL of benzene. Combine the extract with the ester layer and dry the combined solution with anhydrous magnesium sulfate. Decant the dried organic layer into a dry distilling flask. Remove the benzene and unreacted ethyl acetate by distillation with a water bath. Stop heating when the temperature rises to 95 ℃. Distil the residue from a small distilling flask by vacuum distillation, collecting the fraction boiling at 54~55 ℃/931 Pa (7 mmHg). The yield of the product is about 1.8 g, the percentage yield is about 35.4%.

Pure ethyl acetoacetate is a colorless liquid, b.p. 180.4 ℃ (decomposition), d_4^{20} 1.0282, n_D^{20} 1.4194.

(2) *Property experiment of ethyl acetoacetate*

Ethyl acetoacetate has the properties of both the carbonyl group of ketone and the enol because it has tautomer of keto form and the enol form (it contains 93% keto form and 7% enol form at room temperature).

$$\underset{\text{keto form}}{CH_3\overset{O}{\overset{\|}{C}}-CH_2-\overset{O}{\overset{\|}{C}}-OC_2H_5} \rightleftharpoons \underset{\text{enol form}}{CH_3\overset{OH}{\overset{|}{C}}=CH-\overset{O}{\overset{\|}{C}}-OC_2H_5}$$

There are two coordination centers in the structures. It can form the chelates with some metallic ions. They can be detected qualitatively by this property.

① The reaction of it with 2,4-dinitrophenylhydrazine　　Add 3 drops of the just prepared 2,4-dinitrophenylhydrazine solution to a test tube, and add 2 drops of ethyl acetoacetate, and then warm the mixture gently. After cooling, the yellow precipitation appears.

② The reaction of it with bromine water and ferric chloride　　Add 2 drops of ethyl acetoacetate and 1 drop of 1% ferric chloride solution to a test tube. Observe the color of the solution. And then add several drops of bromine water, shake the tube. Observe the color of the solution again. Allow it to stand for a while, observe the changes of the color. Record these phenomena and explain them.

Notes

1. All apparatus used must be dry in this experiment. During the treatment of metallic sodium, it can not touch water. Cut the sodium quickly to avoid the moisture and oxygen in air.
2. Ethyl acetate must be absolutely anhydrous. But it should contain 1%~2% ethanol. The purification method of ethyl acetate is as follows: wash the ordinary ethyl acetate with saturated calcium chloride solution for several times to remove the partial ethanol in it, then dry it with anhydrous potassium carbonate baked in high temperature oven, and finally distill it in water bath. Collect the fraction boiling at 76~78 ℃. After these process, the required ethyl acetate is obtained (it contains 1%~2% ethanol). If it is analytically pure, use it directly.
3. Add 50% aqueous acetic acid solution to the flask after the sodium reacts completely to avoid the fire hazard.
4. The solid separates while neutralizing the reaction solution with acetic acid solution. The solid disappears gradually until a clear liquid is obtained when the acetic acid is added continuously while stirring. If a small amount of solid is still undissolved under the condition of weak acidity, add a little water to dissolve it. Pay attention to avoid the excessive acetic acid solution being added, otherwise the solubility of ethyl acetoacetate in water would be increased. In addition, if the acidity is excessively high, the by-product, "the dehydration acetic acid", will produce and the yield will be reduced.
5. The vacuum distillation is recommended to collect the ethyl acetoacetate because it is decomposed easily during the distillation at atmospheric pressure and "the dehydration acetic acid" is produced.

"The dehydration acetic acid" dissolves in the ester generally. It will separate as the brown solid when excessive ethyl acetate is distilled off gradually, especially ethyl acetoacetate is distilled off partially under the reduced pressure.

$$\underset{\text{enol form}}{\underset{\displaystyle}{\begin{array}{c}CH_3-C\\H-C\\C-OC_2H_5\\\parallel\\O\end{array}}\!\!\!\!\!\!\!\!\!\!\!\!\!\!\overset{O-H}{}} \qquad \underset{\text{keto form}}{\underset{\displaystyle}{\begin{array}{c}C_2H_5O\\C=O\\CH-C-CH_3\\H\parallel\\O\end{array}}} \longrightarrow \underset{\text{"the dehydration acetic acid"}}{\underset{\displaystyle}{\begin{array}{c}CH_3-C\overset{O}{}C=O\\\!\!\!\!\!CH-CCH_3\\\parallel\\O\end{array}}}+2C_2H_5OH$$

6. Calculate the percentage yield by the used amount of sodium because ethyl acetate is excess.

Chapter 3 Semimicro Synthetic Experiment of Typical Organic Compounds

> **Questions**
>
> 1. What is the effect on the reaction if the used apparatus has not been dried thoroughly in this experiment?
> 2. What are the purposes of adding 50% aqueous acetic acid solution and saturated sodium chloride solution to the reaction solution during the final treatment in this experiment?
> 3. What is the tautomerism? How to prove that ethyl acetoacetate is an equilibrium mixture of two tautomers, enol form and keto form, by experiments?
> 4. What is the condensation agent used in this experiment? What is the rate of the mole for it and the reactants? Which starting material should be used as the standard to calculate the percentage yield?

3.8 Oxidation and reduction reaction

Experiment 28 Preparation of cyclohexanone

Reaction equation

$$\text{C}_6\text{H}_{11}\text{OH} \xrightarrow[\text{H}_2\text{SO}_4]{\text{NaCr}_2\text{O}_7} \text{C}_6\text{H}_{10}\text{O}$$

Equipment Apparatus for: extraction/separation, distillation.

Materials	
cyclohexanol (FW 100.16)	5.00 g (5.2 mL 50 mmol)
sodium dichromate dihydrate (FW 298)	5.25 g (17.6 mmol)
sodium chloride	
concentrated sulfuric acid	
anhydrous magnesium sulfate	
diethyl ether	

Procedure

In a 100 mL beaker, dissolve 5.25 g of sodium dichromate dihydrate in 30 mL of water. Add carefully 4.3 mL of concentrated sulfuric acid with stirring, and cool the deep orange-red solution to 30 ℃.

Place 5.00 g of cyclohexanol in a 100 mL 2-neck flask and add the dichromate solution in one portion to the flask. Swirl the mixture to insure thorough mixing and observe its temperature with a thermometer. The mixture rapidly becomes warm. When the temperature reaches 55 ℃, cool the flask in a cold water bath, and regulate the amount of cooling so that the temperature remains between 55 ℃ and 60 ℃. The temperature falls after 0.5 h. Remove the cold water bath, allow the flask to stand for 1 h, occasional shaking it. The reaction solution becomes greenish black.

Add 30 mL of water and two boiling stones to the flask and assemble an apparatus for distillation. Distill the mixture until about 25 mL of distillate, consisting of water and an upper layer of cyclohexanone, has been collected (cyclohexanone and water can form an azeotrope, the boiling point is 95 ℃). Saturate the aqueous layer with sodium chloride, separate the organic layer, and extract the aqueous layer twice with 2×15 mL portions of diethyl ether. Combine the solvent extracts with the cyclohexanone layer and dry it over anhydrous magnesium sulfate. Filter the dried solution into a distilling flask of suitable size and carefully distill the ether from a water bath. Remove the water-cooled condenser and attach an air-cooled condenser. Distill the residual cyclohexanone at atmospheric pressure and collect the fraction boiling at 151~155 ℃. The yield is 3.0~3.50 g, the percentage yield is 61%~71%.

Pure cyclohexanone is a colorless liquid, b. p. 155.7 ℃, d_4^{20} 0.9478, n_D^{20} 1.4507.

Notes
1. It is substantively a simple steam distillation for distilling the mixture of the product and water. Cyclohexanone and water can form an azeotrope containing cyclohexanone 38.4%, the boiling point is 95 ℃.
2. The distillation yield of the water should not be too much. Otherwise a small amount of cyclohexanone will dissolve in water and result in the loss of it, in spite of using the salting out. The solubility of cyclohexanone in water is 2.4 g/100 mL at 31 ℃. Add sodium chloride to the distillate in order to decrease the solubility of cyclohexanone in water and be favorable to separate the layers.

Questions

1. Why should the mixture of sodium dichromate and concentrated sulfuric acid be cooled below 30 ℃ before use?
2. What is the action of the salting-out?
3. Whether the potassium permanganate can be used to oxidate the cyclohexanol under the condition of the alkaline? What product will be obtained?

Experiment 29 Preparation of adipic acid

Adipic acid is one of the main raw materials for manufacture of nylon 66. It can be prepared either from cyclohexanone via oxidation reaction using potassium permanganate as an oxidizer or from cyclohexanol via oxidation reaction using nitric acid or potassium permanganate as an oxidizer.

Reaction equation

Method 1

$$\text{cyclohexanone} \xrightarrow{KMnO_4} HOOC(CH_2)_4COOH$$

Method 2

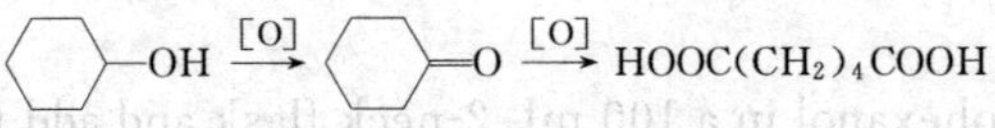

$$\text{cyclohexanol} \xrightarrow{[O]} \text{cyclohexanone} \xrightarrow{[O]} HOOC(CH_2)_4COOH$$

Equipment Mechanical stirrer; apparatus for: reflux, suction filtration, recrystallization.

Materials	
Method 1	
cyclohexanone (FW 98.15)	2 mL (1.9 g, 19.3 mmol)
potassium permanganate (FW 158.04)	6.3 g (40 mmol)
sodium hydroxide solution (0.3 mol/L)	
sodium bisulfite	
concentrated hydrochloric acid	
Method 2	
cyclohexanol (FW 100.16)	2.6 mL (2.5 g, 25 mmol)
potassium permanganate (FW 158.04)	12 g (76 mmol)
sodium carbonate solution (10%)	
concentrated sulfuric acid	

Chapter 3 Semimicro Synthetic Experiment of Typical Organic Compounds

Method 1

Set up a 100 mL 3-neck flask with a mechanical stirrer, thermometer and reflux condenser. Place 6.3 g of potassium permanganate, 50 mL of 0.3 mol/L sodium hydroxide solution and 2 mL of cyclohexanone in the flask. *Note*: *If the reaction temperature rises above 45 ℃ by itself, should cool appropriately the flask in a cold water bath.* Keep the temperature at approximately 45 ℃ for 25 minutes. Then heat the flask over an asbestos gauze until the mixture boils for 5 minutes to complete the oxidation and coagulate the manganese dioxide. Test for residual potassium permanganate by placing a drop of the reaction mixture from the tip of a stirring rod on a piece of filter paper. Unreacted potassium permanganate will appear as a purple ring around the brown manganese dioxide. If unreacted permanganate remains, decompose it by adding small portions of solid sodium bisulfite until the spot test is negative.

Suction-filter the mixture, thoroughly wash the manganese dioxide on the filter with water, and concentrate the filtrate to about 10 mL by transferring it to a beaker, adding a boiling stone, and boiling on a hot plate or over a flame. Acidify the solution with concentrated hydrochloric acid, first by adding acid to pH 1 to 2 (pH paper) and then an additional 2 mL of concentrated hydrochloric acid. Allow the solution to cool to room temperature and collect the crystallized adipic acid by suction filtration. The crude adipic acid can be recrystallized from water and decolorized by charcoal. The white crystals are obtained. The yield is 1.5 g, the percentage is 53%, the melting point is 153~152 ℃.

The pure adipic acid is a white prismatic crystal, m.p. 153 ℃.

Method 2

Fit a 250 mL 3-neck flask with a mechanical stirrer and a thermometer. Add 2.6 mL of cyclohexanol and sodium carbonate solution (dissolve 3.8 g of sodium carbonate in 35 mL of warm water). Add 12 g of powdered potassium permanganate in batches to the flask while stirring and controlling the temperature above 30 ℃. The whole process needs about 2.5 h. After adding, continuously stir the mixture until the reaction temperature no longer rises. Then heat the mixture with stirring for 30 minutes on warm water (50 ℃). A large number of manganese dioxide has been produced in the process of reaction.

Suction-filter the mixture, wash the residue with 10 mL of 10% sodium carbonate solution. Add concentrated sulfuric acid slowly until the solution is strongly acidic and adipic acid has been precipitated. Cool the solution thoroughly collect the crystals of adipic acid by suction filtration and dry it in air. The yield is about 2.2 g, the percentage yield is 60.2%, and the melting point is 153 ℃.

Notes

Method 1

This reaction is itself exothermic. The temperature of the reaction will exceed 45 ℃ in the initial stage of the reaction. If the temperature does not rise to 45 ℃ by itself after 5 minutes, carefully heat the mixture to 40 ℃ in order to make the reaction start.

Method 2

1. If the water in sodium carbonate solution is less, it will affect the stirring effect and cause potassium permanganate can not sufficiently react.
2. If the reaction does not start immediately after adding potassium permanganate, carefully

heat mixture in a water bath. When the temperature rises to 30 ℃, remove the warm water bath at once. The exothermic reaction can carry on automatically.

Questions

1. Why the temperature of the oxidation reaction must be controlled strictly?
2. What is the function of adding sodium carbonate solution to the flask in method 2?

Experiment 30 Preparation of benzoic acid

Benzoic acid and sodium benzoate are important food preservative. Benzoic acid can be used as the pharmaceutical and dye intermediates, also used in the manufacture of plasticizers, initiator in polyester polymerization, spice, etc., in addition, it can be used as the antirusting agent of the steel equipment.

Reaction equations

$$\text{C}_6\text{H}_5\text{CH}_3 + 2\text{KMnO}_4 \longrightarrow \text{C}_6\text{H}_5\text{COOK} + \text{KOH} + 2\text{MnO}_2 + \text{H}_2\text{O}$$

$$\text{C}_6\text{H}_5\text{COOK} + \text{HCl} \longrightarrow \text{C}_6\text{H}_5\text{COOH} + \text{KCl}$$

Equipment   Apparatus for: reflux, suction filtuation, recrystallization.

Materials	
toluene (FW 92.15)	2.7 mL (2.3 g, 25 mmol)
potassium permanganate (FW 158.04)	8.5 g (53.8 mmol)
concentrated hydrochloric acid	
sodium bisulfite	
Congo-red test paper	

Procedure

In a 250 mL round-bottomed flask, place 2.7 mL of toluene and 10 mL of water. Attach a reflux condenser to the flask and heat the flask over an asbestos gauge until the mixture boils. Add in small portions 8.5 g of potassium permanganate from the top of the condenser. Finally wash potassium permanganate adhering to the internal wall of the condenser into the flask with 25 mL of water. Continue the boiling and shake the flask occasionally until the toluene layer almost disappears and the refluxing liquid appears no longer the droplets of oil (require about 4~5 h).

Filter the hot reaction mixture under reduced pressure and wash the residue of manganese dioxide with a small mount of hot water (benzoic acid dissolves in hot water, not cold water). Combine the filtrate and the washings, cool it an ice-water bath, and then acidify it with concentrated hydrochloric acid until the Congo-red test paper becomes blue and benzoic acid separates out.

After cooling thoroughly, filter benzoic acid under reduced pressure, wash it with a little cold water, and allow them to dry thoroughly. The yield of the crude product is about 1.7 g. The crude product can be purified by recrystallization from water.

Pure benzoic acid is a colorless needle-shape crystal, m.p. 122.4 ℃.

Chapter 3 Semimicro Synthetic Experiment of Typical Organic Compounds

Notes
1. If the solution is colored purple by residual permanganate, add a small amount of solid sodium bisulfite and filter again under reduced pressure.

$$KMnO_4 + NaHSO_3 \longrightarrow MnO_2 \downarrow + Na_2SO_4 + K_2SO_4 + H_2O$$

2. If the obtained benzoic acid is not colorless, recrystallize it from appropriate amount of hot water and decolor it with the active charcoal. In 100 g of water the solubility of benzoic acid is 0.18 g at 4 ℃, 0.27 g at 18 ℃, and 2.2 g at 75 ℃.

Questions
1. Are there any other ways to prepare benzoic acid?
2. After the reaction has finished, if the filtrate is purple, why should add the sodium bisulfite to the solution?

Experiment 31 Preparation of *p*-toluidine

Reaction equation

$$p\text{-}CH_3C_6H_4NO_2 \xrightarrow[H^+]{Fe} p\text{-}CH_3C_6H_4NH_2$$

Equipment Mechanical stirrer; apparatus for: reaction with reflux, suction filtration, extraction/separation, distillation.

Materials	
p-nitrotoluene (FW137.14)	4.50 g (32.8 mmol)
reductive ferrum powder (FW 55.84)	7.00 g (125.4 mmol)
ammonium chloride	0.90 g
sodium bicarbonate	
hydrochloric acid (5%)	
sodium hydroxide solution (20%)	
zinc powder	
benzene	

Procedure

　　Place 7.00 g of reductive ferrum powder, 0.90 g of ammonium chloride and 20 mL of water in a 50 mL 3-neck flask equipped with a reflux condenser and a mechanical stirrer. Heat the mixture while stirring and boil it gently over a small flame for 15 minutes. Cool the flask slightly, and then add 4.50 g *p*-nitrotoluene to the flask. Continue the reflux while stirring for 1 h. After the reaction is finished, cool the mixture to room temperature and neutralize it with 5% sodium bicarbonate solution. Add appropriate amount of benzene to the mixture with stirring. Remove the residue of ferrum powder by suction filtration and wash the residue with a little benzene. Pour the filtrate into a separatory funnel, separate the benzene layer, extract the aqueous phase three times with benzene, and then combine the extracts. Extract again the extracts of benzene three times with 5% hydrochloric acid and combine the extracts of hydrochloric acid. Add 20% sodium hydroxide solution to the extract of hydrochloric acid with stirring. The crude product separates. Collect the obtained crude product by suction filtration, wash it with a little water and re-extract the aqueous phase with a little benzene. Combine the extract with the crude product. Distill off the benzene from a water bath. Add a small amount of zinc powder to the residue and distill the residue over an asbestos gauge. Collect the fraction boiling at 198~201 ℃. The yield is about 2.50 g, m.p. 44~45 ℃.

Pure p-toluidine is a colorless sheet-shape crystal, m.p. 44~45 ℃, b.p. 200.3 ℃. p-Toluidine becomes black easily in air and light because of the oxidation.

Notes

1. In this experiment, ferrum-hydrochloric acid is used as the reducer, in which the hydrochloric acid is obtained by the hydrolysis of ammonium chloride.

$$NH_4Cl + H_2O \rightleftharpoons NH_4OH + HCl$$

2. Control the pH value between 7 and 8 while adding sodium bicarbonate to the reaction mixture to avoid the formation of gelatinous ferric hydroxide under the condition of stronger alkaline, which will make the separation difficult.
3. The ferrum residue is the active ferrum mud, containing 44.7% divalent ferrum (calculated by FeO), the black particles. It will liberate heat vigorously in air, thus it should be poured into the waste tank containing water in time.
4. p-Toluidine can be also purified by the recrystallization from ethanol-water except distillation.

Questions

1. Before the reduction reaction, why is the ferrum powder pretreated?
2. During the final treatment, why are the sodium bicarbonate solution and benzene added first, and then extract the layer of benzene with 5% hydrochloric acid?

Experiment 32 Preparation of benzhydrol (diphenyl carbinol)

Reaction equations

Method 1

$$C_6H_5COC_6H_5 \xrightarrow[CH_3OH]{NaBH_4} C_6H_5CH(OH)C_6H_5$$

Method 2

$$C_6H_5COC_6H_5 \xrightarrow{Zn, NaOH} C_6H_5CH(OH)C_6H_5$$

Equipment

Method 1   Apparatus for: reflux, suction filtration, recrystallization.
Method 2   Magnetic stirrer; apparatus for: reflux, suction filtration, recrystallization.

Materials

Method 1
benzophenone (FW 182.2)	0.92 g (5.05 mmol)
sodium borohydride (FW 37.8)	0.12 g (3.2 mmol)
methanol (FW 32.04)	4 mL
petroleum ether (b.p. 60~90 ℃)	

Method 2
benzophenone (FW 182.2)	1.83 g (10.04 mmol)
zinc powder (FW 65.39)	1.97 g (30.1 mmol)
sodium hydroxide (FW 40)	1.97 g (49.3 mmol)
ethanol (95%)	
concentrated hydrochloric acid	
petroleum ether (b.p. 60~90 ℃)	

224

Procedure

Method 1

Dissolve 0.92 g of benzophenone in 4 mL of methanol in a dry, standard-taper ground Erlenmeyer flask. Measure rapidly 0.12 g of sodium borohydride and add it to the flask. Add two boiling stones and attach speedily a reflux condenser to the flask. The temperature of the reaction rises spontaneously until the solution boils. Allow the mixture to stand for 20 minutes at room temperature and shake the contents frequently. Add 1.5 mL of water to the mixture and heat it in a water bath until boil. Keep the boil for 5 minutes. After cooling slightly the flask, place the flask in an ice-water bath. In a few seconds, copious crystals of benzhydrol separate. Collect the crystals by suction filtration and wash them with small amount of cold water and dry them in air. The crude product may be recrystallized from petroleum ether. The percentage yield is 70% ~ 80%, the melting point is 67~68 ℃.

Pure benzhydrol (diphenyl carbinol) is a white, needle-shape crystal, m.p. 69 ℃.

Method 2

In a dry 50 mL round-bottomed flask place 1.97 g of sodium hydroxide, 1.83 g of benzophenone, 1.97 g of zinc powder and 20 mL of 95% ethanol one after another. Shake the contents to dissolve gradually the sodium hydroxide and benzophenone. Attach a reflux condenser to the flask and place the flask in a water bath. Stir the solution. Filter the solution by suction filtration and wash the residue with a small amount of 95% ethanol. Pour the filter liquor into a beaker containing a mixture of 90 mL ice-water and 4 mL concentrated hydrochloric acid. The white precipitate appears immediately. Collect the crude product by suction filtration, wash it with a small amount of water and dry it in air. The crude product is recrystallized from petroleum ether. The white needle-shape crystal is obtained. The yield is 1.4~1.6 g. The percentage yield is about 80%. The m.p. is 67~68 ℃.

Notes

1. Sodium borohydride is strongly basic, deliquescent, and caustic. Weigh it quickly, handle it carefully and do not permit it to touch the skin.
2. In the method 2, control the pH value of the solution is in the range of 5~6 while acidification, otherwise the product separates difficultly.

Questions

1. What is the difference in the reductive character between lithium aluminum hydride ($LiAlH_4$) and sodium borohydride ($NaBH_4$).
2. In the method 1, why is 1.5 mL of water added to the flask, and heat the mixture to boil, and then again cool it until crystallization after the reaction has finished?

3.9 Diazo reaction

The reaction of primary aromatic amines with sodium nitrite results in the formation of a diazonium salt, in the presence of excess and cold aqueous solution of inorganic acid (hydrochloric acid or sulfuric acid is used usually). This process is called diazo reaction.

$$\text{ArNH}_2 + 2\text{HX} + \text{NaNO}_2 \xrightarrow{0\sim5\ ℃} \text{ArN}_2^+ \text{X}^- + \text{NaX} + 2\text{H}_2\text{O}$$

A diazonium salt can be prepared by dissolving 1 mol of primary aromatic amine in about 2.5~3 mol of dilute hydrochloric acid in a flask, cooling the solution to 0~5 ℃ in an ice-salt bath, and adding dropwise an aqueous solution of sodium nitrite while keeping the temperature and stirring constantly until the reaction solution makes potassium iodide-starch test paper change blue.

Pay attention to the following points in the preparation of the diazonium salt:

① Control the reaction temperature strictly because the diazotization is an exothermal reaction, most of diazonium salts are unstable and easily decompose at room temperature. The reaction temperature is generally maintained at 0~5 ℃. The diazonium salts of primary aromatic amines with strong electron-withdrawing group (such as nitro- and sulfo-) on the aromatic ring are relatively stable and can be prepared at higher temperature.

② Maintain the acidity of the solution. In the preparation of diazonium salt, the used amount of the acid is 0.5~1 mol more than the theoretical amount. If the amount of the acid is insufficient, the formed diazonium salt will be coupled with the unreacted aromatic amine.

③ Control the amount of sodium nitrite because the diazonium salt will be oxidated by the excess nitrous acid and the yield of it will be reduced. So toward the end of the addition, the solution should be tested constantly for excess nitrous acid using potassium iodide-starch test paper until the test paper changes blue. An excess of nitrous acid can be eliminated by adding a litter area to the solution.

The prepared diazonium salt solution should not be placed too long time, use it for the synthesis of the next step in time. If the reaction continues, it is unnecessary to separate the diazonium salt from the solution.

Diazonium salts have broad application. Its chemical reactions have two types: One type is that the diazonio group (N_2^+) can be replaced by many different groups (such as —H, —OH, —F, —Cl, —Br, —CN, —NO$_2$, —SH, etc.) under the different condition while liberating nitrogen, and the corresponding aromatic compound is yield, such as the **Sandmeyer reaction**. Other type is the coupling reaction, which are widely used for preparing dyes. The coupling reaction is usually carried out in a weak acid or weak alkaline medium.

Experiment 33 Preparation of *p*-chlorotoluene

Reaction equations

$$2\text{CuSO}_4 + 2\text{NaCl} + \text{NaHSO}_3 + 2\text{NaOH} \longrightarrow 2\text{CuCl}\downarrow + 2\text{Na}_2\text{SO}_4 + \text{NaHSO}_4 + \text{H}_2\text{O}$$

$$\text{H}_3\text{C}-\text{C}_6\text{H}_4-\text{NH}_2 + \text{NaNO}_2 \xrightarrow[0\sim5\ ℃]{\text{HCl}} [\text{H}_3\text{C}-\text{C}_6\text{H}_4-\overset{+}{\text{N}}\equiv\text{N}]\text{Cl}^- + \text{NaCl} + \text{H}_2\text{O}$$

$$[\text{H}_3\text{C}-\text{C}_6\text{H}_4-\overset{+}{\text{N}}\equiv\text{N}]\text{Cl}^- \xrightarrow[\text{HCl}]{\text{CuCl}} \text{H}_3\text{C}-\text{C}_6\text{H}_4-\text{Cl} + \text{N}_2\uparrow$$

Equipment Apparatus for: steam distillation, extraction/separation, distillation at atmospheric pressure.

Materials

p-toluidine (FW 107.16)	2.1 g (19.6 mmol)
sodium nitrite (FW 69)	1.50 g (21.7 mmol)

Chapter 3 Semimicro Synthetic Experiment of Typical Organic Compounds

copper sulfate pentahydate (FW 249.68)	6.00 g (24.0 mmol)
sodium bisulfite (FW 104.06)	1.40 g (13.4 mmol)
sodium chloride	
sodium hydroxide	
sodium bicarbonate	
concentrated sulfuric acid	
concentrated hydrochloric acid	
anhydrous calcium chloride	
petroleum ether (b.p. 60~90 ℃)	
starch-potassium iodide test paper	
Congo-red test paper	

Procedure

(1) *Preparation of cuprous chloride solution*

In a 100 mL round-bottomed flask dissolve 6.00 g of copper sulfate pentahydrate and 1.8 g of sodium chloride in 20 mL of water by heating. In a beaker prepare a solution of 1.4 g of sodium bisulfite and 0.9 g of sodium hydroxide in 10 mL of water, and add this solution with swirling to the hot copper sulfate solution (60~70 ℃). The color of the reaction solution has changed from blue-green to light green or colorless and the white powdery solid has been separated. Cool the flask in an ice-water bath. Decant off the upper liquid and wash the precipitated cuprous chloride twice with water. The cuprous chloride is obtained as a white powder. Dissolve the cuprous chloride by adding 10 mL of cold concentrated hydrochloric acid. Cork the flask to minimize oxidation and place it an ice-water bath until it is to be used.

(2) *Preparation of the diazonium salt solution*

In a 100 mL beaker dissolve 2.10 g of *p*-toluidine in 6.0 mL of water and 6 mL of concentrated hydrochloric acid by heating. After cooling the flask slightly, place it in an ice-salt bath with stirring until the mixture becomes pappy and control the temperature of the reaction mixture below 5 ℃. Add the solution dissolved 1.50 g of sodium nitrite in 4.0 mL of water dropwise to the cold suspension of *p*-toluidine hydrochloride by a dropper at such a rate that the temperature of the reaction mixture dose not rise above 5 ℃ all the time. Add a small piece of ice at a pinch to the reaction solution to prevent the temperature from rising. After the vast majority of sodium nitrite solution (85%~90%) has been added, test the reaction solution with a starch-potassium iodide test paper. If the test paper is colored dark blue immediately, it indicates that sodium nitrite is already right amount, need not to add it again. Continue to stir the reaction solution for a while to complete the reaction. Then put the beaker in an ice bath until it is to be used.

(3) *Preparation of p-chlorotoluene*

Pour the cold diazonium salt solution of *p*-toluidine slowly with shaking into the hydrochloric acid solution of cuprous chloride which has been cooled to 0 ℃. A orange-red double salt of the diazonium chloride and cuprous chloride, $[CH_3-C_6H_4-N\equiv N]^+ CuCl_2^-$, is precipitated in the flask after a while and nitrogen is evolved slowly. After adding, allow

227

the mixture to stand at room temperature for 30 minutes, then warm it gradually to 50~60 ℃ with a water bath to decompose the cuprous complex until the nitrogen is evolved no longer.

Arrange an assembly for steam distillation. When the cuprous chloride complex has decomposed completely, distill off p-chlorotoluene by steam distillation. Separate the organic layer, extract the aqueous layer with 2×10 mL portions of petroleum ether (b. p. 60~90 ℃). Combine the extracts with the organic phase, wash them with 2 mL of 5% sodium hydroxide solution, 2 mL of water, 2 mL of concentrated sulfuric acid, 2 mL of water, 2 mL of 5% sodium bicarbonate solution and water one after another until the solution is neutral. Dry the organic layer with anhydrous calcium chloride. After drying, filter the dried liquid into a dry distilling flask, and then distil off the petroleum ether at atmospheric pressure. Then continue to distill the residues and collect the fraction boiling at 158~162 ℃. The yield is 1.50 g, the percentage yield is 60%.

Pure p-chlorotoluene is a colorless and transparent liquid, b. p. 162 ℃, d_4^{20} 1.072, n_D^{20} 1.521.

Notes
1. Sodium bisulfite deteriorates easily. Use the pure sodium bisulfite (the purity is above 90%) in this experiment, otherwise the reduction reaction is incomplete and the yield will be reduced.
2. The quality of cuprous chloride prepared at 60~70 ℃ is better, the granule of cuprous chloride is thick, and it is rinsed easily.
3. If the reaction solution is still blue-green, it indicates that the reduction is incomplete, add sodium bisulfite solution again according to the situation. If the precipitate is tawny, add several drops of hydrochloric acid immediately and shake slightly to change the cuprous hydroxide to cuprous chloride completely. The quantity of adding hydrochloric acid should be controlled because the cuprous chloride dissolves in the acid.
4. When cuprous chloride is washed by water, shake gently, and allow the solid to settle, and then decant off the water layer carefully. Do not shake vigorously, otherwise the settlement is slow because the granule of cuprous chloride is thinner.
5. When cuprous chloride is exposed to air, it is oxidized easily because of the hot and light in air. The diazonium salt decomposes easily when it is placed for a long time. Prepare them at the same time and mix them immediately after preparing. When the diazo reaction closes to the end, the speed of the reaction is slower. So the diazonium salt should be prepared firstly, then the cuprous chloride.
6. When the diazo reaction closes to the end, the speed of reaction is slower. Therefore, after adding one drop of sodium nitrite solution each time, stir the mixture 1~2 minutes and test the reaction solution with a starch-potassium iodide test paper. If the test paper is colored blue, it indicates that nitrous acid is excess (separating iodine meets starch to reveal the blue).

$$2KI + 2HNO_2 + 2HCl \longrightarrow I_2 + 2NO + 2H_2O + 2KCl$$

If the solution of sodium nitrite is added excessively, decompose it with urea.

$$H_2NCONH_2 + 2HNO_2 \longrightarrow CO_2 \uparrow + 2N_2 + 3H_2O$$

7. Maintain the solution to be acidic in the process of the diazotization and make the Congo-

red test paper change blue.
8. When preparing *p*-chlorotoluene, the speed of pouring the diazonium salt solution into the hydrochloric acid solution of cuprous chloride should not be too quickly in order to avoid producing the by-product, azobenzene.
9. The double salt of the diazonium chloride and cuprous chloride is unstable, and it decomposes to produce *p*-chlorotoluene at 15 ℃. The decomposition can be accelerated by heating slightly. But if the temperature of the solution is higher, the tar-like material and *p*-methyl phenol will be produced, and the yield of product will be reduced. If the time is permitted, place the solution of double salt overnight at room temperature, and then heat it to decompose in a water bath. Stir the reaction solution unceasingly to avoid it overflowing because of producing a great amount of nitrogen in the process of the decomposition.
10. The concentrated sulfuric acid can remove the by-product, azobenzene.

Questions

1. What is the diazo-reaction? What use dose it have in organic synthesis?
2. Why dose the diazo-reaction have to be carried out at lower temperature? If the temperature of the reaction is above 5 ℃ or the acidity of the reaction solution is not sufficient, what by-product will form?
3. Can toluene be chlorinated directly to prepare *p*-chlorotoluene?
4. The cuprous chloride is oxidized by nitrous acid in the presence of hydrochloric acid. The evolution of one kind of reddish brown gas may the observed during the oxidation. Try to explain this phenomenon, and express it with a reaction equation. What harm dose this gas have to the human body?
5. Why can the starch-potassium iodine test paper test the presence of nitrous acid? Which reactions have occurred? If sodium nitrite is added excessively, what disadvantage will it have to this reaction?

Experiment 34 Preparation of methyl orange

Reaction equations

$$H_2N-\langle\bigcirc\rangle-SO_3H + NaOH \longrightarrow H_2N-\langle\bigcirc\rangle-SO_3Na + H_2O$$

$$H_2N-\langle\bigcirc\rangle-SO_3Na \xrightarrow[0\sim 5\ ℃]{NaNO_2, HCl} \left[HO_3S-\langle\bigcirc\rangle-\overset{+}{N}\equiv N\right]Cl^-$$

$$\xrightarrow[HAc]{PhNMe_2} \left[HO_3S-\langle\bigcirc\rangle-N=N-\langle\bigcirc\rangle-NHMe_2\right]^+ Ac^-$$

helianthin (red)

$$\xrightarrow{NaOH} NaO_3S-\langle\bigcirc\rangle-N=N-\langle\bigcirc\rangle-NMe_2 + NaAc + H_2O$$

methyl orange (orange)

Equipment Apparatus for: suction filtration, recrystallization.

Materials	
sulfanilic acid dihydrate (FW 209.19)	2.00 g (9.56 mmol)
sodium hydroxide solution 5%	10 mL (263.4 mmol)
sodium nitrite (FW 69)	0.8 g (11.6 mmol)
concentrated hydrochloric acid	
N,N-dimethylaniline (FW 121.18)	

glacial acetic acid	1.3 mL (1.24 g, 10.3 mmol)
sodium hydroxide solution (10%)	
sodium chloride	
ethanol	
diethyl ether	
starch-potassium iodide test paper	
pH indicator paper	

Procedure

(1) *Preparation of the diazonium salt of sulfanilic acid*

In a 100 mL beaker place 2.00 g of sulfanilic acid dihydrate, add 10 mL of 5% sodium hydroxide solution. Warm the mixture in a hot water bath while stirring with a glass rod until it dissolves. Cool the solution to room temperature in an ice-water bath, add 0.8 g of sodium nitrite, and stir until it dissolves. Pour this solution batchwise, with stirring, into a 250 mL beaker containing 13 mL of cold water and 2.5 mL of concentrated hydrochloric acid. Control the reaction temperature below 5 ℃ in the process of the reaction. In a short while the diazonium salt of sulfanilic acid should separate as a finely divided white precipitate. After adding, test the reaction solution with a starch-potassium iodide test paper. Keep the reaction solution in an ice bath for 15 minutes to complete the reaction.

(2) *Coupling*

In a test tube, mix together 1.3 mL of freshly redistilled N,N-dimethlyaniline and 1 mL of glacial acetic acid. Slowly, with stirring, add this solution to the cooled solution of diazotized sulfanilic acid in the 250 mL beaker. After adding, continue to stir the mixture vigorously with a stirring rod for 10 minutes to ensure the completion of the coupling reaction. In a few minutes, a red precipitate of helianthin should form. Next, add 15 mL of 10% aqueous sodium hydroxide solution. Do this slowly and with stirring while still cooling the beaker in an ice bath. Check with litmus or pH indicator paper to make sure the solution is basic. If not, add extra base. The color of product turns to orange. The crude methyl orange is precipitated with fine-granular shape.

Heat the mixture with a boiling water bath for 5 minutes to dissolve most of the newly formed methyl orange. When all of methyl orange is dissolved, cool the mixture slightly, and then cool it in an ice bath. The methyl orange should recrystallize. When the precipitation appears complete, collect the crystals by vacuum filtration with a Büchner funnel, wash them with a small amount of saturated sodium chloride solution, ethanol, and finally with diethyl ether in turn. Recrystallize the crude product from boiling water containing a small amount of sodium hydroxide (about 0.15 g). Usually about 25 mL of hot water will be required for each gram of the crude product to be recrystallized. The orange red slice-shape crystals of methyl orange are obtained. The yield is about 2.5 g, the percentage yield is 75%.

Do not attempt to determine the melting point of this compound because it has no definite melting point.

Dissolve a little methyl orange in water, add a few drops of dilute hydrochloric acid, and then make alkaline again with dilute sodium hydroxide solution. Observe the color changes.

Chapter 3 Semimicro Synthetic Experiment of Typical Organic Compounds

Notes

1. Sulfanilic acid is an amphoteric compound. Its acidity is slightly stronger than the basicity. It exists in the form of acidic inner salt. It can react with base to produce the salt, but hardly react with acid. So it can not dissolve in acid. Because the acidic solution is required in the diazo reaction, the sulfanilic acid should react first with base to produce sodium p-aminobenzene-sulfonate whose solubility in water is bigger.

$$\underset{SO_3^-}{\underset{|}{C_6H_4}}-\overset{+}{N}H_3 + NaOH \longrightarrow \underset{SO_3^-Na^+}{\underset{|}{C_6H_4}}-NH_2 + H_2O$$

2. In the diazo reaction, nitrous acid is produced while acidizing the reaction solution.

$$NaNO_2 + HCl \longrightarrow HNO_2 + NaCl$$

    Meanwhile, p-aminobenzenesulfonic acid will separate from the solution as fine-granular precipitation, and carry on the diazo reaction with the nitrous acid immediately to produce the powdery diazo salt. The reaction solution must be stirred constantly during the reaction to make p-aminobenzensulfonic acid be diazotized completely.

3. Control strictly temperature of the reaction in the range of 0~5 ℃ during the diazo reaction. If the temperature of the aqueous diazonium salt is allowed to rise above 5 ℃, the produced diazonium salt will be easily hydrolyzed to the corresponding phenol, reducing the yield of the desired product. So the prepared diazonium salt is still cooled in an ice-water bath until it is to be used. The prepared diazonium salt of sulfanilic acid is more stable because the strong electron withdrawing sulfo group exists in the *para*-position of the diazo group.

4. In the test with the starch-potassium iodide test paper, if the test paper is not colored blue immediately the extra sodium nitrite solution should be added while stirring thoroughly until the test paper is colored blue just. If the nitrous acid has been excessive, add a small amount of urea to remove it. The nitrous acid has the function of the oxidation and nitrosylation, thus the excessive nitrous acid will give a series of side reactions.

$$2HNO_2 + 2KI + 2HCl \longrightarrow I_2 + 2NO + 2H_2O + 2KCl$$
$$H_2NCONH_2 + 2HNO_2 \longrightarrow CO_2\uparrow + N_2\uparrow + 3H_2O$$

5. The N,N-dimethylaniline is oxidized easily after it is stored for a long time, so use the redistilled one in this experiment.

6. In the coupling reaction, add 10% aqueous sodium hydroxide solution to the mixture solution until the orange precipitate is no longer formed. In this period, the temperature of the reaction mixture is always maintained at 0~5 ℃. Make sure the solution is alkaline by testing it with the pH paper, otherwise the color of the crude methyl orange is poor.

7. The crude product is alkaline. When the temperature of the solution is allowed to rise, the product will deteriorate easily and the color of it will deepen quickly. So control the temperature of dissolving the methyl orange at about 60 ℃. The color of wet methyl orange can also deepen quickly when it is exposed in sunlight. It will dry quickly if washed first with alcohol and then ether.

Questions

1. What is the coupling reaction? Discuss in which medium the coupling reaction can be carried out according to this experiment.
2. Why should sulfanilic acid be changed into the corresponding sodium salt when the diazo salt is prepared in this experiment? Whether the sulfanilic acid can be mixed with hydrochloric acid directly?
3. Why does the N,N-dimethylaniline couple with the diazonium salt at the *para*-position of the ring?
4. Explain the reason that methyl orange changes the color in acidic medium? Express it with reaction equation.

3.10 The Diels-Alder reaction

The Diels-Alder is the 1,4-addition of a conjugated diene to the dienophile to form an adduct that is a six-membered ring containing a double bond. This reaction is not only an important method of synthesizing six-membered organic compound, but also occupies a very important position in theory. The presence of electron-donating groups (such as alkyl or alkoxy) on the diene or of electron-withdrawing group (such as carbonyl, cyano) on the dienophile accelerates the rate of reaction. It is a concerted reaction and does not have any reactive intermediates. It possesses two characters of the reversibility and the stereospecific *cis*-addition.

Experiment 35 The Diels-Alder reaction of cyclopentadiene with maleic anhydride

Reaction equation

Equipment Apparatus for: distillation, fractional distillation, recrystallization, suction filtration.

Materials	
cyclopentadiene (FW 66.1)	2.4 g (3 mL, 36 mmol)
maleic anhydride (FW 98.06)	3 g (31 mmol)
petroleum ether (b.p. 60~90 ℃)	
ethyl acetate	

Procedure

Place 3 g of maleic anhydride in a 50 mL Erlenmeyer flask and dissolve it in 10 mL of ethyl acetate by warming in a hot water bath. Add 10 mL of petroleum ether and then cool the solution in an ice bath. Add 2.4 g of freshly distilled cyclopentadiene to the maleic anhydride solution and swirl the solution to mix in the ice water bath until the white product crystallizes from solution and the initial exothermic reaction is over. Heat the flask in a water bath until the product has dissolved, and then allow the solution to cool slowly and stand undisturbed. The white needle-shape crystals are obtained. Collect the product by suction filtration and wash it with 5 mL the mixed solvent of ethyl acetate and petroleum ether. After drying, weigh it. The yield is about 4.00 g, the percentage yield is 78.7% and the melting point is 164~165 ℃.

This adduct has retained a carbon-carbon double bond, so it can make the solution of potassium permanganate or the solution of bromine in carbon tetrachloride depigmentize.

Notes

1. Cyclopentadiene easily dimerizes to form the dimer at room temperature. Fortunately, a balance between cyclopentadiene and the dimer can be formed at 170 ℃ (the boiling point of the dimmer). Thus, monomeric cyclopentadiene can be obtained by 'cracking' the dimer by heat.

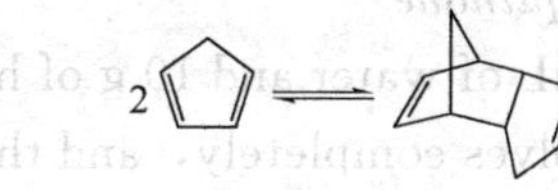

Place 10 g of dicyclopentadiene in a 100 mL round-bottomed flask, then fit a Vigrenx fractionating column 30 cm in length, stillhead with thermometer (100 ℃), condenser, and receiver containing calcium chloride. Cool the receiving flask in an ice bath and heat the distillation flask with an oil bath until the dicyclopentadiene start to boil gently (b. p. 170 ℃) and the monomer begins to distill steadily in the range of 40~42 ℃. Control the temperature of distilling vapor below 45 ℃. Collect the fraction of 40~45 ℃. The freshly distilled cyclopentadiene should be used immediately, otherwise it reverts to the dimer.

2. Maleic anhydride should be recrystallized before use if it is stored for a long time. Place 10 g of maleic anhydride and 15 mL of chloroform in an Erlenmeyer flask. Heat and boil the solution until the solid dissolves, and then filter the hot solution. Allow the filtrate to cool and stand until crystallization is complete. Collect the crystals by suction filtration and dry it in desiccator. The melting point of pure maleic anhydride is 60 ℃.

Question

What should be noticed in the preparation of cyclopentadiene from dicyclopentaodiene? Why?

Experiment 36 The Diels-Alder reaction of cyclopentadiene with 1,4-benzoquinone

Reaction equations

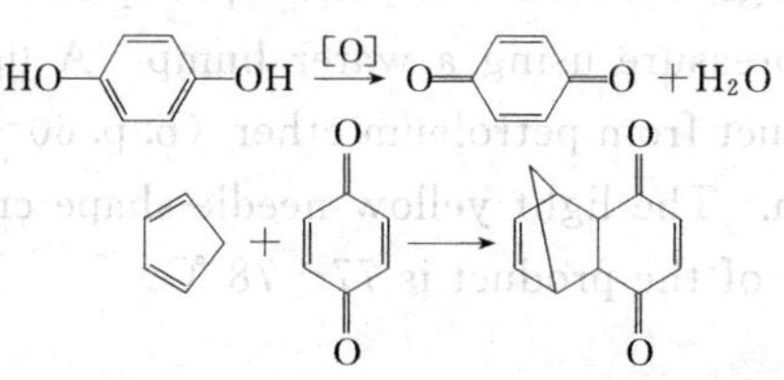

Equipment   Magnetic stirrer, rotary evaporator; apparatus for: extraction/separation, suction filtration, distillation under reduced pressure with a water bump.

Materials	
hydroquinone (FW 110.11)	10 g (90.8 mmol)
sodium dichromate (FW 298)	14 g (47 mmol)
concentrated sulfuric acid	

benzene	
anhydrous calcium chloride	
1,4-benzoquinone (FW 108.10)	1.8 g (17 mmol)
cyclopentadiene (FW 66.1)	
petroleum ether (b. p. 60~90 ℃)	1.2 g (1.5 mL, 18 mmol)
ethanol	

Procedure

(1) *Preparation of 1,4-benzoquinone*

Warm to 50 ℃ a mixture 200 mL of water and 10 g of hydroquinone in a 250 mL beaker. Stir the solution until the solid dissolves completely, and then cool the solution to room temperature. Dropwise add 5.4 mL of concentrated sulfuric acid with stirring. Here the color of the solution becomes yellow. Add in small portions, with swirling, a solution of 14 g sodium dichromate in 6.5 mL of water. The degree of viscosity of the reaction mixture accretes gradually, the color gradually becomes yellowish-green, and a small amount of greenish-black precipitate separates. Cool the reaction mixture to 10 ℃, collect the crystals of quinone with suction and suck them as dry as possible. Extract the aqueous phase with 2× 15 mL portions of benzene, dissolve the crude product with 50 mL of benzene, combine the benzene solution and dry it over anhydrous calcium chloride. An orangeish-yellow transparent solution is obtained after several hours. Filter off the drying agent, evaporate the mass of benzene on the rotary evaporator until a small amount of the bright yellow crystals of quinone separate out, and then cool the residue until the crystals completely separate out. Collect the crystals of quinone with suction. After drying, weigh it. The percentage yield is 70%. Pure 1,4-benzoquinone is a bright yellow crystals, m. p. 115~116 ℃.

(2) *Preparation of cyclopentadien-1,4-benzoquinone adduct*

Place 1.8 g of pure 1,4-benzoquinone and 7 mL of ethanol in a 100 mL Erlenmeyer flask. A yellow suspension liquid is obtained. Cool the flask in an ice bath to 0~5 ℃, then rapidly add 1.5 mL of freshly distilled cyclopentadiene to the liquid. Swirl the flask and place it in an ice bath. Remove the flask from the ice bath after 15 minutes and allow the flask to stand for 45 minutes at room temperature. The solution changes from cloudy to the clear orangeish-yellow and a great amount of yellow precipitate separate concomitantly. Distill ethanol under reduced pressure using a water bump. A light yellow solid is obtained. Recrystallize the crude product from petroleum ether (b. p. 60~90 ℃) at 70 ℃. Collect the crystals by suction filtration. The light yellow needle-shape crystals are obtained. Dry the product. The melting point of the product is 77~78 ℃.

Notes

1. Hydroquinone will become a thick material in the presence of sulfuric acid when it is impure. Thus, filter to remove the insoluble substance. If you use commercial 1,4-benzoquinone, it should be purified by sublimation. It assumes the purple black frequently because of impure.
2. Stir rapidly the solution to raise the yield of product because the solution changes gradually thick.
3. Evade the light to store 1,4-benzoquinone because it changes the color easily. 1,4-Benzoquinone is extremely volatile, therefore it is not suitable for higher temperature in the

course of dissolving it with benzene.

Questions

1. Which configuration is the obtained adduct, *endo* or *exo*?
2. What is the mole rate of 1,4-benzoquinone and cyclopentadiene in this experiment? If the used amount of cyclopentadiene is increased, what product will be obtained? Write the structural formula.

Experiment 37 The Diels-Alder reaction of anthracene with maleic anhydride

Reaction equation

Equipment Apparatus for: reflux, suction filtration, recrystallization.

Materials
anthracene (FW 178.24)	2 g (11.2 mmol)
maleic anhydride (FW 98.06)	1 g (10.2 mmol)
anhydrous xylene	

Procedure

In a dry 50 mL round-bottomed flask dissolve 2 g of pure anthracene and 1 g of maleic anhydride in 25 mL of anhydrous xylene, and then add two boiling stones into the flask. Attach a water-cooled reflux condenser protected with a calcium chloride drying tube, heat the mixture under reflux for 25~30 minutes, swirl the contents occasionally in the course of the reflux to fall the crystals adhering to the wall of the flask into the solution. After the reaction is complete, allow the flask to stand and cool it to room temperature, a great amount of the white solid separates simultaneously. Collect the white crystals with suction filtration. Recrystallize the crude product from anhydrous xylene. Dry the product with a vacuum desiccator containing wax flakes and silica gel. The yield is about 2 g, m.p. 262~263 ℃ (decomposition).

The melting point of the pure product is 263~264 ℃.

Notes

1. Because maleic anhydride and the adduct will hydrolyze to become binary acid when met water, all apparatus and reagents must be dry.
2. Because the wax can absorb the hydrocarbon gas, use it to remove the trace of xylene in the product.
3. The product should be kept in the desiccator, or else the product will hydrolyze to form binary acid because of absorbing the water in air and the determination of the melting point is difficult too.

Questions

1. What influence would it have on this experiment if xylene had not been dried?

2. Why can anthracene and maleic anhydride carry on the Diels-Alder reaction and the reaction occurs in the 9,10-position of anthracene?

3.11 Phase-transfer catalytic reaction

In organic synthesis, homogeneous reaction carries on easily than heterogeneous reaction. However, the he heterogeneous reaction that the two reactants are in different phases is common in organic synthesis. Because the reactants can not close together each other, the yield will be reduced, the reaction rate will be slower, even the reaction hardly occurs. Here, the method of phase transfer catalysis can be used, which makes the hardly miscible reactants in two phase react or the reaction be accelerated by a phase-transfer catalyst. And this process is called phase-transfer catalytic reaction. The phase-transfer catalyst (abbreviated as PTC) should have two basic conditions: ①can transfer a reagent from one phase to another; ②make the transferred reagent active.

Commonly used phase transfer catalysts have three categories: quaternary salt, crown ether and noncyclic polyether.

① Quaternary salt includes quaternary ammonium salt, quaternary phosphonium salt, etc. In this family, the alkyl radical is lipophilic, but the anion is more strongly hydrophilic. The cation volume of alkyl radical should be moderate. If the volume is too big, the solubility of the salt in water will be reduced. On the contrary, it will increase the solubility in water and effect the catalytic reaction.

② Crown ether can make some metal ions dissolve in organic phase by complexing with them. Crown ether is usually limited use because of the expensive price and strong toxicity.

③ The phase-transfer catalytic mechanism of noncyclic polyether is similar to crown ether. It can also complex with some metal ions such as polyethylene glycol (PEG). When PEG appears bending, it looks like crown ether. The PEG whose relative molecular mass is in the range 400~600 has a moderate aperture, a strong ability of complexing with metal ion and a better catalytic effect.

Experiment 38 Preparation of 7,7-dichlorobicyclo [4.1.0] heptane
(7,7-dichloronocarane)

Reaction equation

$$\text{cyclohexene} \xrightarrow[\text{TEBA}]{\text{NaOH, H}_2\text{O, CHCl}_3} \text{7,7-dichlorobicyclo[4.1.0]heptane}$$

Equipment Magnetic stirrer; apparatus for: reflux with stirring, extraction/separation, distillation under reduced pressure.

Materials	
cyclohexene (FW 82.14)	1.62 g (2.0 mL, 20.0 mmol)
chloroform (FW 119.4)	20 mL
benzyl triethyl ammonium chloride (TEBA)	0.3 g
sodium hydroxide	4.00 g (100 mmol)

anhydrous magnesium sulfate
pH indicator paper

Procedure

Set up a 50 mL 3-neck flask with a reflux condenser, thermometer, pressure-equalizing dropping funnel and a magnetic stirrer bar. Place 2.0 mL of freshly redistilled cyclohexene, 0.3 g of phase transfer catalyst (TEBA) and 10 mL of chloroform in the flask. Stir the mixture vigorously with a magnetic stirrer and dropwise add the solution dissolved 4.00 g of sodium hydroxide in 4 mL of water to the mixture from the dropping funnel. The reaction is exothermic during the addition. After adding, heat the mixture to reflux with the water bath for 40 minutes with vigorous stirring. The color of the reaction solution becomes yellow and the solid separates.

Allow the reaction solution to cool to room temperature, add 10 mL of water to the solution to dissolve the solid. Transfer the biphasic solution to a separatory funnel. Separate the layers, saving the chloroform layer. Extract the aqueous layer once with 10 mL of chloroform. Combine this extract with organic layer. Wash the combined organic phase with 3×10 mL portions of water until it is neutral to pH indicator paper. Dry the organic layer with anhydrous magnesium sulfate. Distill off chloroform and unreacted cyclohexene in a water bath. Distill the residue under reduced pressure in a small flask and collect in a weighed bottle the portion boiling at 80～82 ℃/2.13 kPa (16 mmHg). The yield is 1.80 g.

Pure 7,7-dichlorobicyclo [4.1.0] heptane is a colorless and transparent liquid, b.p. 197～198 ℃, n_D^{20} 1.5014.

Notes

1. This experiment belongs to the addition reaction of carbene with alkene. 7,7-Dichlorobicyclo [4.1.0] heptane is synthesized by the reaction of phase-transfer catalysis. Other phase-transfer catalyst can be also used for this experiment, such as tetrabutylammonium bromide, polyethylene glycol-400 (PEG-400), *et al*. The circulation equations of phase-transfer catalysis in this experiment are as follows:

   aqueous phase $\quad PhCH_2\overset{+}{N}Et_3Cl^- + NaOH \rightleftharpoons PhCH_2\overset{+}{N}Et_3OH^- + NaCl$

   $\quad\quad\quad\quad\quad\quad\quad\quad\quad\quad \Big\updownarrow \quad\quad\quad \Big\updownarrow CHCl_3$

   organic phase $\quad PhCH_2\overset{+}{N}Et_3Cl^- + :CCl_2 \rightleftharpoons PhCH_2\overset{+}{N}Et_3CCl_3^- + H_2O$

2. Use the freshly redistilled cyclohexene.
3. The reaction temperature should be controlled between 50～55 ℃. If the temperature is below 50 ℃, the reaction is incompletely. If the temperature is above 60 ℃, the color of reaction solution will deepen, the reaction solution will become thick, the raw material or carbene will volatilize and the yield of the product will decrease.
4. Because this reaction is carried out in two-phase, stir the reactants vigorously to increase the yield of the product.

5. Much floccule often suspends in the reaction solution during stratification. Remove it by the suction filtration with a Büchner funnel.
6. Separating the layers, the aqueous layer should be separated as completely as possible. The organic layer should be dried thoroughly.

> **Questions**
>
> 1. Why can dichlorocarbene carry on the addition reaction with cyclohexene in the presence of water in this experiment?
> 2. What is the function of phase-transfer catalyst in this experiment?
> 3. What is the purpose of stirring vigorously the mixture during adding sodium hydroxide solution?

Experiment 39   Preparation of *p*-methyl phenylthioacetic acid

Reaction equation

$$\text{HS-}\underset{\text{CH}_3}{\text{C}_6\text{H}_4}\text{-} + \text{ClCH}_2\text{COOH} \xrightarrow[\text{PEG-400}]{\text{NaOH, CH}_3\text{CN}} \text{HOOCCH}_2\text{S-}\underset{\text{CH}_3}{\text{C}_6\text{H}_4}\text{-}$$

Equipment   Mechanical stirrer; apparatus for: reflux with stirring, distillation, suction filtration, recrystallization.

Materials

p-thiocresol (FW 124.18)	2.48 g (20 mmol)
chloroacetic acid (FW 94.5)	1.89 g (20 mmol)
sodium hydroxide (FW 40)	2.0 g (50 mmol)
polyethylene glycol-400	
acetonitrile	
concentrated hydrochloric acid	
petroleum ether (b. p. 60~90 ℃)	
Congo-red test paper	

Procedure

Place 2.48 g of *p*-thiocresol, 1.89 g of chloroacetic acid, 2.0 g of sodium hydroxide, 100 mL of acetonitrile and 0.4 g of polyethylene glycol-400 (PEG-400) in a dry 250 mL 3-neck flask equipped with a reflux condenser and a mechanical stirrer. Heat the mixture to reflux for 3 hours with stirring in an oil bath at about 125 ℃. Then distill off the acetonitrile thoroughly in a boiling water bath, Cool the flask and add 40 mL of water to the flask to dissolve the sodium salt. Transfer the solution to a beaker, and then add concentrated hydrochloric acid dropwise to the solution with stirring until the Congo-red test paper becomes blue. The product separates immediately. Filter off the product by suction and wash the filter cake with a little cold water.

After drying, recrystallize the crude product from petroleum ether (b. p. 60~90 ℃) and dry it again, Weigh the product, determine its melting point, and calculate the yield.

The melting point of pure *p*-methyl phenylthioacetic is 93.5~94.5 ℃.

Notes
1. *p*-Thiocresol is toxic, irritant and stenchy, measure it in the hood.
2. Distill off acetonitrile in water bath and place it in the recovery bottle.

3.12 Ultrasonic radiation reaction

Ultrasonic wave, a new technology to activate and promote chemical reactions, has been developed after middle period of 1980s. This technology which is applied and researched in the field of chemistry and chemical engineering, has formed an emerging cross-discipline, Ultrasonic chemistry. Because of the research and development of ultrasonic chemistry, this technology has begun to move from the laboratory to the industrialization. The sound wave whose frequency is in the range of 20~1000 kHz is called the ultrasonic wave. The used ultrasonic wave frequency in organic synthesis is usual at 20~80 kHz. The synthetic chemistry mainly uses the cavitation effect of ultrasonic wave. A high energy environment of high temperature and high pressure is produced in the micro-area of medium and the extremely short time, accompanied by powerful shock wave and the micro-jet flow, as well as discharge and radiation, etc. It creates an extreme physical environment for promoting and opening the pass of chemical reactions. Ultrasonic wave can not only increase reaction rate, initiate the reaction easily, reduce the harsh reaction conditions, but also can change the reaction pathway and selectivity. Once this technology is used to the reaction with important economic value, it will produce huge application prospect.

3.12.1 The Grignard reaction

Magnesium metal can react with alkyl and aryl halides in anhydrous diethyl ether or tetrahydrofuran to yield organomagnesium halides. They are called Grignard reagents which are invented by Victor Grignard, a French chemist. Grignard reagents react with the unsaturated compounds easily, especial the substances containing double bong which is between carbon atom and other atom, by nucleophilic addition reaction. This process is called Grignard reaction. Grignard reagents are extraordinarily useful and versatile compounds in organic synthesis and used to synthesize various types of compounds such as alkanes, alcohols, aldehydes, carboxylic acids, etc.

In the synthesis of Grignard reagents, the traditional way requires absolutely anhydrous diethyl ether and a small amount of iodine, as a inductor. While under ultrasonic radiation, this reaction does not require the anhydrous diethyl ether of special treatment, the rate of the reaction is quick, the reaction have no induction periods and the yield is high.

Experiment 40 Preparation of triphenylmethanol (triphenylcarbinol)

Reaction equations

Method 1 The reaction of benzophenone with phenylmagnesium bromide

$$\text{Ph-Br} \xrightarrow{\text{Mg, Et}_2\text{O}}_{)))} \text{Ph-MgBr} \xrightarrow[\text{Et}_2\text{O, }))))]{\text{Ph-CO-Ph}} \text{Ph}_3\text{C-OMgBr} \xrightarrow{\text{H}^+,\ \text{H}_2\text{O}} \text{Ph}_3\text{C-OH}$$

Method 2 The reaction of ethyl benzoate with phenylmagnesium bromide

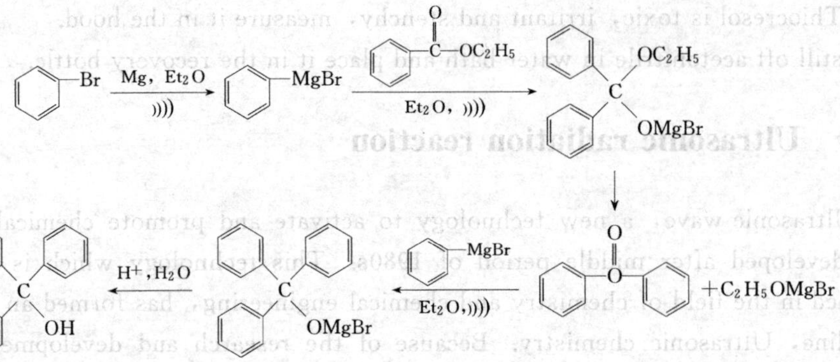

Ph-Br $\xrightarrow{\text{Mg, Et}_2\text{O},)))}$ Ph-MgBr $\xrightarrow{\text{PhCO}_2\text{C}_2\text{H}_5,\ \text{Et}_2\text{O},)))}$ Ph-C(OC$_2$H$_5$)(Ph)(OMgBr)

$\downarrow$

Ph$_3$C-OH $\xleftarrow{\text{H}^+,\text{H}_2\text{O}}$ Ph$_3$C-OMgBr $\xleftarrow{\text{PhMgBr},\ \text{Et}_2\text{O},)))}$ Ph-C(=O)-Ph + C$_2$H$_5$OMgBr

Secondary reaction

Ph-MgBr + Ph-Br $\longrightarrow$ Ph-Ph

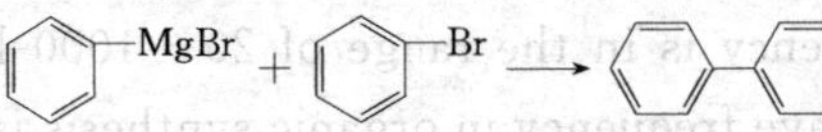

Equipment Mechanical stirrer; apparatus for: stirring with addition and reflux, distillation, suction filtration, recrystallization.
Instrument Ultrasonic wave cleaning instrument.

Materials	
Method 1	
bromobenzene (FW 157.02)	2.7 mL (4.04 g, 25.7 mmol)
magnesium turnings (FW 24.3)	0.7 g (28.8 mmol)
benzophenone (FW 182.21)	4.5 g (25.0 mmol)
iodine crystal	a small piece
diethyl ether (dry)	
sulfuric acid (20%)	
ethanol (95%)	
petroleum ether (b. p. 90~120 ℃)	
Method 2	
bromobenzene (FW 157.02)	1.7 mL (2.54 g, 16.2 mmol)
magnesium turnings (FW 24.3)	0.4 g (16.5 mmol)
ethyl benzoate (FW 150.18)	0.9 mL (0.95 g, 6.3 mmol)
iodine crystal	a small piece
diethyl ether (dry)	
sulfuric acid (20%)	
ethanol (95%)	
petroleum ether (b. p. 90~120 ℃)	

Procedure

Method 1 *The reaction of benzophenone with phenylmagnesium bromide*

  All apparatus must be thoroughly dried in a hot (>120 ℃) oven before use. Fix a 250 mL 3-neck flask in the ultrasonic wave cleaning instrument. Add water to the trough of the cleaning instrument (about 5~8 cm in height). Add 0.7 g of magnesium turnings and 5 mL of anhydrous diethyl ether (the reagent bottle is opened just) to the flask provided with a reflux condenser and a pressure-equalizing dropping funnel. Add about 1 mL of the liquid mixture containing 2.7 mL of bromobenzene and 10 mL of anhydrous diethyl ether to the magnesium from the dropping funnel. Ultrasonic wave radiates the mixture in the flask about 1~2 minutes, then stop the radiation and add a small crystal of iodine to the flask. The reaction is initiated at once (if the reaction does not start immediately, warm the flask gently in a bath of warm water). After the reaction has started, the iodine color disappears gradually, the turbidity appears and the solution in the flask spontaneously boils. When the initial vigorous reaction has slowed down, add dropwise the remainder of bromobenzene and diethyl ether to the previously activated reaction mixture, at such a rate that the ether refluxes gently without external heating. Radiate the contents in the flask intermittently by ultrasonic wave during dropping the bromobenzene solution. After all of the bromobenzene solution has been added (about 40 minutes), continuously radiate the contents in the flask by Ultrasonic wave about 5 minutes in order to react completely. The gray phenylmagnesium bromide, Grignard reagent, is obtained.

Add very slowly the mixed solution containing 4.5 g of benzophenone and 13 mL of anhydrous diethyl ether to the Grignard reagent while radiating intermittently by ultrasonic wave, and add at intervals extra anhydrous diethyl ether to insure good mixing and effective cooling. The addition product of phenylmagnesium bromide with benzophenone separates as a white precipitate. After all of the benzophenone has been added, continue the radiation by ultrasonic wave about 10 minutes to react completely (**caution**: The temperature of the water in the trough of ultrasonic wave cleaning instrument should not surpass 25 ℃ during the radiation).

Remove the ultrasonic wave cleaning instrument, put and fix the flask in an ice-water bath, and then add 20% sulfuric acid solution while stirring by a mechanical stirrer in order to decompose the addition product into triphenylmethanol. Separate the ether layer and distill off diethyl ether in a water bath. Add 10 mL of petroleum ether (b. p. 90~120 ℃) to the residue in the flask and stir the mixture about 10 minutes. The white crystals separate during the stir. Collect the crude product with suction filtration and wash it with a little cold solvent twice. Recrystallize the crude product from the mixed solvent of petroleum ether (b. p. 90~120 ℃) and 95% ethanol. Cool the solution to room temperature, collect the crystals with suction filtration and wash them with a little cold solvent. After drying, weigh the product, determine its melting point, and calculate the yield.

Pure triphenylmethanol is a white and lamellar crystal, m. p. 164.2 ℃.

Method 2   The reaction of ethyl benzonate with phenylmagnesium bromide

All apparatus must be thoroughly dried in a hot (>120 ℃) oven before use. Fix a 50 mL 3-neck flask in the ultrasonic wave cleaning instrument. Add water to the trough of the cleaning instrument (about 5~8 cm in height). Add 0.4 g of magnesium turnings and 2 mL of anhydrous diethyl ether (the reagent bottle is opened just) to the flask provided with a reflux condenser and a pressure-equalizing dropping funnel. Add about 1 mL of the liquid mixture containing 1.7 mL of bromobenzene and 7 mL of anhydrous diethyl ether to the magnesium from the dropping funnel. Ultrasonic wave radiates the mixture in the flask about 1~2 minutes, then stop the radiation and add a small crystal of iodine to the flask. The reaction is initiated at once (if the reaction does not start immediately, warm the flask gently in a bath of warm water). After the reaction has started, the iodine color disappears gradually, the turbidity appears and the solution in the flask spontaneously boils. When the initial vigorous reaction has slowed down, add dropwise the remainder of bromobenzene and diethyl ether to the previously activated reaction mixture, at such a rate that the ether refluxes gently without external heating. Radiate the contents in the flask by ultrasonic wave intermittently during dropping the bromobenzene solution. After all of the bromobenen solution has been added (about 40 minutes), continuously radiate the contents by ultrasonic wave for 5 minutes in order to react completely. The gray phenylmagnesium bromide, Grigard reagent, is obtained.

Add very slowly the mixed solution containing 0.9 mL of ethyl benzoate and 2 mL of anhydrous diethyl ether to the Grignard reagent while radiating by ultrasonic wave intermittently, and add at intervals extra anhydrous diethyl ether to insure good mixing and effective cooling. After all of the benzophenone has been added, continue the radiation by ultrasonic wave for 10 minutes to react completely (**caution**: The temperature of the water in the trough of ultrasonic wave cleaning instrument should not surpass 25 ℃ during the radiation).

Remove the ultrasonic wave cleaning instrument, put and fix the flask in an ice-water bath, then add 20% sulfuric acid solution (about 15 mL) while stirring in order to decompose the addition product into triphenylmethanol. Separate the ether layer and distill off diethyl ether in a water bath. Add 5 mL of petroleum ether (b. p. 90~120 ℃) to the residue in the flask and stir the mixture about 10 minutes. The white crystals separate during the stir. Collect the crude product with suction filtration and wash them with a little cold solvent twice. Recrystallize the crude product from the mixed solvent of petroleum ether (b. p. 90~120 ℃) and 95% ethanol. Cool the solution to room temperature, collect the crystals with suction filtration and wash them with a little cold solvent. After drying, weigh the product, determine its melting point and calculate the yield.

Notes

1. The temperature of the water in the trough of ultrasonic wave cleaning instrument should not surpass 25 ℃ in the course of ultrasonic wave radiation, otherwise the ultrasonic cavitation effect will be weakened, the yield of the product will be reduced and diethyl ether is also volatilized easily.
2. Use the opened just bottle absolute diethyl ether in this experiment, the absolute diethyl ether need not be treated specially. All the apparatuses must be dried thoroughly before use.
3. Ultrasonic wave radiation is not used at the initial stage of the reaction in order to keep the local concentration of halohydrocarbon be higher in reaction solution. This will be favorable for initiating the reaction. But if the concentration of halohydrocarbon is always higher in the course of whole reaction, the coupled secondary reaction occurs easily. So ultrasonic wave radiation is kept intermittently after the reaction is initiated, and the speed of dropping halohydrocarbon should be very slow.

$$RMgBr + RBr \longrightarrow R-R + MgBr_2$$

4. The by-product can be removed because it dissolves in petroleum ether easily.

Questions

1. What applications does Grignard reaction have in organic synthesis?
2. What secondary reactions may occur in this reaction?

3. What is the effect on this experiment if the adding rate of bromobenzene is too fast or it is added once?
4. What characters of ultrasonic wave are used to accelerate the reaction speed in synthetic chemistry?
5. Compared with the classical method, what advantages does this synthetic method have?

Experiment 41 Preparation of 2-methyl-1-phenyl-2-propanol

2-Methyl-1-phenyl-2propanol has the slightly sweet, delicate fragrance of the medicinal herbs and the smell of fresh flower such as roses. It is applied widely in chemical industry of daily use and edible essence.

Reaction equation

$$\text{PhCH}_2\text{Cl} \xrightarrow{\text{Mg, Et}_2\text{O}} \text{PhCH}_2\text{MgCl} \xrightarrow{\text{CH}_3\text{COCH}_3} \text{PhCH}_2-\underset{\underset{\text{CH}_3}{|}}{\overset{\overset{\text{CH}_3}{|}}{\text{C}}}-\text{OMgCl} \xrightarrow[\text{H}_2\text{O}]{\text{NH}_4\text{Cl}} \text{PhCH}_2-\underset{\underset{\text{CH}_3}{|}}{\overset{\overset{\text{CH}_3}{|}}{\text{C}}}-\text{OH}$$

Equipment Mechanical stirrer; apparatus for: stirring with addition and reflux, distillation at atmospheric pressure, distillation under reduced pressure.
Instrument Ultrasonic wave cleaning instrument.

Materials	
acetone (dry) (FW 58.08)	2.0 mL (1.58 g, 27.2 mmol)
benzyl chloride (FW 126.59)	3.2 mL (3.52 g, 27.8 mmol)
magnesium turnings (FW 24.3)	0.7 g (28.8 mmol)
iodine crystal	a small piece
diethyl ether (dry)	
saturated ammonium chloride	
saturated sodium bicarbonate	
anhydrous potassium carbonate	

Procedure

All apparatus must be thoroughly dried in hot ($> 120\ ℃$) oven before use. Fix a 250 mL 3-neck flask in the ultrasonic wave cleaning instrument. Add water to the trough of the cleaning instrument (about 5~8 cm in height). Add 0.7 g of magnesium turnings and 5 mL of anhydrous diethyl ether to the flask provided with a reflux condenser and a pressure-equalizing addition funnel. Add about 1 mL of the liquid mixture containing 3.2 mL of benzyl chloride and 10 mL of diethyl ether to the magnesium from the dropping funnel. Ultrasonic wave radiates the mixture in the flask about 1~2 minutes, then stop the radiation and add a small crystal of iodine to the flask. At this time the reaction is initiated at once. After the reaction has started, the bubbles are generated on the surface of magnesium, the solution in the flask spontaneously boils, the iodine color disappears gradually and the turbidity appears. When the initial vigorous reaction had slowed down, add dropwise the remainder of benzyl chloride and dicey ether to the previously activated reaction mixture, at such a rate that the ether refluxes gently without external heating. Radiate intermittently the contents in the flask by ultrasonic wave during dropping the benzyl chloride solution. After all of the benzyl chloride solution has been added (about 40 minutes), continuously radiate the contents by ultrasonic wave about 5 minutes in order to react completely. The gray benzyl magnesium chloride, Grignard reagent, is obtained.

Add very slowly the mixed solution containing 2.0 mL of acetone and 6.0 mL of anhydrous diethyl ether to the Grignard reagent while radiating intermittently by ultrasonic wave and add at intervals extra anhydrous diethyl ether to insure good mixing and effective cooling. The addition product of benzyl magnesium chloride with acetone separates as a white precipitate. After all of the acetone has been added (about 20 minutes), continue the radiation by ultrasonic wave about 10 minutes to react completely (**caution**: The temperature of the water in the trough of ultrasonic wave cleaning instrument should not surpass 25 ℃ during the radiation).

Remove the ultrasonic wave cleaning instrument, put and fix the flask in an ice-water bath, then add 25 mL of saturated ammonium chloride solution while stirring by a mechanical stirrer in order to decompose the addition product into 2-methyl-1-phenyl-2-propanol. Separate the ether layer, and extract the aqueous layer with diethyl ether twice. Combine all the ether layers and wash them with saturated sodium bicarbonate solution and water respectively. Dry the ether solution over anhydrous potassium carbonate. Distill off diethyl ether with a hot water bath. Transfer the residue to a small distillation set, and distil the liquid under reduced pressure using a vacuum pump, collecting the product boiling at 88~90 ℃/533 Pa (4 mmHg).

Pure 2-methyl-1-phenyl-2-propanol is a colorless oily liquid, b.p. 88~90 ℃/400~533 Pa.

Note

See the experiment 40.

3.12.2 Aldol condensation reaction

In the presence of dilute base or acid, there is an intermolecular condensation reaction involving two molecules of carbonyl compound with α hydrogen atom, aldehyde or ketone. The product is a β-hydroxyaldehyde when an aldehyde is used or a β-hydroxyketone when a ketone is the carbonyl compound. When the temperature of the reaction is raised, the resulting compound loses water spontaneously to produce an α,β-unsaturated aldehyde or α,β-unsaturated ketone. These reactions are called **aldol condensations**. It is an important method to preparing α,β-unsaturated carbonyl compounds and an important reaction to lengthen the carbochain in organic synthesis too.

The aldol condensation involves the self-condensation and crossed aldol condensation. Crossed aldol condensations of an aromatic aldehyde having no α hydrogen and an aldehyde or ketone having α hydrogen produce α,β-unsaturated aldehydes or α,β-unsaturated ketones. These crossed aldol condensation reactions are called **Claisen-Schmidt reactions**.

Experiment 42 Preparation of benzalacetophenone

Reaction equation

$$PhCHO + CH_3COPh \xrightarrow[\text{))))}, 25\sim30\ ℃]{NaOH\ (10\%),\ EtOH} PhCH=CHCOPh + H_2O$$

Chapter 3 Semimicro Synthetic Experiment of Typical Organic Compounds

Equipment Apparatus for: suction filtration, recrystallization.
Instruments Ultrasonic wave cleaning instrument.

Materials	
acetophenone (FW 120.16)	1.00 mL (1.03 g, 8.57 mmol)
benzaldehyde (FW 106.13)	0.8 mL (0.83 g, 7.82 mmol)
sodium hydroxide solution (10%)	
ethanol (95%)	
pH indicator paper	

Procedure

In a 50 mL Erlenmeyer flask place 2.1 mL of 10% sodium hydroxide solution, 2.5 mL of ethanol. Cool the mixture to room temperature, shake well, and add 0.8 mL of freshly redistilled benzaldehyde. Fix the flask in the ultrasonic wave cleaning instrument whose trough has been added some water in advance. The water surface in the cleaning trough should be slightly higher than the liquid level in the flask. Turn on the ultrasonic wave cleaning instrument, and radiate the mixture in the flask by ultrasonic wave for about 30~35 minutes. Control the temperature of the water in the trough at 25~30 ℃ during the radiation. Then chill the reaction mixture in an ice-water bath for thirty minutes or longer to crystallize completely. Collect the product with suction filtration and wash it thoroughly with ice water, until the washings are neutral to pH indicator paper. Recrystallize the crude product from 95% ethanol. The yield is about 85%, m.p. 55~57 ℃.

Notes

1. If the reaction temperature is higher than 30 ℃ or less than 15 ℃, it is unfavorable to the reaction.
2. Benzoic acid may be produced by the automatic oxidation of benzaldehyde if it is stored for a long time. Benzaldehyde should be redistilled before use.
3. Because of the lower melting point of the product, the oil beads may appear during the recrystallization, add the solvent again until the solution is homogeneous.
4. The sodium hydroxide should not be excessive in order to avoid producing a lot of polymer and reducing the yield of the product.
5. Determine the temperature of water in the trough with a thermometer after turning off the ultrasonic wave cleaning instrument.

Questions

Outline the merit of synthesizing benzalacetophenone by the method of the ultrasonic wave radiation through consulting other related experimental books or the literature.

3.13 Leuckart Reaction

The α-amine alcohol is formed by the reaction of ammonia with aldehyde or ketone, which is unstable, dehydrates to form an imine. The imine is then reduced to form the amine product through the catalytic hydrogenation. This is an important method of synthesizing the amine from carbonyl compound.

$$\underset{\alpha\text{-amino alcohol}}{\overset{NH_2}{\underset{OH}{-C-}}} \xrightarrow{-H_2O} \underset{\text{imine}}{\overset{NH}{-C=}} \xrightarrow{H_2/Ni} \underset{\text{amine}}{\overset{NH_2}{\underset{H}{-C-}}}$$

(from: $\overset{O}{\underset{}{C}} + NH_3 \longrightarrow$)

If the formic acid is used as the reducing agent in this experiment instead of the hydrogen gas and a nickel catalyst, the process of this reductive amination is called the **Leuckart reaction**. The formic acid, or formate ion, is acting as a reducing agent because its hydrogen atom is being transferred to the imine as a hydride ion.

$$\overset{NH}{\underset{}{-C=}} + H - \overset{O}{\underset{}{C}} - \ddot{\underset{}{O}}^- \xrightarrow{-CO_2} \overset{:\ddot{N}H^-}{\underset{H}{-C-}} \xrightleftharpoons{H_2O} \overset{NH_2}{\underset{H}{-C-}} + OH^-$$

Because the hydride may be introduced from either side of the imine molecule, the product of the Leuckart reaction is racemic. If we are to obtain one of the enantiomers in optically pure form, the raceme must be resolved.

Actually, because of the presence of the formate ion, the free α-phenylethylamine is not formed directly above the reaction, but rather the hydride transfer process takes place with the formation of α-phenylethylformamide. The formamide must be hydrolyzed by the acid and neutralized by the base in a subsequent step, then the amine is obtained.

Experiment 43   Preparation of (±)-α-phenylethylamine

In this experiment (±)-α-phenylethylamine will be prepared by the Leuckart Reaction using acetophenone and ammonium formate as starting materials.

Reaction equations

$$Ph-\overset{O}{\underset{}{C}}-CH_3 + 2HCOONH_4 \longrightarrow Ph-\overset{NHCHO}{\underset{}{CHCH_3}} + 2H_2O + CO_2 + NH_3$$

$$Ph-\overset{NHCHO}{\underset{}{CHCH_3}} + H_2O + HCl \longrightarrow Ph-\overset{NH_3Cl^-}{\underset{}{CHCH_3}} + HCOOH$$

$$Ph-\overset{NH_3Cl^-}{\underset{}{CHCH_3}} + NaOH \longrightarrow Ph-\overset{NH_2}{\underset{}{CHCH_3}} + NaCl + H_2O$$

(±)-α-phenylethylamine

Equipment   Apparatus for: reflux, extraction/separation, distillation steam distillation.

Materials	
acetophenone (FW 120.16)	11.7 mL (12.0 g, 100 mmol)
ammonium formate (FW 63.03)	20.3 g (322 mmol)
benzene	
concentrated hydrochloric acid	
sodium hydroxide	
ethanol (95%)	
ferric chloride solution (1%)	
pH indicator paper	

Chapter 3 Semimicro Synthetic Experiment of Typical Organic Compounds

Procedure

Assemble a 100 mL 3-neck flask, fitted with a thermometer (immerge it in the reaction mixture), a distillation head and a condenser as for a simple distillation. Place 11.7 mL of acetophenone, 20.3 g of ammonium formate and two boiling stones into the flask. Heat slowly the reaction mixture using a heating mantle. During this period of heating, the mixture will melt into two phases, and the distillation will begin. When the temperature of the reaction mixture reaches to 150~155 ℃, the mixture will become homogeneous, and the reaction will take place. Continue gentle heating until the temperature of the reaction mixture reaches 185 ℃. Do not heat the mixture beyond this temperature. About 1~1.5 hours of heating are usually required to reach this stage. A little acetophenone has been distilled. Stop the heating and transfer the distillate to a separatory funnel. Remove the acetophenone layer by and return it to the reaction flask, resume the heating, maintain the temperature at 180~185 ℃ for 1~1.5 hours.

Allow the reaction mixture to cool, and then add 10 mL of water to the flask with shaking to ensure complete mixing. Transfer this solution to a separatory funnel, rinsing the reaction flask with 10 mL of water. Combine the washings with the solution (what does the water wash off?). Separate the organic layer [the crude (±)-α-phenylethylamine] and return it to the original reaction flask. Extract the aqueous layer with 2×10 mL portions of benzene. Save the benzene extracts and discard the aqueous layer. Combine the benzene extracts with the α-phenylethylamine, add 12 mL of concentrated hydrochloric acid and two boiling stones. Heat the mixture until all of the benzene has distilled. Reflux the solution for an additional 30 minutes.

Cool the reaction solution to room temperature. Extract it with 2×10 mL portions of benzene (what is washed off?). Transfer the aqueous layer to a 250 mL round-bottomed flask which has been equipped for steam distillation. Add slowly the solution dissolved 10 g of sodium hydroxide in 10 mL of water and begin the steam distillation. Continue the steam distillation until the distillate, which is initially basic, is no longer basic when tested with pH indicator paper. About 80 mL distillate will be obtained.

Extract the distillate, which contains the free amine, with 3×15 mL portions of benzene. Combine the extracts and dry it with some sodium hydroxide pellets. Remove the solvent with a water bath using a simple distillation. Transfer the residue to a small flask and distil it under reduced pressure, collect the fraction boiling at 82~83 ℃/2.4 kPa (18 mmHg). (±)-α-Phenylethylamine is obtained, the yield is about 5 g, the percentage yield is about 41%.

The pure (±)-α-phenylethylamine is a colorless liquid, b. p. 187.4 ℃, n_D^{20} 1.5238.

Notes
1. Do not heat the reaction mixture beyond 185 ℃, otherwise some solid ammonium carbonate may form in the condenser. During the period of distillation, some solid ammonium carbonate will form in the distillation head and the condenser. If it is more, the condenser becomes plugged with this solid. At the moment, stop heating, wash off the solid with water, and then resume the heating.
2. Before the beginning of the steam distillation, make sure that the ground glass joints are well greased to prevent freezing of the joints caused by the basic solution.

3. Store the α-phenylethylamine in a tightly stoppered flask because it readily absorbs atmospheric carbon dioxide.
4. Because α-phenylethylamine has stronger corrosion behavior, do not determine the refractive index.

Questions

1. Why is the α-phenylethylamine produced by the Leuckart reaction only a raceme?
2. Why is the solution made basic before the steam distillation of the reaction mixture?
3. Why is the reaction temperature controlled strictly to not surpass 185 ℃ ?
4. Write the mechanism to prepare the (±)-α-phenylethylamine by Leuckart reaction.

3.14 Resolution of racemic compound

Raceme is a mixture mixed by equivalent enantiomers. Their physical and chemical properties (except chiral condition) are general the same besides the opticity. It is impossible to separate the raceme by distillation, extraction, recrystallization, etc..

At present, there are many methods for the resolution of raceme. To change the raceme into diastereomers, a general method in the laboratory is that make an optical compound react with a given raceme. Then the diastereomers are separated and refined by the fractional crystallization according to the different solubility in a selected solvent. Finally, remove the resolving agent and obtain the pure optical isomers.

An optical alkaloid, such as (−)-ephedrine, (−)-brucine, etc., is generally used to resolve the acidic raceme. And the basic raceme is resolved by an optical acid, such as tartaric acid, camphor-β-sulfonic acid, etc..

Experiment 44 Resolution of (±)-α-phenylethylamine

(±)-α-Phenylethylamine belongs to basic raceme. It may be resolved by an acidic resolving agent such as tartaric acid. In this experiment the diastereoisomeric salts, (−)-α-phenylethylamine-(+)-tartrate salt and (+)-α-phenylethylamine-(+)-tartrate salt, are formed by the reaction tartaric acid with (±)-α-phenylethylamine. The two diastereoisomeric salts have sufficiently different solubilities in methanol. The former have a lower solubility than the latter in methanol. So the method of the fractional crystallization may be used, the (−)-α-phenylethylamine-(+)-tartrate salt first separates from the solution as crystals. The pure (−)-α-phenylethylamine can be obtained by treating the salt with purification and basification. The pure (+)-α-phenylethylamine can be also obtained by treating the (+)-α-phenylethylamine-(+)-tartrate salt in mother liquor with the similar method.

Reaction equations

$$\underset{\underset{Ph}{|}}{\overset{\overset{CH_3}{|}}{H}}\!\!-\!\!NH_2 + H_2N\!\!-\!\!\underset{\underset{Ph}{|}}{\overset{\overset{CH_3}{|}}{H}} + \underset{\underset{COOH}{|}}{\overset{\overset{COOH}{|}}{H}}\!\!-\!\!\overset{OH}{\underset{H}{|}} \longrightarrow B(-)A(+) + B(+)A(+)$$

$$\underbrace{B(-) \qquad B(+)}_{(\pm)\text{-}\alpha\text{-phenylethylamine}} \qquad \underbrace{A(+)}_{(+)\text{-tartaric acid}} \qquad \underbrace{}_{\text{diastereomeric salts}}$$

Chapter 3 Semimicro Synthetic Experiment of Typical Organic Compounds

$$\begin{array}{c}\text{CH}_3\\ \text{H}\!\!-\!\!\!\overset{+}{\text{NH}_3}\\ \text{Ph}\end{array} \quad \begin{array}{c}\text{COO}^-\\ \text{H}\!\!-\!\!\text{OH}\\ \text{HO}\!\!-\!\!\text{H}\\ \text{COOH}\end{array} \xrightarrow{2\text{NaOH}} \begin{array}{c}\text{CH}_3\\ \text{H}\!\!-\!\!\text{NH}_2\\ \text{Ph}\end{array} + \begin{array}{c}\text{COONa}\\ \text{H}\!\!-\!\!\text{OH}\\ \text{HO}\!\!-\!\!\text{H}\\ \text{COONa}\end{array}$$

$\underbrace{\text{B}(-) \qquad \text{A}(+)}_{\text{less soluble salt in CH}_3\text{OH}}$ $(-)$-α-phenylethylamine
$[\alpha]_D^{22} = -40.3°$

$$\begin{array}{c}\text{CH}_3\\ \text{H}_3\overset{+}{\text{N}}\!\!-\!\!\text{H}\\ \text{Ph}\end{array} \quad \begin{array}{c}\text{COO}^-\\ \text{H}\!\!-\!\!\text{OH}\\ \text{HO}\!\!-\!\!\text{H}\\ \text{COOH}\end{array} \xrightarrow{2\text{NaOH}} \begin{array}{c}\text{CH}_3\\ \text{H}_2\text{N}\!\!-\!\!\text{H}\\ \text{Ph}\end{array} + \begin{array}{c}\text{COONa}\\ \text{H}\!\!-\!\!\text{OH}\\ \text{HO}\!\!-\!\!\text{H}\\ \text{COONa}\end{array}$$

$\underbrace{\text{B}(+) \qquad \text{A}(+)}_{\text{more soluble salt in CH}_3\text{OH}}$ $(+)$-α-phenylethylamine
$[\alpha]_D^{22} = +40.3°$

Equipment Apparatus for: reflux, suction filtration, recrystallization, extraction/separation, distillation under reduced pressure.
Instrument Polarimeter.

Materials	
$(\pm)$-α-phenylethylamine (FW 121)	3 g (3.25 mL, 25 mmol)
L-$(+)$-tartaric acid (FW 150.09)	3.8 g (25 mmol)
methanol	
diethyl ether	
sodium hydroxide solution (50%)	

Procedure

(1) *Fractional crystallization*

Place 3.8 g of L-$(+)$-tartaric acid, 50 mL of methanol and two boiling stones in a 100 mL Erlenmeyer flask. Fit the flask with a reflux condenser. Heat this mixture with a water bath until the solid dissolves completely. Cool the flask slightly, slowly add 3 g of racemic α-phenylethylamine to the warm solution with a Pasteur pipette while swirling the flask to mix well. After dropping, stopper the flask and allow it to stand quietly for at least 24 hr at room temperature. The white prismatic crystals should form as the solution cools. Filter the crystals by suction filtration and wash them with a very small amount of cold methanol. Save the obtained solution because it can be processed for eventual recovery of the $(+)$-amine. Spread the crystals on paper to dry. After drying, weigh the product, and calculate the percentage yield of $(-)$-amine $(+)$-hydrogen tartrate.

Pure $(-)$-α-phenylethylamine-$(+)$-hydrogen tartrate is a white prismatic crystals, m.p. 179~182 ℃ (decompose), $[\alpha]_D^{22} = 13°$ (H_2O, 8%).

(2) *Isolation of amine*

① The recovery of $(-)$-α-phenylethylamine Dissolve the obtained crystalline amine-tartrate salt in 10 mL of water, add 2 mL of 50% sodium hydroxide solution. Swirl the mixture until all of the crystals have dissolved. Extract this mixture with 3×10 mL portions of diethyl ether using a separatory funnel. Combine the extracts from each extraction in a stoppered Erlenmeyer flask and dry the ether layer over some sodium hydroxide pellets or anhydrous magnesium sulfate. Decant the dried ether solution into a clean, dry distilling flask. Remove the ether by simple distillation using a water bath. Transfer the residue to a small distilling flask and distil it under reduced pressure (combine the crude products obtained by several groups because the crude product obtained by one group is a little), collect the fraction boiling at 81~81.5 ℃/2.4 kPa (18 mmHg). Weigh the weight of amine collected, calculate the yield, determine the specific rotation, and calculate the optical purity.

The reported specific rotation of pure (−)-α-phenylethylamine is $[\alpha]_D^{22}=-40.3°$.

② The recovery of (+)-α-phenylethylamine The mother liquor removed (−)-amine-(+)-tartrate salt, which contains mostly the (+)-amine-(+)-tartrate salt, contain methanol. Remove the methanol by simple distillation using a water bath. The white solid, that is the (+)-amine-(+)-tartrate salt, is obtained. May treat it using the same operation method in the method① to yield the (+)-amine. Hydrolyze this salt with water and 50% sodium hydroxide solution, isolate the free amine by extraction with diethyl ether, dry the ether layer over some sodium hydroxide pellets, distill off the ether using a water bath. Distill the liquid residue by vacuum distillation. Collect the fraction boiling at 85~86 ℃/2.8 kPa (21 mmHg). Weigh the collected (+)-amine, calculate the yield, determine the specific rotation, and calculate the optical purity.

The reported specific rotation of pure (+)-α-phenylethylamine is $[\alpha]_D^{22}=+40.3°$.

Notes
1. When preparing (−)-α-phenylethylamine-(+)-hydrogen tartrate, if the obtained crystals are not prismatic crystals, but needle-shape crystals or the mixture of prismatic and needles, the mixture should be heated with a water bath while shaking until most of the solid has dissolved. The needle-shape crystals dissolve easily and usually a small amount of the prismatic crystals will remain to seed the solution. Allow the solution to cool slowly at room temperature and form prismatic crystals. The α-phenylethylamine obtained by the needle-shape crystals do not have a sufficiently high optical purity.
2. The optical purity of an optical active substance may be calculated by the following expression.

$$\text{optical purity} = \frac{\text{the measured specific rotation}}{\text{the specitic rotation of the pure enatiomer}} \times 100\%$$

3. Do not inhale the vapor of methanol as it is toxic, or it will blind the both eyes.

Questions

1. After (±)-α-phenylethylamine is added to the methanol solution of L-(+)-tartaric acid, the prismatic crystals separate out. Whether the filtrate has the optical rotation?
2. What is the key step in the resolution experiment? How is the reaction condition controlled in order to resolve the enantiomers completely?

3.15 Photochemical reaction

The chemical reaction which can be occurred due to the molecule excited by light, is called photochemical reaction.

Ultraviolet light and visible light can cause the photochemical reaction generally, the wavelength of light is in the range of 200~700 nm. The substances which contain unsaturated bonds, such as alkenes, aldehydes, ketones, etc. generally can carry on the photochemical reactions.

When ultraviolet light and visible light irradiate an organic molecule, it can cause the electron transition. That is, an electron from one of the bonding orbitals or nonbonding orbitals is excited to an antibonding orbital, usually to the one of higher energy. So the original **ground-state** molecule changes into an excited state molecule.

The majority of organic molecules in the ground-state is in singlet state (the ground-state singlet state is recorded as S_0, at this time all the electrons are paired). When such a molecule absorbs ultraviolet light of the appropriate wavelength, an electron is excited. Because the electron must retain its spin value during this transition (because a change of spin is quantum-mechaically forbidden process during an electronic transition), a **singlet state** is produced (the first excited state of a molecule is recorded as S_1). But the excited state singlet S_1 is very unstable, so the excited electron may undergo a change of spin to give a thermodynamically stable **triplet state** (the excited triplet state is recorded T_1, at this time the two electrons are spin parallel). The conversion from the first excited singlet state to the triplet state is called **intersystem crossing** (it is recorded as ISC). The excited state singlet S_1 may return to the ground state S_0 by reemission of the absorbed photon of energy. This process is called **fluorescence**.

The triplet state may return to the ground state by emitting **phosphorescence** (its wavelength is longer than that of the fluorescence), or a process called a **radiationless transition** to dissipate the excess energy. The two ways of the transition $T_1 \rightarrow S_0$ would require a change of spin for the electron, and this is a forbidden process. Hence, the triplet excited state usually has a longer lifetime than that of the singlet state. Many photochemical reactions are carried out when the molecules of the reactant are at the triplet state. So the triplet state is very important in the photochemistry.

Experiment 45 Preparation of benzopinacol

The photoreduction of benzophenone is one of the most thoroughly studied photochemical reactions. If benzophenone is dissolved in a "hydrogen-donor" solvent, such as 2-propanol, and exposed to ultraviolet light, an insolube dimeric product, benzpinacol, will form.

Experiments show that the photoreduction of benzophenone is a reaction of the n-π* triplet state (T_1) of benzophenone. When the 2-propanol solution of benzophenone is exposed to ultraviolet light in the range of 300~350 nm, only the π electron of the carbonyl π bond in benzophenone can absorb the photon of energy, but 2-propanol can not. A nonbonding electron from the highest-energy occupied orbital in the carbonyl π bond happens the n-π* transition to the triplet state (T_1) via the singlet state (S_1) and intersystem crossing. Because the triplet state has a longer half life period and considerable energy (314~334.7 kJ/mol), it can abstract a hydrogen atom on the *sec*-carbon from 2-propanol. Then the C—H bond on the *sec*-carbon happens homolytic cleavage. The diphenylhydroxymethyl radicals and isopropyl radicals are formed. Two of diphenylhydroxymethyl radicals, once formed, may couple to form benzpinacol. The complete mechanism for photoreduction is outlined in the steps that follow.

Reaction equation

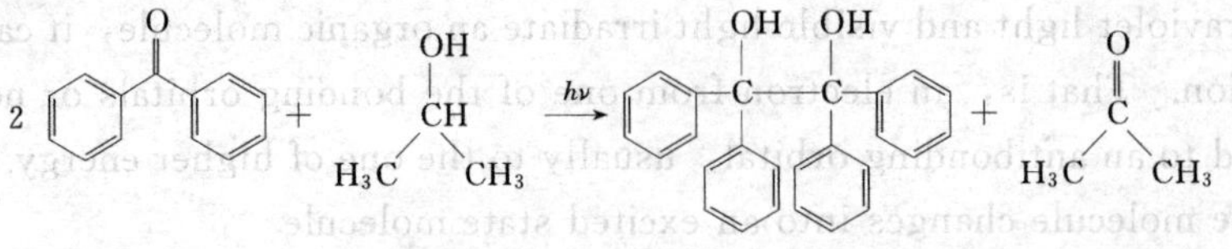

$$2 \text{ } Ph-CO-Ph + (CH_3)_2CHOH \xrightarrow{h\nu} Ph_2C(OH)-C(OH)Ph_2 + (CH_3)_2C=O$$

Mechanism

(reaction scheme showing benzophenone excitation from $S_0 \to S_1 \to T_1$, hydrogen abstraction from isopropanol by the triplet diradical, generating diphenylhydroxymethyl radical and 2-hydroxy-2-propyl radical; the triplet ketone abstracts a second H from isopropanol, and two diphenylhydroxymethyl radicals couple to form benzpinacol)

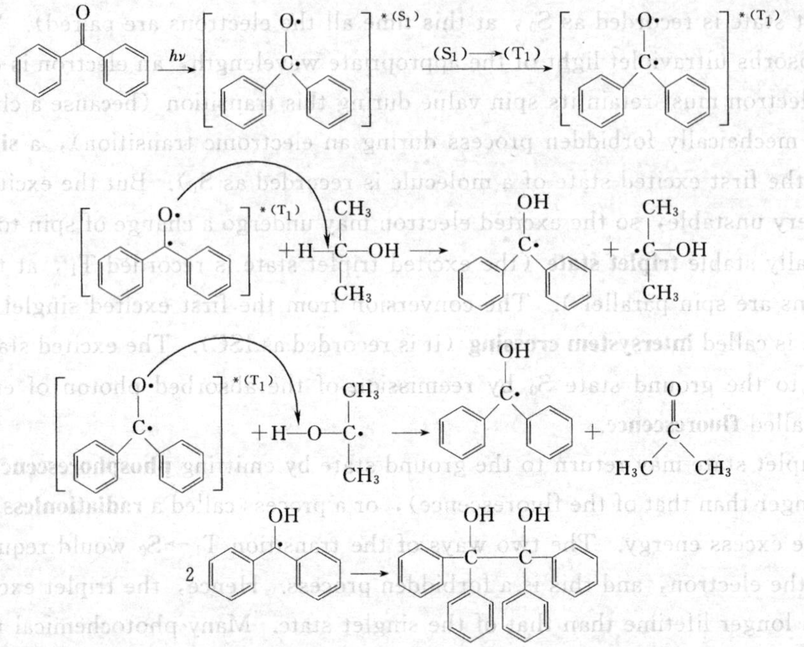

Equipment Apparatus for: recrystallization, suction filtuation.

Materials

benzophenone (FW 182.21)	0.5 g (2.7 mmol)
isopropanol (FW 60.11)	
glacial acetic acid (FW 60.05)	

Procedure

Add 0.5 g of benzophenone and 3 mL of isopropanol to a 10 mL test tube, dissolve the solid in a warm water bath. Add one drop of glacial acetic acid to the tube, shake thoroughly, and then add sufficient isopropanol to fill the tube in order to exclude the air in the tube. Stopper the test tube tightly with a rubber stopper, place it in a beaker on a windowsill where they will receive direct sunlight for a week. A large amount of colorless crystals have been separated from the solution. Collect the colorless crystals with suction filtration and the crude product from glacial acetic acid. After drying, weigh it, determine its melting point and calculate the yield.

Pure benzpinacol is a colorless and needle-shape crystal, m. p. 188~190 ℃ (decomposition).

Notes

1. In order to eliminate the traces of alkali from the glassware, add one drop of glacial acetic into the test tube. If there were the traces of alkali, benzopinacol would be decomposed

Chapter 3 Semimicro Synthetic Experiment of Typical Organic Compounds

into diphenyl carbinol and benzophenone.
The reaction equalations are as follow.

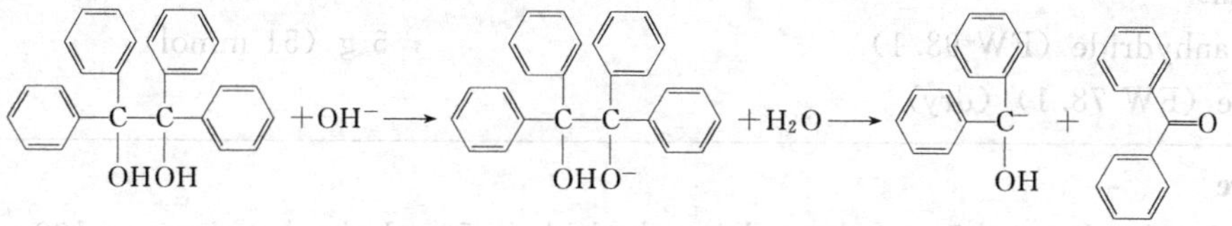

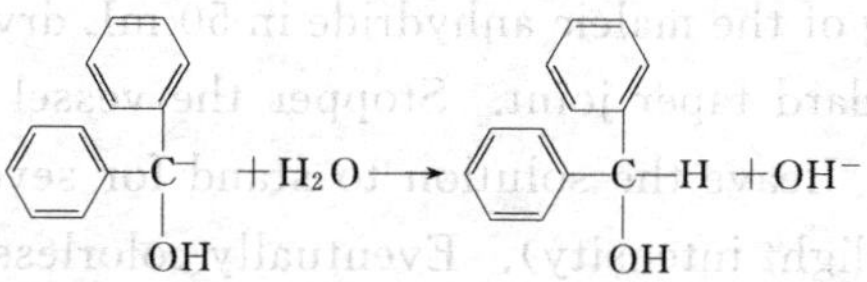

$$Ph_2C(OH)-C(OH)Ph_2 + OH^- \longrightarrow Ph_2C(OH)-C(O^-)Ph_2 + H_2O \longrightarrow Ph_2C(OH)^- + Ph_2C=O$$

$$Ph_2C(OH)^- + H_2O \longrightarrow Ph_2CH(OH) + OH^-$$

2. In the photochemical reaction benzophenone can produce radicals, but because of the presence of oxygen, the radicals will be consumed, the reaction rate will be reduced, furthermore the oxygen make the reaction be complicated. So the test tube should be filled with the solvent to exclude the oxygen from air in the test tube.
3. Because the photochemical reactions are carried out mainly at the thin layer solution near the wall of the container, shaker the test tube frequently to prevent crystallizing on the wall of the test tube. The extent of the reaction relates to the beaming time.
4. Select a smaller reaction container in order to raise the yield, because hither concentration can promote the bimolecular reaction.

Questions

1. State the reduction mechanism of benzophenone in this photochemical reaction.
2. If the mouth of the test tube is not stoppered, what is the influence on the reaction?
3. If glacial acetic acid is not added before the reaction, what is the influence on the experimental result? Write out the relevant equations.
4. A reaction similar to the one here described occurs when benzophenone is treated with the metal magnesium.

$$2Ph_2C=O \xrightarrow{Mg} Ph_2\underset{|}{C}(OH)-\underset{|}{C}(OH)Ph_2$$

Compare the mechanism of this reaction with the photoreduction mechanism. What are the differences?
5. Benzopinacol will rearrange under the acid catalysis. Try to write the rearrangement product and the reaction mechanism.

Experiment 46 Addition of benzene to butenedioic anhydride
(maleic anhydride)

Reaction equation

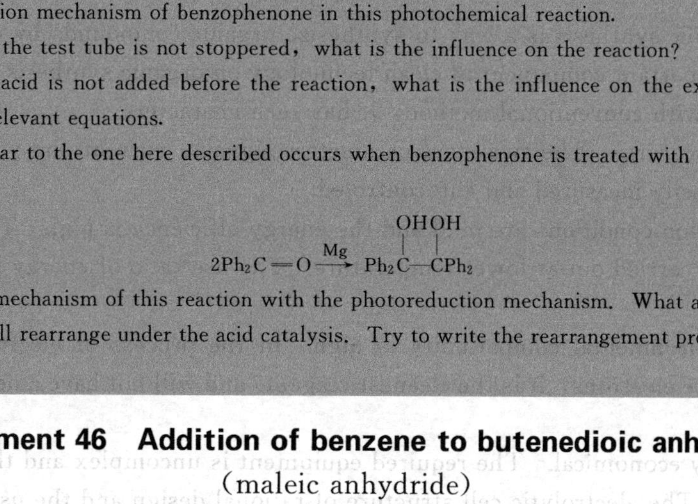

Equipment   Apparatus for: suction filtration.

Materials	
maleic anhydride (FW 98.1)	5 g (51 mmol)
benzene (FW 78.1) (dry)	

Procedure

In a hood, dissolve 5 g of the maleic anhydride in 50 mL dry benzene in a 100 mL Erlenmeyer flask that has a standard taper joint. Stopper the vessel securely, and place it on a south facing window ledge. Leave the solution to stand for several days in bright sunshine (length of time depends on light intensity). Eventually colorless crystals of the photoadduct will appear. Return the solution to a hood and collect the crystals of the product by suction filtration. Dry the crystals at the pump for a few minutes and then in air for about 15 minute. Record the yield, m.p. (it is over 300 ℃).

> **Question**
>
> 1. What sort of reactions are involved in (a) the formation of the intermediate 1 : 1 adduct and (b) the formation of the final 1 : 2 adduct?
> 2. If the experiment is carried out using benzene that is not completely dry, solid with mp 135 ℃ crystallizes out of solution within a few hours. What do you think this solid might be?

3.16 Organic electrochemical reaction

Electroorganic synthesis is a way to synthesize organic compounds by electrolytic reaction. It is an important component of clean technology on organic synthesis in green chemistry. Compared with conventional method, it has such characters.

① Can autocontrol. The two main parameters signals, current and voltage in electrochemisty, are easily measured and autocontroled.

② The reacion conditions are mild and the energy efficiency is high. The electrochemical reactions are carried out at lower temperature. The use ratio of energy is high due to no Carnot cycle.

③ The environmental compatibility is high. In the process of electrochemistry, the main reagents are electrons, it is the cleanest reagents and will not have a negative impact on the enviroment.

④ It is very economical. The required equipment is uncomplex and the cost of equipment is lower. The electrolytic cell structure of rational design and the use of advanced electrode materials can achieve the requirements of zero-emission. At present, the application of electrochemical method in organic synthesis has attrated widely attention. The electrochemistry method has been recognized increasingly in the field of chemistry and chemical industry.

Experiment 47   Preparation of iodoform

Iodoform, also called yellow iodine, is a bright yellow crystal. It can be used as preservatives and antiseptics.

Chapter 3 Semimicro Synthetic Experiment of Typical Organic Compounds

Reaction equations
Main reactions

cathode: $2H^+ + 2e \longrightarrow H_2$

anode: $2I^- - 2e \longrightarrow I_2$

$$I_2 + 2OH^- \rightleftharpoons IO^- + I^- + H_2O$$

$$CH_3COCH_3 + 3IO^- \longrightarrow CHI_3 \downarrow + CH_3COO^- + 2OH^-$$

Secondary reaction

$$3IO^- \longrightarrow IO_3^- + 2I^-$$

Equipment Magnetic stirrer; apparatus for: suction filtration, recrystallization.

Materials	
Potassium iodide (FW 166.00)	2.2 g (13.3 mmol)
acetone (FW 58.08)	0.5 mL (0.4 g, 6.9 mmol)
ethanol (FW 46.07)	

Procedure

Make use of a 50 mL beaker as electrolyzer and two carbon sticks of 1# battery are used as electrodes. The electrodes are fixed vertical on a hard cardboard or a plexiglass plate. Add 40 mL of water, 2.2 g of potassium iodide and 0.5 mL of acetone to the beaker. Stir the mixture until the solution is clear. Switch on the power source of electrodes (connect 6V direct current) and electrolyze at room temperature. A great amount of yellow precipitation is produced during the reaction. Change the electrodes to stop the reaction after about 1 h to accelerate the reaction. Switch off the power of electrodes to stop the reaction after 2.5 h, and continue stirring for 1~2 minutes again. Collect the yellow precipitate of iodoform with suction filtration and wash it with a little cold water twice, dry it in the air. Recrystallize the crude product from ethanol (the apparatus of the recrystallization sees the Figure 2-45). After drying, weigh, determine its melting point, calculate the yield.

Pure iodoform is a bright yellow crystal, m.p. 119 ℃. It can sublimate, but can not dissolve in water.

Notes

1. In order to reduce the loss of current passing through the medium, the two electrodes should be as close as possible each other.
2. The superficial area of electrode is bigger, and the reaction is faster. Guarantee the area of the electrodes immersing in the reaction solution.
3. Pure iodoform is a yellow crystal, but when the graphite electrode is used, the crystal is gray green because of the mixed graphite, so it should be purified.

Questions

1. Why does the pH of the electrolyte rise gradually during the electrolysis?
2. Except recrystallization, what methods can be also used to purify iodoform?

3.17 Isolation of natural product

The organic compounds derived from the inner bodies of animals and plants is called natural products. There are varieties of natural products and they widely exist in nature. Most of the extracts from natural products have specially physiological activity. Some of them can be used as spices and dyes, and some even have magical pharmaceutical effect. The isolation and identification of natural products are a very active research field in organic chemistry. In the course of the research, the first problem to be solved is the isolation, purification and identification of natural products. Along with the development of the modern chromatography and spectrum technology, the isolation and identification of natural substances have become more favorable and convenient.

Experiment 48 Isolation of caffeine from tea

Tea contains a variety of alkaloids, tannins, pigments, vitamins and proteins and other substances. Caffeine is one of the alkaloids in tea. The caffeine content of tea is *ca*. 1%~5%. In addition, it also contains a small amount of theophylline and theobromine. They belong to the purine derivatives, their strutural formulas are as follows.

purine   caffeine   theobromine   theophylline

The chemical name of caffeine is 1,3,7-trimethyl-2,6-dioxopurine. It is a shiny, colorless and needle crystal and contains one crystal water. It can lose the crystal water and sublimate at 100 ℃. The sublimination is quite significant at 120 ℃ and very quick at 178 ℃. Caffeine sublimates to be a needle-like crystal. The melting point of anhydrous caffeine is 235 ℃. It is a weakly basic substance and has bitter taste. It can dissolve in hot water (about 80 ℃), ethanol, diethyl ether, acetone, dichloromethane and chloroform, but undissolve in petroleum ether.

The chemical name of theobromine is 3,7-dimethyl-2,6-dioxopurine. The theobromine content of tea is *ca*. 0.05%. It is a colorless and needle crystal with bitter taste. The melting point is 342~343 ℃. It sublimates at 290 ℃, dissolves in hot water, but hardly dissolves in cold water, ethanol and undissolves in ether.

The chemical name of theophylline is 1,3-dimethyl-2,6-dioxopurine, and it is the isomer of theobromine. It is a white powder crystal with bitter taste. The melting point is 273 ℃. Dissolve in the boiling water and chloroform easily, in cold water and ethanol slightly.

The alkaloids in tea have a certain degree of pharmic function on the human body. Caffeine can make the central nervous system excited, relieve fatigue, has cardiotonic effect. The function of theophylline is similar to that of caffeine. It has fewer central nervous system effects but is a stronger myocardial stimulant than that of caffeine. The function of theobromine is also similar to that of caffeine. It has a fewer effect on the central nervous system than caffeine and theophylline, but the cardiotonic function is between them.

Caffeine is often used as the incitant of heart, respiratory organ and nervous system in medicine. It is also one component of the APC which is a drug for treating cold. The exces-

Chapter 3 Semimicro Synthetic Experiment of Typical Organic Compounds

sive use of caffeine will increase the resistance and have a mild addiction.

Caffeine is identified not only by measuring the melting point and spectrum, but also further confirmed by preparing a derivative salt of caffeine, caffeine salicylate. As a weakly basic compound, caffeine can react with salicylic acid to give the caffeine salicylate, whose melting point is 137 ℃.

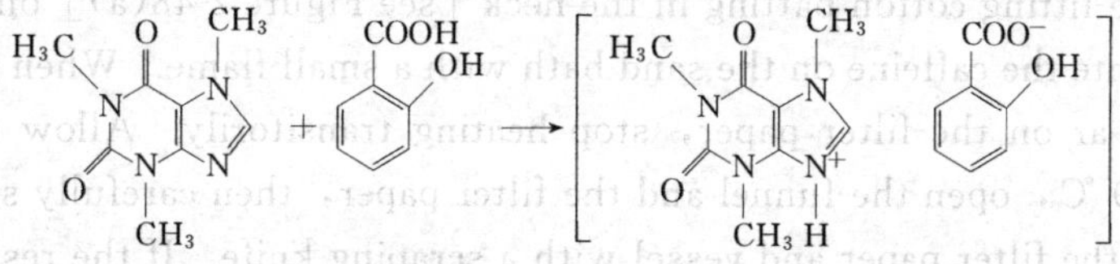

caffeine   salicylic acid   caffeine salicylate

Equipment   Apparatus for: Soxhlet extractor, sublimation, distillation, suction filtration, extraction/separation.

Materials	
Method 1	
tea	3 g
ethanol (95%)	60 mL
quicklime powder	
Method 2	
tea	10 g
calcium carbonate	10 g
chloroform	
salicylic acid	
toluene	
petroleum ether (b. p. 60～90 ℃)	

Procedure

(1) *Isolation of caffeine from tea*

Method 1 Add 20 mL of 95% ethanol and two boiling stones to the round-bottomed flask. Place 3 g of tea leaves and 40 mL of 95% ethanol in the filter paper thimble of Soxhlet extractor (see Figure 2-39). Arrange the apparatus for continues extraction with a water bath. Reflux continuously and extract until the color of the extract is very light. When the condensed liquid just is siphoned back to the flask from the Soxhlet, stop heating immediately.

Allow the flask to cool slightly, and arrange the apparatus for distillation. Distill out and recover most ethanol in the extract, which can be reused afterwards. Pour the warm concentrated solution in the flask into an evaporating dish, rinse the flask with small amounts of recovered ethanol twice, and then combine them with the concentrated solution. Add 1～1.5 g of quicklime powder to the evaporating dish and stir the mixture continuously with a stirring rod to form the tea sand. Evaporate the tea sand to dryness on a steam bath (stir constantly and crush the blocks). Then move the evaporating dish on an asbestos gauge and bake the tea sand with a small flame for a moment. Make sure the moisture in the mixture is removed thoroughly. After cooling, wipe the powder off the wall of the evaporating dish wish a filter paper to avoid it polluting the caffeine as sublimating.

Cover a large round filter paper which has been pierced a lot of small holes and the holes stings are upward on the evaporating dish. Cover a glass funnel of the proper size which has been placed loose-fitting cotton batting in the neck [see Figure 2-48(a)] on the filter paper. Carefully sublimate the caffeine on the sand bath with a small flame. When the white needle-like crystals appear on the filter paper, stop heating transitorily. Allow the sand bath to cool to about 100 ℃, open the funnel and the filter paper, then carefully scrape off the caffeine adhered to the filter paper and vessel with a scraping knife. If the residue is green colored, resublime the residue completely with a larger flame after stirring until the color of the residue is brown. Combine all the obtained caffeine. Weigh the product, determine its melting point, and calculate the content of caffeine in tea leaves.

Method 2 Place 10 g of tea leaves packed by a piece of gauge in a 250 mL beaker containing 100 mL of water and 10 g of calcium carbonate powder. Boil the mixture for 30 minutes to produce a concentrated extract. Place the tea bag in an empty beaker to cool for a few minutes and then squeeze as much extract as possible from them into the original beaker. Filter the warm mixture solution with suction. After the filtrate is cooled, extract it with 15 mL of chloroform twice. Combine the extracts and distill off the chloroform with a water bath. The residue is used for the sublimation of caffeine. The sublimating method will follow the procedure mentioned in the method 1.

(2) *Preparation of caffeine salicylate*

Mix 40mg of caffeine with 30mg of salicylic acid and 2.5 mL of toluene in a test tube. Warm the test tube in a water bath while shaking until all the solid dissolves, then carefully add 1.5 mL petroleum ether (b. p. 60~90 ℃). Cool the mixture in an ice bath to obtain the crystals of caffeine salicylate. If no crystals are separated, rub the wall of the test tube with a stirring rod. Collect the caffeine salicylate by suction filtration. After drying, determine the melting point. The melting point of the pure caffeine salicylate is 137 ℃.

Notes
1. The size of the filter paper thimble should be moderate. It must be not only tightly close to the wall of the Soxhlet extractor, but also be convenient to use. Its height is not higher than the siphon tube of the Soxhlet extractor.
2. While the mixture of the extract and quicklime is baked on the asbestos gauge with a small flame, make sure the solvent in the mixture be removed thoroughly, otherwise there will be small water drops on the inner wall of the funnel at the beginning of sublimation. If so, move the flame away, wipe water drops off quickly with filter paper, and continue the sublimation.
3. Control the temperature of the sand bath in the course of the sublimation. If the temperature were too high, the product would be impure and lose by carbonization.
4. Tannin does not refer to a single homogeneous compound. Tannins are phenolic compounds having molecular weights between 500 and 3000. Tannins in tea are usually divided in two classes: those that can be hydrolyzed and those that can not. Tannins of the first type are esters and they represent the structures in which some of the hydroxyl

groups in glucose have been esterified by digalloyl groups. They can be hydrolyzed to yield gallic acid and glucose in hot water. The nonhydrolyzable tannins are condensation polymers of catechin and glucose. The gallic acid hydrolyzed by tannins can react with caffeine to give an insoluble salt.

5. Calcium carbonate, a base, is added to the crude caffeine, the calcium salts of acidic substances such as tannins are formed. So the caffeine can be purified by sublimation.
6. Tannins in tea can be removed as insoluble calcium salts by adding calcium carbonate, so that caffeine is remained behind in water as a soluble alkaloid.

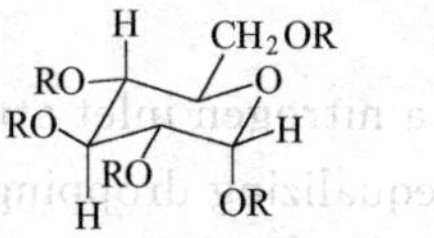

glucose if R=H
a tannis if some R=digalloyl

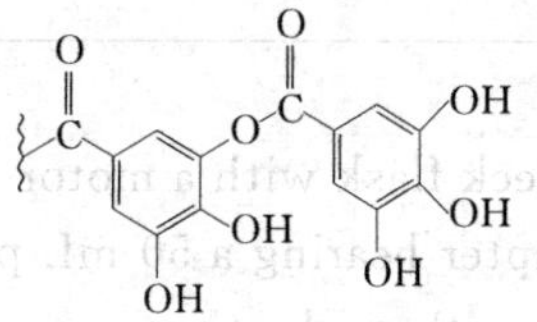

a digalloyl group

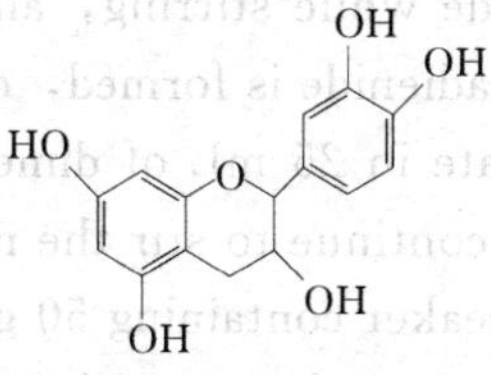

catechin

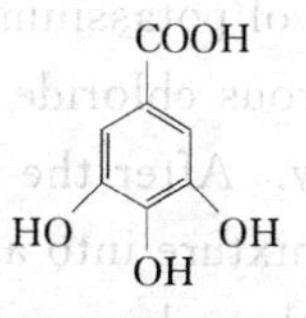

gallic acid

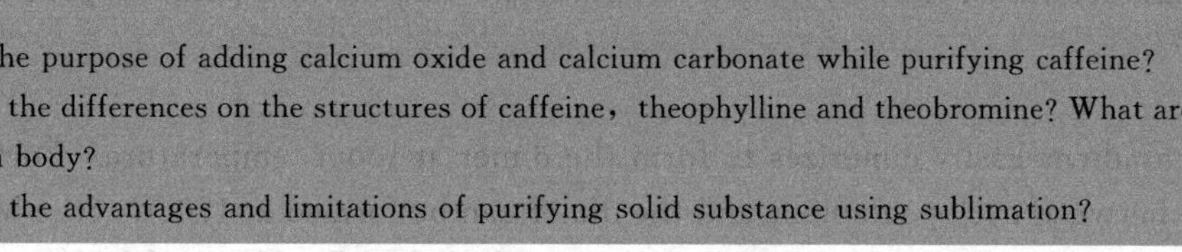

caffeine + gallic acid → caffeine gallate

Questions

1. What is the purpose of adding calcium oxide and calcium carbonate while purifying caffeine?
2. What are the differences on the structures of caffeine, theophylline and theobromine? What are the effects on human body?
3. What are the advantages and limitations of purifying solid substance using sublimation?

3.18 Preparation of organometallic compound

Experiment 49 Preparation of ferrocene

Reaction equations

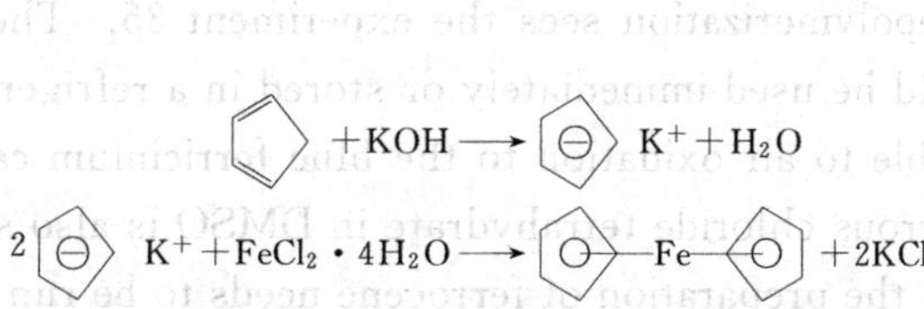

$$\text{C}_5\text{H}_6 + \text{KOH} \longrightarrow \text{C}_5\text{H}_5^- \text{K}^+ + \text{H}_2\text{O}$$

$$2\,\text{C}_5\text{H}_5^- \text{K}^+ + \text{FeCl}_2 \cdot 4\text{H}_2\text{O} \longrightarrow (\text{C}_5\text{H}_5)_2\text{Fe} + 2\text{KCl}$$

Equipment Motor stirrer; apparatus for: stirring under inert atmosphere, suction filtration.

Materials

cyclopentadiene (FW 66.18)	2.6 mL (2.1 g, 31.7 mmol)
potassium hydroxide (FW 56.11)	1.3 g (23.2 mmol)
ferrous chloride tetrahydrate	
dimethyl sulfoxide	
nitrogen	
hydrochloric acid (2mol/L)	
pH indicator paper	

Procedure

Fit a 100 mL 3-neck flask with a motor stirrer, a nitrogen inlet attached to the nitrogen line and a Claisen adapter bearing a 50 mL pressure-equalizing dropping funnel in an adapter and a gas bubbler in another adapter.

In the flask place 1.3 g of potassium hydroxide, followed by 2.6 mL of freshly distilled cyclopentadiene and 30 mL of dimethyl sulfoxide while stirring, and then bubble nitrogen. After the black solution of potassium cyclopentadienide is formed, drop the freshly prepared solution of 3.5 g of ferrous chloride tetrahydrate in 25 mL of dimethyl sulfoxide (DMSO) while stirring vigorously. After the addition, continue to stir the mixture for 20 minutes.

Pour the reaction mixture into a 250 mL beaker containing 50 g of ice and 50 g of water and stir the mixture with a glass rod. Acidify the solution with 2 mol/L hydrochloric acid until the pH is 3~5. After the yellow solid has separated completely, collect it with suction filtration and wash it with 4×10 mL portions of water. Dry the crude product with suction, the yield of crude ferrocene is about 2.2 g.

The crude product can be purified by sublimation. Place the crude product in a clean and dry 400 mL beaker, cover the beaker with a watch glass and stopper the mouth of the beaker with cotton. Heat the beaker gently at 100 °C while cooling the watch glass with a piece of humid cloth. Yellow crystals are obtained. The melting point of pure ferrocene is 173~174 °C.

Notes

1. Cyclopentadiene easily dimerizes to form the dimer at room temperature. Fortunately, a balance between cyclopentadiene and the dimer can be formed at 170 °C (the boiling point of the dimmer). Thus, monomeric cyclopentadiene can be obtained by "cracking" the dimer by heat.

The method of the depolymerization sees the experiment 35. The freshly depolymerized cyclopentadiene should be used immediately or stored in a refrigeratory.

2. Ferrocene is susceptible to air oxidation to the blue ferricinium cation $[Fe^{3+}(C_5H_5)_2]$. Furthermore, the ferrous chloride tetrahydrate in DMSO is also susceptible to air oxidation to Fe^{3+}. Thus, the preparation of ferrocene needs to be run under nitrogen in order to seclude air.

3. If the ferrous chloride tetrahydrate has become brown, wash it with ethanol or diethyl e-

ther until it is light green. Grind it into powder and dissolve it in DMSO before use. If there are many brown iron chlorides in the ferrous chloride, the experiment effect will be reduced.

4. Use the ground potassium hydroxide. The grinding operation should be quick to prevent potassium hydroxide from absorbing moisture in air.

> **Questions**
> 1. Talk about the functions of ferrocene through consulting the literatures.
> 2. Ferrocene occurs more easily the electrophilic substitution reaction than benzene, but it can not be nitrated by the mixed acid of concentrated nitric acid and concentrated sulfuric acid. Why?

3.19 The Wittig reaction

The Wittig reaction involves the reaction of a phosphorus ylid with a ketone or aldehyde to give an alkene. Advantages of the Wittig synthesis are that the position of the carbon-carbon double bond which is formed by the reaction of a carbonyl compound with a phosphorus ylid is confirmable, the secondary reactions of rearrangement and dislocation can not occur generally, the conjugated alkenes can prepared easily by α,β-unsaturated aldehydes or ketones, the reaction conditions are mild and the yield of the product is higher. Therefore, Wittig reaction has been applied widely as a general method of synthesizing alkenes in organic synthesis.

The ylid (Wittig reagent) is prepared by reaction of a nucleophilic triphenylphosphine with an alkyl halide, followed by subsequent treating the produced phosphonium salt with a strong base to eliminate the α-H on the alkyl group.

$$(C_6H_5)_3P + BrCHR^3R^4 \longrightarrow (C_6H_5)_3\overset{\oplus}{P}CHR^3R^4 Br^{\ominus} \xrightarrow[-LiBr]{C_6H_5Li} $$

$$(C_6H_5)_3\overset{\oplus}{P}-\overset{\ominus}{C}R^3R^4 \longleftrightarrow (C_6H_5)_3P=CR^3R^4$$
$$\text{ylid}$$

The ylid is a kind of nucleophilic reagent. It can attack the carbonyl carbon of the aldehydes and ketones to produce the dipolar intermediate called a betaine. The betaine, in either one or two steps, eliminates the triphenylphosphine oxide to yield the desired alkene.

$$(C_6H_5)_3\overset{\oplus}{P}-\overset{\ominus}{C}R^3R^4 + O=C\begin{smallmatrix}R^1\\R^2\end{smallmatrix} \longrightarrow (C_6H_5)_3\overset{\oplus}{P}-CR^3R^4$$
$$\overset{\ominus}{O}-CR^1R^2$$
ylid   carbonyl compound   betaine

$$(C_6H_5)_3P-CR^3R^4 \longrightarrow (C_6H_5)_3P=O + R^1R^2C=CR^3R^4$$
$$\,|\qquad|$$
$$O-CR^1R^2$$
oxaphosphetane   triphenylphosphine oxide   alkene

The Wittig reagent has been limited in the actual application to a certain extent. First of all, the Wittig reagent whose α-carbon is bonded to the hydrogen atom or alkyl group is active and can react with oxygen, water halogen acid, alcohol, etc. So the reaction must be carried out in the absence of them. Secondly, the Wittig reagent is relatively stable when the electron withdrawing groups, such as —COR, —CN, —COOR, —CHO, etc., are connected to α-carbon, it can react with aldehydes. But it react with ketones slowly, even no reaction. In addition, the usually used strong base, organolithium reagent, is not easily treated and dangerous. The treatment should be carried out in inert gas. The phosphine reagents are largely toxic and expensive.

In order to increase the nucleophilicity of α-carbon, many improvements of Wittig reaction have been done. Among them, the most effective one is the Horner-Emmons method. That is, replace the phenyl groups on phosphorus atom in Wittig reagent with oxygen atom or oxethyl group. Compared with Wittig reagent, the Horner-Emmons reagent has many advantages. It is not very sensitive to alkaline medium, react with oxygen and water very slowly. The operation of the experiment is more convenient, finally produced diethyl phosphate can dissolve in water and the produced alkene can be separated easily from the mixture.

Experiment 50 Preparation of (E)-1,2-diphenylethene (stilbene)

Reaction equations

$$PCl_3 + 3C_2H_5OH + 3C_6H_5N(CH_3)_2 \longrightarrow (C_2H_5O)_3P + 3C_6H_5N(CH_3)_2 \cdot HCl$$

$$(C_2H_5O)_3P + C_6H_5CH_2Cl \longrightarrow (C_2H_5O)_2\overset{O}{\underset{\|}{P}}CH_2C_6H_5 + C_2H_5Cl$$

$$(C_2H_5O)_2\overset{O}{\underset{\|}{P}}CH_2C_6H_5 + CH_3ONa \xrightarrow{DMF} \left[(C_2H_5O)_2\overset{O}{\underset{\|}{P}}\overset{\ominus}{C}HC_6H_5\right]Na^{\oplus} \xrightarrow{C_6H_5CHO}$$

$$\underset{C_6H_5}{\overset{H}{\diagdown}}C=C\underset{H}{\overset{C_6H_5}{\diagup}} + (C_2H_5O)_2\overset{O}{\underset{\|}{P}}ONa$$

Equipment   Mechanical stirrer; apparatus for: reflux, suction filtration, distillation at atmospheric pressure, distillation under reduced pressure, recrystallization.

Materials	
(1) absolute ethanol (FW 46.07)	4.6 g (5.8 mL, 100 mmol)
N,N-dimethylaniline (FW 121.18)	12.1 g (12.7 mL, 100 mmol)
phosphorus trichloride (FW 137.5)	4.5 g (2.9 mL, 33 mmol)
petroleum ether (b.p. 30~60 ℃)	
(2) benzyl chloride (FW 126.6)	6.0 mL (6.60 g, 52 mmol)
triethyl phosphite (FW 166.2)	9.0 mL (8.70 g, 52 mmol)
(3) sodium methoxide (FW 54)	3.00 g (55.5 mmol)
benzaldehyde (FW 106.1)	5.2 mL (5.40 g, 51 mmol)
N,N-dimethylformamide (DMF)	
methanol	
isopropanol	

Procedure

(1) *Preparation of triethyl phosphite*

Set up a 100 mL 3-neck flask with a mechanical stirrer, pressure-equalizing dropping funnel and reflux condenser protected with a calcium chloride drying tube. Place 4.6 g of absolute ethanol, 12.1 g of freshly redistilled N,N-dimethylaniline and 30 mL of dry petroleum ether (b. p. 30~60 ℃) in the flask. From the funnel, add a solution of 4.5 g of the freshly redistilled phosphorus trichloride in 15 mL dry petroleum ether (b. p. 30~60 ℃) dropwise to the flask with stirring at such a rate as to maintain gentle boil. Because this reaction is exothermal, cool the flask with a cold water bath if necessary. The white precipitate appears in the flask during adding the solution of phosphorus trichloride (What compound is it?). When the addition is complete, continue to stir the mixture for 30 minutes, and then heat the mixture under reflux with a water bath for 1 h while stirring. Cool the mixture to room temperature, filter off the precipitate by suction and wash it twice with dry petroleum ether. Transfer the filtrate to a distilling flask and distill off the petroleum ether in a water bath. Distil the residue from a small distilling flask under reduced pressure. Collect the fraction boiling at 57~58 ℃/2.13 kPa (16 mmHg). The yield of the product is about 4.5 g and the percentage yield is about 82%.

The boiling point of pure triethyl phosphite is 156.5 ℃.

(2) *Preparation of diethyl benzyl phosphonate*

In a dry 50 mL 2-neck flask provided with a 300 ℃ thermometer and a reflux condenser with calcium chloride drying tube, place 6 mL of benzyl chloride, 9 mL of triethyl phosphite and two boiling stones. Heat the flask gently over an asbestos gauze with a small flame. When the temperature of the solution reaches 130~140 ℃, the evolution of ethyl chloride begins. The solution boils at about 165 ℃. Continue to heat for 1.5~2 h. The internal temperature continues to rise and attains above 200 ℃ by the end of the hour. Stop heating and allow the reaction solution to cool to room temperature to be used.

(3) *Preparation of (E)-1,2-diphenylethene*

In a dry 100 mL 3-neck flask provided with a pressure-equalizing dropping funnel, thermometer and a calcium chloride drying tube, place 3.00 g of sodium methoxide and 25 mL of dried N,N-dimethylformamide. Cool the flask in an ice-water bath, then add the prepared diethyl benzyl phosphonate (the reaction of forming ylid is exothermal). When the temperature of the reaction mixture in the flask is below 20 ℃, add drop by drop 5.2 mL of freshly redistilled benzaldehyde from the funnel at such a rate as to maintain the reaction temperature between 30~40 ℃. When the addition is complete, remove the ice-water bath, and allow the flask to stand 10 minutes at room temperature to finish the reaction completely. Add 10 mL of water to the flask with stirring. The product separates as the crystals at the moment. Cool the flask in an ice-water bath with stirring for a while to crystallize completely. Collect the white crystals with suction, wash them with a little cold mixed solvent of equal volume methanol and water, and then dry them. Recrystallize the crude product form isopropanol (about 30~50 mL). The yield is about 5~6 g, m. p. 124~126 ℃.

The melting point of pure (E)-1,2-diphenylthene is 125 ℃.

Notes

1. Phosphorus trichloride is highly toxic and irritant. This experiment should be carried out in the hood to prevent its vapor being inhaled. If any of phosphorus trichloride is spilled on your skin, wash off it immediately with soap and water. N,N-dimethylaniline is toxic, avoid inhaling the vapor. If any of N,N-dimethylaniline is spilled on your skin, wash off it immediately with 5 % acetic acid, followed by soap and water.

2. Phosphorus trichloride meets water to decompose, therefore all apparatus, solvent and chemicals must be rigorously dried before use. Phosphorus trichloride and N,N-dimethylaniline should be redistilled before use. The commercial absolute ethanol should be soaked by 5A molecular sieves before use.
3. During preparing triethyl phosphite, use the dry Büchner funnel and filter flask when filtering the white precipitate with suction and suck the precipitate as dry as possible.
4. Benzyl chloride is a lachrymator. Wear goggles while operating. Triethyl phosphite has an unpleasant smell.
5. When preparing diethyl benzyl phosphonate, if the time of heating reactants is short of one hour and the reaction temperature is below 220 ℃, the yield will be reduced. The by-product, ethyl chloride is harmful to the human body, lead off it outside.
6. Sodium methoxide is extremely hygrometric, so avoid it being exposed to air. Measure it rapidly and cover the lid of the reagent bottle in time.
7. The amount of the used isopropanol is decided by the drying degree of the crude product.

Preparation of sodium methoxide

Set up a 100 mL 2-neck flask with a reflux condenser protected with a calcium chloride drying tube. With the help of a rubber stopper, connect the end of the drying tube to a rubber tube which has been introduced to the air drain. Add 15 mL of anhydrous methanol to the flask and stopper the flask with a ground glass stopper. Take a small piece of metallic sodium with a tweezers and cut off the any oxide at the surface. Measure 6 g of clean sodium rapidly, cut it into small pieces, and then put them directly in a 100 mL beaker containing petroleum ether. Add one piece of sodium to the flask, hydrogen evolves immediately. To keep the reaction under control one piece of sodium should be allowed to react completely before another is added. This operation should be carried on in the hood, no flame is nearby, can not add many pieces of the sodium once. After all of the sodium has reacted, remove the excess methanol by distillation with a hot water bath at atmospheric pressure, and then under reduce pressure with a water pump using an oil bath maintained at 150 ℃. The dry white solid of sodium methoxide is obtained. Place it in a dry wide mouth bottle, stopper the bottle tightly with a rubber stopper and store it in a vacuum drier to be used.

Questions

1. Which characteristics does the Wittig reaction have in preparing olefin? Write the reaction mechanism.
2. What methods can be also taken to prepare olefin with aldehyde or ketone?
3. What impurity does benzaldehyde which is stored for a long time contain? What is the effect of using it on the Wittig reaction?

3.20 Hofmann degradation reaction

Experiment 51 Preparation of anthranilic acid (2-aminobenzoic acid)

Reaction equations

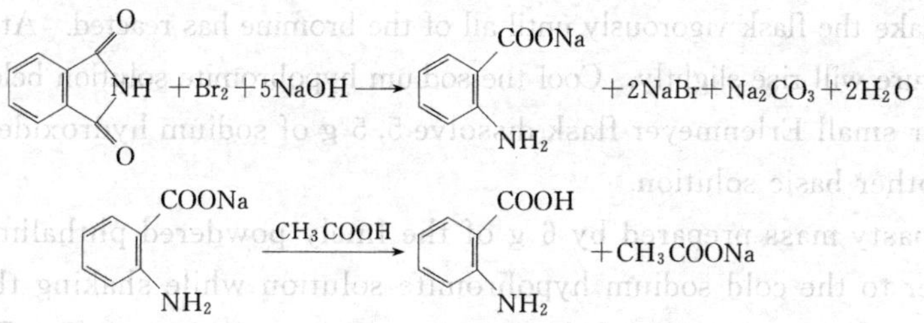

Chapter 3 Semimicro Synthetic Experiment of Typical Organic Compounds

$$\text{Phth-NH} + Br_2 + 5NaOH \longrightarrow \text{o-}C_6H_4(COONa)(NH_2) + 2NaBr + Na_2CO_3 + 2H_2O$$

$$\text{o-}C_6H_4(COONa)(NH_2) + CH_3COOH \longrightarrow \text{o-}C_6H_4(COOH)(NH_2) + CH_3COONa$$

Reaction mechanism

$$\text{Phth-NH} + Br_2 + 5NaOH \longrightarrow \text{o-}C_6H_4(COONa)(NH_2) + 2NaBr + Na_2CO_3 + 2H_2O$$

via intermediates:
- o-C_6H_4(COONa)(CONH_2) →[NaOBr/NaOH]
- o-C_6H_4(COONa)(CONHBr) →[−HBr]
- o-C_6H_4(COONa)(CON:) →[H_2O] o-C_6H_4(COONa)(N=C=O) → o-C_6H_4(COONa)(NH_2)

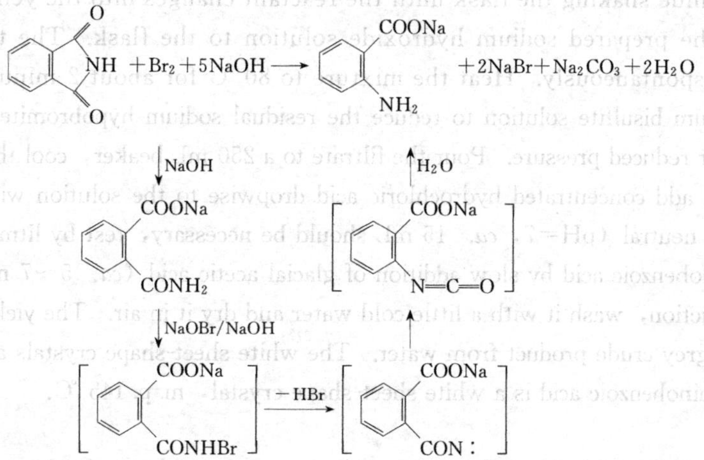

Equipment Apparatus for: reflux, suction filtration, recrystallization.

Materials	
phthalic anhydride (FW 148.12)	10 g (67.5 mmol)
ammonia water (concentrated)	
bromine (FW 159.8)	2.1 mL (6.55 g, 41 mmol)
sodium hydroxide	
concentrated hydrochloric acid	
sodium bisulfite solution (saturated)	
glacial acetic acid	
pH indicator paper	

Procedure

(1) *Preparation of phthalimide*

Place 10 g of phthalic anhydride and 10 mL of concentrated ammonia water in a 100 mL 2-neck flask. Fit the flask with an air condenser and a 360 ℃ thermometer. First heat the flask over an asbestos gauge, then heat the flask directly with a small flame while shaking occasionally, until the reaction temperature rises gradually to 300 ℃. Push the solid formed by sublimation in the condenser into a porcelain dish. Pout the hot reaction product into a porcelain dish. After cooling, place the concretionary solid in a mortar and grinds it to powder to be used The yield is about 8 g, m. p. 232~234 ℃.

(2) *Preparation of anthranilic acid*

In a 125 mL Erlenmeyer flask dissolve 7.5 g of sodium hydroxide in 30 mL of water and cool the solution to $-5 \sim 0$ ℃ in an ice-salt bath. Add 2.1 mL of bromine in one portion to the basic solution and shake the flask vigorously until all of the bromine has reacted. At the moment the reaction temperature will rise slightly. Cool the sodium hypobromite solution below 0 ℃ to be used.

In another small Erlenmeyer flask dissolve 5.5 g of sodium hydroxide in 20 mL of water to prepare another basic solution.

Add the pasty mass prepared by 6 g of the finely powdered phthalimide and a small amount of water to the cold sodium hypobromite solution while shaking the flask vigorously and maintaining the temperature of the reaction mixture at about 0 ℃. Remove the ice-salt bath and continue shaking the flask until the reactant changes into the yellow clear solution. Rapidly add the prepared sodium hydroxide solution to the flask. The temperature of the mixture rises spontaneously. Heat the mixture to 80 ℃ for about 2 minutes. Add 2 mL of saturated sodium bisulfite solution to reduce the residual sodium hypobromite. Cool the solution and filter under reduced pressure. Pour the filtrate to a 250 mL beaker, cool the solution in an ice-water bath and add concentrated hydrochloric acid dropwise to the solution with stirring until the solution is just neutral (pH=7, ca. 15 mL should be necessary, test by litmus paper). Precipitate the 2-aminobenzoic acid by slow addition of glacial acetic acid (ca. $5 \sim 7$ mL), filter the precipitate with suction, wash it with a little cold water and dry it in air. The yield is about 4 g. Recrystallise the grey crude product from water. The white sheet-shape crystals are obtained.

Pure 2-aminobenzoic acid is a white sheet-shape crystal, m. p. 145 ℃.

Notes
1. Bromine is highly corrosive and irritant. Always wear goggles and rubber gloves and measure out in the hood. Do not inhale the bromine vapor.
2. Anthranilic acid can dissolve both in the alkali and in the acid, therefore the excessive hydrochloric acid will make the product dissolved. If too much acid is added, neutralize the mixture by sodium hydroxide solution.
3. The isoelectric point of anthranilic acid is about $3 \sim 4$. In order to isolate anthranilic acid completely, add the right amount acetic acid to the solution.

Questions
1. Which applications does anthranilic acid have in the synthesis and the analysis?
2. If the used bromine and the sodium hydroxide are insufficient or more excessive, what is the influence on this reaction respectively?
3. When the basic solution of anthranilic acid is just neutral with adding hydrochloric acid to the solution, why should the right amount of acetic acid be added rather than the hydrochloric acid to isolate the product completely?

3.21 Multistep synthesis

3.21.1 The multistep synthesis of 4-iodophenoxyacetic acid from phenol
Multistep synthesis route 1 The preparation of 4-iodophenoxyacetic acid

$$\text{C}_6\text{H}_5\text{OH} \xrightarrow{\text{ClCH}_2\text{COOH/OH}^-} \text{C}_6\text{H}_5\text{OCH}_2\text{COOH} \xrightarrow{\text{ICl}} \text{4-I-C}_6\text{H}_4\text{OCH}_2\text{COOH}$$

Experiment 52   Experiment 53

Chapter 3 Semimicro Synthetic Experiment of Typical Organic Compounds

Experiment 52 Preparation of phenoxyacetic acid

Phenoxyacetic acid is a white flake crystal or needle-shape crystal. It can be used for the synthesis of dye, medicine and pesticide. It can be directly used as a plant growth regulator which is harmless to humans and animals. Phenoxyacetic acid has been widely applied. It can be prepared by the Williamson reaction of phenol with chloroacetic acid in aqueous alkali.

Overall reaction equation

$$\text{C}_6\text{H}_5\text{OH} \xrightarrow{\text{ClCH}_2\text{COOH/OH}^-} \text{C}_6\text{H}_5\text{OCH}_2\text{COOH}$$

Stepwise reaction equations

$$\text{C}_6\text{H}_5\text{OH} + \text{NaOH} \longrightarrow \text{C}_6\text{H}_5\text{ONa} + \text{H}_2\text{O}$$

$$2\text{ClCH}_2\text{COOH} + \text{Na}_2\text{CO}_3 \longrightarrow 2\text{ClCH}_2\text{COONa} + \text{H}_2\text{O} + \text{CO}_2$$

$$\text{C}_6\text{H}_5\text{ONa} + \text{ClCH}_2\text{COONa} \longrightarrow \text{C}_6\text{H}_5\text{OCH}_2\text{COONa} + \text{NaCl}$$

$$\text{C}_6\text{H}_5\text{OCH}_2\text{COONa} + \text{HCl} \longrightarrow \text{C}_6\text{H}_5\text{OCH}_2\text{COOH} + \text{NaCl}$$

Equipment Motor stirrer; apparatus for: reflux with stirring, extraction/separation, suction filtration.

Materials	
phenol (FW 94.11)	2.8 g (30 mmol)
sodium hydroxide (FW 40)	1.3 g (32.5 mmol)
chloroacetic acid (FW 94.50)	3.1 g (32.8 mmol)
sodium carbonate (FW 105.99)	2 g (18.9 mmol)
hydrochloric acid (20%)	
diethyl ether	
sodium chloride solution (15%)	

Procedure

(1) *Preparation of sodium chloroacetate solution*

Add 3.1 g of chloroacetic acid and 10 mL of 15% sodium chloride solution to a 100 mL beaker by turn, then add about 2 g of sodium carbonate to the beaker while stirring, at such a rate that the temperature of the reaction solution does not surpass 40 ℃. After all of the sodium carbonate has been added, the pH of the solution is 7～8. If not, adjust that until 7～8 with saturated sodium carbonate solution.

(2) *Preparation of sodium phenate solution*

Fit a 100 mL 3-neck flask with a motor stirrer, reflux condenser and thermometer. Dissolve 2.8 g of phenol and 1.3 g of sodium hydroxide in 7.5 mL of water in the flask with stirring. Then cool the solution to the room temperature.

(3) *Preparation of phenoxyacetic acid*

Pour the prepared sodium chloroacetate solution into the flask containing sodium phenate solution. Heat the flask gently over an asbestos gauze with a small flame while stirring for 2 h and keep the temperature of the reaction in the rang of 100~110 ℃.

After the reaction has finished, pour the hot reaction mixture into a 250 mL beaker, add 30 mL of water to the beaker, stir uniformly, adjust the pH of the solution between 1~2 with concentrated hydrochloric acid. In this process, the crude phenoxyacetic acid separates in white crystals. After cooling, collect the white crystals with suction filtration and wash them with 5 mL of cold water. Place the crude product into a 250 mL beaker, add 30 mL of water and sodium carbonate while stirring until the crude product is dissolved. Transfer the solution to a separatory funnel, and then extract the aqueous solution with 10 mL diethyl ether to remove the unreacted phenol. Separate the aqueous layer and acidify it with 20% hydrochloric acid until the pH of the solution is 1~2. Allow the contents to cool, collect the white crystals with suction filtration and wash them with a little cold water twice. After drying, the pure phenoxyacetic acid is obtained. Weigh it, determine the melting point of it and calculate the yield.

Pure phenoxyacetic acid is a white and needle-shape crystal, m. p. 98~99 ℃.

Notes
1. Monochloroacetic acid and phenol is corrosive, avoid our skin being touched.
2. Install and fix the motor stirrer carefully to avoid damaging the glassware.
3. Preparing the solution of sodium chloroacetate, the use of the aqueous sodium chloride solution is favourable to inhibit the hydrolysis of sodium chloroacetate. When the temperature of the solution is over 40 ℃ during neutralization, the sodium chloroacetate hydrolyzes easily.
4. When the reaction of preparing phenoxyacetic acid has just started, the pH value of the reaction mixture is 12. The pH value will fall gradually along with the reaction until it is 7~8. At this time, the reaction finishes.
5. Diethyl ether is used to remove the unreacted phenol.

Questions
1. When the ether, phenoxyacetic acid, is prepared by sodium phenate and monochloroacetic acid, why should the monochloroacetic acid be made into sodium chloroacetate? Whether the ether can be prepared by phenol and monochloroacetic acid directly?
2. Why is the aqueous sodium chloride solution added when monochloroacetic acid is neutralized by sodium carbonate?
3. Why does the pH value of the solution change in the course of preparing phenoxyacetic acid? Why is the reaction complete when the pH value of the reaction solution is 7~8?
4. The infrared spectrum of the phenoxyacetic acid is determined by the nujor mull. Please interpret it.
5. Give other methods of preparing phenoxyacetic acid according to the literature.

Experiment 53　Preparation of 4-iodophenoxyacetic acid

4-Iodophenoxyacetic acid is one kind of the plant growth regulators. It can improve the fruiting rate of the plant and prevent the premature fruit dropping.

Chapter 3 Semimicro Synthetic Experiment of Typical Organic Compounds

Reaction equation

$$\underset{}{\text{C}_6\text{H}_5\text{OCH}_2\text{COOH}} \xrightarrow{\text{ICl}} \text{4-I-C}_6\text{H}_4\text{-OCH}_2\text{COOH}$$

Equipment Motor stirrer; apparatus for: stir with heating and addition, suction filtration, recrystallzation.

Materials	
phenoxyacetic acid (FW 152.16)	0.5 g (3.3 mmol)
iodine chloride (FW 162.35)	0.58 g (3.6 mmol)
concentrated hydrochloric acid (37%)	2 mL

Procedure

Place 0.5 g of phenoxyacetic acid and 38 mL of water in a 100 mL 3-neck flask fitted with a motor stirrer, a pressure-equalizing dropping funnel and a thermometer. Stir and dissolve the phenoxyacetic acid in a hot water bath. Add dropwise the solution containing 0.58 g of iodine chloride and 2 mL of concentrated hydrochloric acid from the dropping funnel to the flask under the condition of 40 ℃ water bath. After the addition, raise the temperature of water bath to 80 ℃, continue to stir and heat for 1 h. Cool the solution to room temperature, light pink needle-shape crystals separate from the solution. Collect the crystals with suction filtration and wash them with a small amount of cold water twice. Recrystallize the crude product from water or 50% ethanol. The pure product is obtained as white and needle-shape crystals. After drying, weigh it, determine the melting point of it and calculate the percentage yield.

Pure 4-iodophenoxyacetic acid is a white and needle-shape crystal, m. p. 154~156 ℃.

Notes

1. Iodine chloride and concentrated hydrochloric acid are corrosive and volatile. The operation involving these reagents should be careful. Measure them in a hood.
2. Iodine chloride decomposes easily when it is exposed to light. It has the strong corrosiveness to the rubber. So store it in a brown glass-stoppered bottle.

Question

In the molecule of phenoxyacetic acid, the —OCH$_2$COOH group has the activation on the benzene ring. In the iodation of phenoxyacetic acid, the two isomers, α-iodophenoxyacetic acid and 4-iodophenoxyacetic acid, can be produced theoretically, but why the latter is the main product according to the experiment?

3.21.2 The multostep synthesis of ethyl 3-nitrobenzoate from toluene

Multistep synthesis route 2 The preparation of ethyl 3-nitrobenzoate

$$\text{C}_6\text{H}_5\text{CH}_3 \xrightarrow{[O]} \text{C}_6\text{H}_5\text{COOH} \xrightarrow{\text{EtOH/H}^+} \text{C}_6\text{H}_5\text{COOEt} \xrightarrow[\text{con. H}_2\text{SO}_4]{\text{con. HNO}_3} \text{3-NO}_2\text{-C}_6\text{H}_4\text{COOEt}$$

Experiment 30   Experiment 20   Experiment 54

Experiment 54 Preparation of ethyl 3-nitrobenzoate

The indroduction of a nitro-group into an aromatic ring by means of nitric and sulfuric acids, is an example of aromatic electrophilic substitution by the nitronium ion, NO_2^+. The ethoxycarbonyl group of ethyl benzoate deactivates the ring and directs substitution to the *m*-position.

Reaction equation

$$\text{C}_6\text{H}_5\text{COOEt} \xrightarrow[\text{con. H}_2\text{SO}_4]{\text{con. HNO}_3} \text{3-NO}_2\text{-C}_6\text{H}_4\text{COOEt}$$

Equipment   Apparatus for: suction filtuation.

Materials	
ethyl benzoate (FW 150.18)	1.8 mL (1.88 g, 12.5 mmol)
concentrated sulfuric acid (FW 98.08)	5.5 mL (10.13 g, 101.2 mmol)
concentrated nitric acid (FW 63.0)	1.5 mL (2.12 g, 23.5 mmol)

Procedure

Place 1.8 mL of ethyl benzoate in a 25 mL Erlenmeyer flask and add 4 mL of the concentrated sulfuric acid with shaking and cool the mixture in an ice bath. In another flask place the nitric acid and add the remaining 1.5 mL of sulfuric acid, shaking and cooling in ice. Using a Pasteur pipet, add the nitric acid solution to the ethyl benzoate solution with shaking, maintaining the temperature between 0~10 ℃ by means of the ice bath. The addition takes about 30 min and after this allows the solution to stand for a further 10 min at room temperature. Pour the solution onto ice and stir until the precipitate becomes granular. Filter the ethyl 3-nitrobenzoate with suction and wash well with water. Recrystallize from ethanol to obtain an almost colorless solid. Record the yield and melting point of your product.

The pure ethyl 3-nitrobenzoate is a colorless solid, m. p. 40~43 ℃, b. p. 297~298 ℃.

Question

Write a mechanistic equation for the nitration of toluene by nitric/sulfuric acids.

3.21.3   The multistep synthesis of *p*-bromoaniline and *p*-nitroaniline from aniline

Multistep synthesis route 3   The preparation of *p*-bromoaniline

$$\underset{\text{aniline}}{\text{C}_6\text{H}_5\text{NH}_2} \xrightarrow[-\text{H}_2\text{O}]{\text{CH}_3\text{COOH}} \underset{\text{Experiment 55}}{\text{C}_6\text{H}_5\text{NHCOCH}_3} \xrightarrow[\text{HAc}]{\text{Br}_2} \underset{\text{Experiment 56}}{\text{4-Br-C}_6\text{H}_4\text{NHCOCH}_3} \xrightarrow{\text{H}^+} \text{4-Br-C}_6\text{H}_4\text{NH}_3^+\text{Cl}^- \xrightarrow[\text{Experiment 57}]{\text{NaOH}} \text{4-Br-C}_6\text{H}_4\text{NH}_2$$

In this experiment, the aniline is first converted to acetanilide which is then brominated in the 4-position. After bromination, the amide group is hydrolyzed back to the amine to furnish the desired *p*-bromoaniline.

Aniline may be acetylated by means of acetyl chloride, acetic anhydride, or glacial acetic acid (with removal of the water formed in the reaction). Acetyl chloride reacts vigorously. Acetyl anhydride is preferred for a laboratory synthesis because its rate of hydrolysis is low enough to allow the acetylation of amine to be carried out in aqueous solution. The procedure with glacial acetic acid is an economical one but requires a relatively long period of heating.

Acetylation is often used to "protect" a primary or a secondary amine functional group.

Chapter 3 Semimicro Synthetic Experiment of Typical Organic Compounds

Acylated amines are less susceptible to oxidation, less reactive in aromatic substitution reactions, and less prone to participate in many of the typical reaction of free amines, since they are less basic. At the end of a synthetic sequence, the amino group can be regenerated readily by hydrolysis in acid or base.

Multistep synthesis route 4 The preparation of *p*-nitroaniline

$$\underset{\text{}}{C_6H_5NH_2} \xrightarrow[-H_2O]{CH_3COOH} \underset{\text{Experiment 55}}{C_6H_5NHCOCH_3} \xrightarrow[H_2SO_4]{HNO_3} \underset{\text{Experiment 58}}{p\text{-}O_2N\text{-}C_6H_4\text{-}NHCOCH_3} \xrightarrow[H_2O]{H_2SO_4} p\text{-}O_2N\text{-}C_6H_4\text{-}NH_3^+SO_4H^- \xrightarrow[-Na_2SO_4]{NaOH} \underset{\text{Experiment 59}}{p\text{-}O_2N\text{-}C_6H_4\text{-}NH_2}$$

The sequence of reaction, beginnng with aniline, is as shown. The conversion of aniline to acetanilide, the first step, was perpformed in Experimented 55.

The NH_2 group activates the benzene ring so strongly for electrophilic substitution that direct nitration of aniline can not be controlled. For the preparation of *p*-nitroaniline, thererfore, the reactivity is moderated by acetylation of the amine, the acetyl group is then removed by acid hydrolysis.

p-Nitroaniline is used to prepare *p*-phenylenediamine and also for the production of a special type of azo dye, such as Para Red (an ingrain color). It is used also as an intermediate for laboratory syntheses leading to compounds that cannot be obtained readily.

Experiment 55   Preparation of acetanilide

Acetanilide is one of the oldest synthetic medicinals (1886) and was used for many years as an antipyretic and analgesic drug.

Reaction equation

$$C_6H_5NH_2 \xrightarrow[-H_2O]{CH_3COOH} C_6H_5NHCOCH_3$$

Equipment   Apparatus for: fractional distillation, filtration with suction.

Materials	
aniline (FW 93.1)	2.5 mL (2.55 g, 27.4 mmol)
glacial acetic acid (FW 60.1)	3.8 mL (3.99 g, 66.4 mmol)
zinc powder	0.03 g

Procedure

Place 2.5 mL of freshly distilled aniline in a 10 mL dry round-bottomed flask, add 3.8 mL of glacial acetic acid and a small amount of zinc powder (about 0.03 g) to the flask while swirling the flask. Provide the flask with a fractionating column with a thermometer and connect a distillation adapter. For the receiver use a 5 mL round-bottomed flask to collect the water formed in the reaction and some acetic acid. Heat the flask gently on an asbestos gauge and boil it gently for 10 minutes, then increase the heating slightly. Control the flame to keep the temperature of distilling vapor being about 100~105 ℃ for 1 h. The reaction has finished when the water formed in the reaction and residual acetic acid have been distilled off and the temperature of distilling vapor falls off continually.

Pour the hot reaction mixture into about 20 mL of cold water in a 50 mL beaker while stirring. The white precipitate of acetanilide forms. Stir the aqueous mixture vigorously to avoid the formation of large lumps of the product. After cooling, collect the acetanilide with suction, wash it with a little cold water to remove the residual acid solution, and recrystallize the moist product from water. Collect the product by suction filtration and dry it. Record the weight, measure the melting point and calculate the yield of acetanilide.

Pure acetanilide is a colorless sheet-shape crystal, m. p. 114 ℃.

Notes
1. Aniline is oxidized easily. After it is stored for a long time, it will contain impurity. Then the quality of acetanilide will be affected. Use the redistilled aniline in this experiment.
2. In order to prevent the aniline from being oxidized during the reaction, zinc powder should be added.
3. If the reaction solution is cooled directly, the product is adhered to the flask easily. Pour the hot solution into cold water.
4. The solubility of acetanilide in 100 mL of water under different temperature (g/℃) is, 0.46/20, 0.56/25, 0.84/50, 3.45/80, 5.5/100.
5. If the crude product is colored, add appropriate amount of active charcoal to discolor during the recrystallization.

Questions

1. What measures should be taken to raise the yield in this experiment? Why is the fractional apparatus used?
2. Why must the temperature of distilling vapor on the top of the fractional column be controlled at 100~105 ℃?
3. What are the commonly used acetylation reagents? Please compare their acetylation capacity.
4. What is the use of the acetylation on aniline?

Experiment 56　Preparation of *p*-bromoacetanilide

Reaction equation

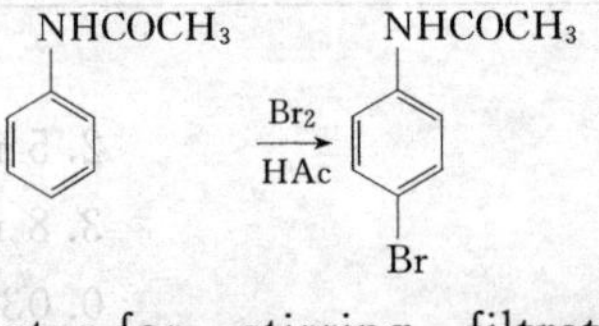

$$\text{C}_6\text{H}_5\text{NHCOCH}_3 \xrightarrow[\text{HAc}]{\text{Br}_2} p\text{-Br-C}_6\text{H}_4\text{NHCOCH}_3$$

Equipment　Motor stirrer; apparatus for: stirring, filtration with suction.

Materials	
acetanilide (FW 135.2)	2.70 g (20 mmol)
bromine (FW 159.8)	3.2 g (1.0 mL, 20 mmol)
glacial acetic acid	
sodium bisulfite	
ethanol	

Procedure

Set up a 50 mL 3-neck flask with a motor stirrer, a thermometer and pressure-equalizing dropping funnel. Connect the top of the funnel to a gas absorption trap to absorb hydrogen bromide that is evolved during the reaction. Place 2.70 g of acetanilide and 6 mL glacial acetic acid in the flask. Heat the flask on a warm water bath while stirring to dissolve acetanilide. Control the temperature of the bath at 45 ℃, add the solution containing 1.0 mL of bromine and 1.2 mL of glacial acetic acid from the dropping funnel while stirring at such a rate that the bromine color just disappears. After all the solution has been added, continue to stir the mixture at 45 ℃ for 1 h while heating the flask with water bath, and then at 60 ℃ for a while until the bromine vapor does not liberated.

Pour 40 mL of cold water with stirring into the flask. The crude p-bromoacetanilide precipitates. The mixture is sometimes colored by free bromine, which may be removed by adding a small amount of sodium bisulfite to decolor the yellow color justly. Cool the mixture to room temperature, collect the product by suction, wash it thoroughly with cold water, press it as dry as possible on the Büchner funnel, and recrystallized the crude product from 95% ethanol. After drying, record the weight, determine the melting point and calculate the percentage yield of p-bromoacetanilide.

Pure p-bromoacetanilide is a needle-shape crystal, m. p. 167 ℃.

Notes
1. The solution of bromine and glacial acetic acid must be prepared in the hood. Bromine is highly toxic and corrosive. It causes severe burns on contact with the skin. Once contact with the skin, treat bromine burns immediately with a liberal quantity of glycerol. Follow this by a thorough washing with water, dry the skin, and apply a healing ointment of fish liver oil. Always wear gloves and measure out in the hood when manipulating bromine.
2. Under the reaction conditions, p-bromoacetanilide (95%) and o-bromoacetanilide (5%) can be obtained. According to the difference of solubility between both isomers in methanol or the mixture of ethanol and water, the crude product may be purified by recrystallization to remove o-isomer which has bigger solubility in the same solvent.
3. The rate of adding the solution of bromine and glacial acetic acid should not be too fast, or else the reaction is vigorous, which will lead to liberate a portion of unreacted bromine together with hydrogen bromide which formed in the reaction and may give disubstitution product.

Questions

1. Why is the p-isomer main product in the monobromide product of acetanilide?
2. What is the effect of the temperature of the reaction on the result during the bromination?
3. What is the purpose of using the sodium bisulfite in the final treatment of the reaction mixture?
4. What impurities may exist in the product? How to remove them?
5. Interpret the infrared spectrum of p-bromoacetanilide and indicate the positions of the absorbing peaks about N—H and C=O bands.

Experiment 57 Preparation of p-bromoaniline

Reaction equation

NHCOCH$_3$-C$_6$H$_4$-Br $\xrightarrow{HCl}$ NH$_3$Cl$^+$-C$_6$H$_4$-Br $\xrightarrow{NaOH}$ NH$_2$-C$_6$H$_4$-Br

Equipment Apparatus for: reflux, filtration with suction.

Materials	
p-bromoacetanilide (FW 214.1)	2.6 g (12 mmol)
concentrated hydrochloric acid	3.4 mL
sodium hydroxide solution (20%)	
ethanol (95%)	
pH indicator paper	

Procedure

In a 50 mL 3-neck flask provided with a reflux condenser and pressure-equalizing dropping funnel, place 2.6 g of pure p-bromoacetanilide, 6 mL of 95% ethanol and several boiling stones. Heat the mixture to boil until all of the p-bromoacetanilide dissolve and then add dropwise 3.4 mL of concentrated hydrochloric acid from the dropping funnel to the flask. After all of the hydrochloric acid has been added, reflux the reaction mixture for 30 minutes. Dilute it with 10 mL of water and change the reflux apparatus into distillation apparatus. Distill the mixture on an asbestos gauze and collect about 8 mL of distillate, which consists of ethanol and water. Pour the residual mixture of p-bromoaniline hydrochloride into a beaker containing about 20 mL of ice-water. Place the beak in an ice-water bath, to cool the solution and cautiously make it alkaline with stirring by the addition of 20% sodium hydroxide (use pH paper). Collect the white precipitate with suction, wash it with a little water and dry it in air. Record the weight, the m.p. and yield of p-bromoaniline.

Pure p-bromoaniline is a colorless crystal, m.p. 65~66 ℃.

Notes

1. In order to avoid the reaction to be vigorous, the rate of adding concentrated hydrochloric acid should not be too fast.
2. The ethyl acetate, ethanol and water formed in the reaction can be removed by azeotropic distillation.
3. The crude product may be recrystallized from a mixture of water and ethanol.

Experiment 58 Preparation of p-nitroacetanilide

Reaction equation

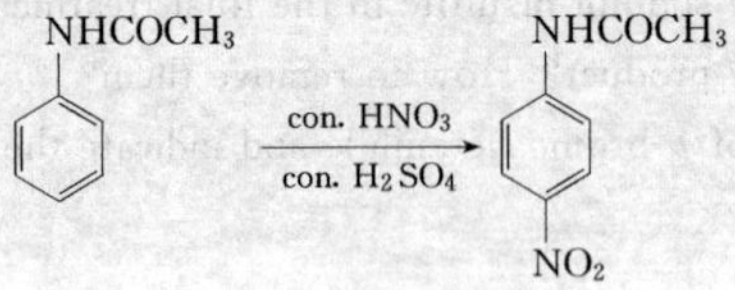

$$\underset{}{\text{C}_6\text{H}_5\text{NHCOCH}_3} \xrightarrow[\text{con. H}_2\text{SO}_4]{\text{con. HNO}_3} \underset{}{\text{p-O}_2\text{N-C}_6\text{H}_4\text{-NHCOCH}_3}$$

Equipment Motor stirrer; apparatus for: suction filtration, recrystallization

Materials	
acetanilide (FW 135.17)	6.75 g (50 mmol)
concentrated sulfuric acid (FW 98.08)	17.5 mL (32.22 g, 322 mmol)
concentrated nitric acid (FW 63.0)	4 mL (5.64 g, 62.7 mmol)
anhydrous sodium carbonate	

Chapter 3 Semimicro Synthetic Experiment of Typical Organic Compounds

Procedure

Place 17.5 mL of concentrated sulfuric acid in a 100 mL beaker. Add 6.75 g of dry acetanilide in small portions with stirring to the beaker at room temperature. Stir the mixture until all but traces of the acetanilide dissolve (a small amount of remaining solid will subsequently dissolve). Place the beaker in an ice-salt bath. Cool sulfuric acid solution of acetanilide to 0 ℃. Add 4 mL of concentrated nitric acid dropwise from a Pasteur pipet while stirring the reaction mixture using a motor stirrer to obtain well mixing in the viscous solution and do not permit the nitration temperature to rise above 5 ℃. After all of the concentrated nitric acid has been added, continue to stir the mixture for about 30 minutes at 5 ℃. Pour the solution slowly with stirring into a mixture of 25 mL of water and 25 g of crushed ice in a 250 mL beaker. The crude p-nitroacetanilide forms. Allow the crude product to stand for 10 minutes. After cooling, collect the product with suction, press it firmly on the filter, and then wash the filter cake with 60 mL of ice-water (30 mL×2) to remove the nitric and sulfuric acids. Press the material as dry as possible.

Transfer the crude product to a 500 mL beaker containing about 50 mL of water. Add sodium carbonate in small portions with stirring until the mixture is distinctly alkaline at the end of the mixing (pH≈10). Add about 50 mL of water to the beaker. Heat the mixture to boil. At the moment, p-nitroacetanilide does not hydrolyze, but o-nitro-acetanilide hydrolyzes to the o-nitroaniline. Cool the mixture slightly (should not be below 50 ℃). Collect the product by vacuum filtration with a Büchner funnel, press it firmly on the filter, wash the filter cake thoroughly with hot water (about 60~70 ℃) to remove the alkaline liquor of o-nitroaniline. Press the product as dry as possible. The crude moist p-nitroacetanilide can be used directly in the next step or allow it to dry in air on a watch glass. The white powder solid is obtained. The yield is about 8 g.

Pure p-nitroacetanilide is colorless crystal, m. p. 215.6 ℃.

Notes

1. This experiment may have the secondary reactions as follows:

   PhNHCOCH$_3$ + H$_2$O $\xrightarrow{H_2SO_4}$ PhNH$_2$ + CH$_3$COOH

   PhNHCOCH$_3$ $\xrightarrow{HNO_3, H_2SO_4}$ o-O$_2$N-C$_6$H$_4$-NHCOCH$_3$ + H$_2$O

   2 o-O$_2$N-C$_6$H$_4$-NHCOCH$_3$ + Na$_2$CO$_3$ + H$_2$O ⟶ 2 o-O$_2$N-C$_6$H$_4$-NH$_2$ + 2CH$_3$COONa + CO$_2$

2. To avoid the hydrolysis of acetanilide, the temperature of dissolving acetanilide can not be risen above 25 ℃. Require 25 minutes to dissolve acetanilide completely in this condition.

3. Do not permit the nitration temperature to rise above 5 ℃ because p-nitroacetanilide is the major product for the reaction of acetanilide with nitro-sulfuric acid below 5 ℃, and yet 25% of o-nitroacetanilide forms at 40 ℃.

4. Heating the alkaline liquor of nitroacetanilide (pH=10), o-isomer is hydrolyzed easily to o-nitroaniline and dissolved in the alkaline liquor, so o-nitroaniline can be removed by vacuum filtration.

5. According to the difference of solubility between of o-nitroacetanilide and p-nitroacetanilide in ethanol, the crude product may be purified by recrystallization from ethanol to

remove *o*-isomer which has bigger solubility in ethanol.
6. The crude, moist *p*-nitroacetanilide can be used directly for hydrolysis to *p*-nitroaniline.

Experiment 59 Preparation of *p*-nitroaniline

Reaction equation

$$\underset{NO_2}{\underset{|}{C_6H_4}}-NHCOCH_3 \xrightarrow[H_2O]{H_2SO_4} \underset{NO_2}{\underset{|}{C_6H_4}}-\overset{+}{N}H_3SO_4H^- \xrightarrow{NaOH} \underset{NO_2}{\underset{|}{C_6H_4}}-NH_2 + Na_2SO_4$$

Equipment Apparatus for: reflux, suction filtuation, recrystallization.

Materials
70% sulfuric acid (FW 98)
20% aqueous sodium hydroxide (FW 40)

Procedure

Transfer the moist, crude *p*-nitroacetanilide to a 100 mL round bottomed flask. Add 37.5 mL of 70% aqueous sulfuric acid solution and two boiling stones into the flask. Connect the flask to a reflux condenser. Heat the mixture slowly on an asbestos gauge and boil gently for 25 minutes. The material gradually dissolves and an orange-colored transparent solution is formed. After cooling, pour the mixture into 100 mL of ice-water. If the precipitate separates, filter off the precipitate with suction (it may be *o*-nitroaniline which does not dissolve in aqueous sulfuric acid). The filter liquor is the solution of *p*-nitroaniline sulfate. After the solution is cooled thoroughly, add 20% aqueous sodium hydroxide slowly with thorough stirring into the solution to precipitate *p*-nitroaniline (pH≈8). Cool the product to room temperature. Collect the orange-yellow precipitated *p*-nitroaniline on a Büchner funnel with vacuum filtration and wash it with cold water to remove the alkaline solution. Dry it in air. The yield is about 4 g.

To obtain pure *p*-nitroaniline recrystallize the product from a large amount of hot water. Pure *p*-nitroaniline is a yellow needle-shape crystal, m. p. 147.5 ℃.

Note

The solubility of *p*-nitroaniline in 100 mL of water is 0.08 g at 18.5 ℃ and 2.2 g at 100 ℃.

Questions

1. Can *p*-nitroaniline be prepared directly by nitration of aniline? Why?
2. How can *o*-nitroacetanilide be removed from the crude *p*-nitroacetanilide in this experiment?
3. During the acidic hydrolysis of *p*-nitroacetanilide, why is the solution transparent? What principle can be taken to precipitate the product?

3.21.4 The multostep synthesis of *p*-aminobenzoic acid from *p*-toluidine

Multistep synthesis route 5   The preparation of *p*-aminobenzoic acid

$$\underset{CH_3}{\underset{|}{C_6H_4}}-NO_2 \xrightarrow{Fe, H^+} \underset{CH_3}{\underset{|}{C_6H_4}}-NH_2 \xrightarrow{(CH_3CO)_2O} \underset{CH_3}{\underset{|}{C_6H_4}}-NHCOCH_3 \xrightarrow[CH_3COONa]{KMnO_4} \underset{COO^-}{\underset{|}{C_6H_4}}-NHCOCH_3 \xrightarrow{H^+} \underset{COOH}{\underset{|}{C_6H_4}}-NH_2$$

Experiment 31 　　　　Experiment 60 　　　　Experiment 61

Chapter 3 Semimicro Synthetic Experiment of Typical Organic Compounds

p-Aminobenzoic acid (PABA) is a member of the group of substances associated with the vitamin B complex. It is present as the central unit of folic acid, vitamin B_{10}, which is made up of a pteridine unit, a p-aminobenzoic unit and a glutamic acid unit. P-Aminobenzoic acid is required for folic acid synthesis by some bacteria, and the sulfa drugs are thought to interfere with this synthesis. Without folic acid, the bacteria cannot synthesize the nucleic acid required for growth. As a result, bacterial growth is arrested until the body's immune system can respond and kill the bacteria.

$$H_2N-C_6H_4-COOH$$
p-aminobenzoic acid
(PABA)

pteridine residue — PBAB residue — glutamic residue

Because PABA can absorb the ultraviolet component of solar radiation, it also finds an important application in sunscreen preparations.

The present synthesis method of p-aminobenzoic acid is simpler for small-scale laboratory preparations and starts from p-toluidine ($CH_3C_6H_4NH_2$). This is acetylated, and the resulting N-acetyl-p-toluidine is oxidized by potassium permanganate. This furnishes p-acetamidebenzoic acid, which is hydrolyzed by heating with hydrochloric acid, to give p-aminobenzoic acid (PBAB).

Experiment 60 Preparation of *N*-acetyl-*p*-toluidine

Reaction equation

$$H_3C-C_6H_4-NH_2 + (CH_3CO)_2O \longrightarrow H_3C-C_6H_4-NHCOCH_3$$

Equipment Apparatus for: filtration with suction, reflux.

Materials	
p-toluidine (FW 107.16)	2.00 g (18.7 mmol)
acetic anhydride (FW 102.09)	2.4 mL (2.59 g, 25.4 mmol)

Procedure

Place 2.00 g p-toluidine, 2.4 mL of acetic anhydride and several boiling stones in a 10 mL round bottomed flask provided with a reflux condenser. The reaction should start immediately and be exothermic, and the entire solid is dissolved. Reflux the solution for 10 minutes. Pour the hot reaction mixture with stirring into 50 mL of cold water. A light yellow solid should appear at this point. After cooling thoroughly, filter the solid by suction, wash it three times with cold water and allow it to dry. The weight of the crude product is 2.60 g, the percentage yield is 93% and the m.p is 147~149 ℃. The crude N-acetyl-p-toluidine is recrystallized from the mixing solvent of ethanol and water.

Notes

1. All apparatus must be dry in this experiment.

2. Acetic anhydride should be redistilled before use. Collect the fraction boiling at 138~139 ℃.

Experiment 61 Preparation of *p*-aminobenzoic acid

Reaction equations

$$CH_3CONH-C_6H_4-CH_3 \xrightarrow[OH^-]{KMnO_4} CH_3CONH-C_6H_4-COO^- \xrightarrow{H^+}$$

$$CH_3CONH-C_6H_4-COOH \xrightarrow[\Delta]{H^+} H_2N-C_6H_4-COOH$$

Equipment Motor stirrer; apparatus for: filtration with suction, reflux.

Materials	
N-acetyl-*p*-toluidine (FW 149.17)	2.60 g (17.4 mmol)
potassium permanganate (FW 158.04)	8.00 g (50.6 mmol)
sodium acetate (FW 82)	2.00 g (24.4 mmol)
sulfuric acid (20%)	
hydrochloric acid (1:1)	
sodium hydroxide solution (20%)	
glacial acetic acid	
litmus paper	

Procedure

Place 2.60 g of the previously prepared *N*-acetyl-toluidine in a 250 mL beaker, along with 2.00 g of sodium acetate, 8.00 g of potassium permanganate and 60 mL of water. Stir the mixture with a motor stirrer at room temperature and heat gently the reaction mixture for 30 minutes. The mixture is dark brown and a large amount of brown precipitate forms. Filter off the precipitated manganese dioxide from the hot solution and wash the manganese dioxide with a small amount of hot water. Cool the colorless filtrate to room temperature and acidify it with 20% sulfuric acid solution (pH=1~2). A large amount of white solid should form at this point. Collect the *p*-acetamidebenzoic acid with suction and press it as dry as possible. The melting point of the pure *p*-acetamidebenzoic acid is 250~252 ℃.

Place the crude *p*-acetamidebenzoic acid from the preceding step, 30 mL of the dilute hydrochloric acid solution (1 volume of concentrated acid to 1 volume of water) and two boiling stones in a 50 mL round-bottomed flask. Add a reflux condenser, and boil the mixture gently for 30 minutes. Cool the reaction mixture, transfer it to a 200 mL beaker, add 20 mL of ice-water, and make the solution just alkaline to litmus paper with 20% sodium hydroxide solution. For each 30 mL of the final solution, add 1 mL of glacial acetic acid, chill the solution in an ice bath, and initiate the crystallization. If necessary, induce crystallization by scratching the inside of the beaker with a glass rod or adding a small seed crystal. Collect the product with suction and allow it to dry in air. The product is a light yellow needle-shape crystal and the weight is 1.50 g. The melting point of the pure *p*-aminobenzoic acid is 186~187 ℃.

Notes

1. To avoid the reaction is too vigorous, the flame must be controlled and the rate of the stirring mixture should be moderate during the oxidation reaction with potassium permanganate serving as the oxidizing agent (this reaction is exothermic).

Chapter 3 Semimicro Synthetic Experiment of Typical Organic Compounds

2. If the filtrate shows the presence of excess permanganate by its purple color, add a small amount of sodium bisulfite until the purple color disappears.
3. Be careful to adjust the acid-base degree in each step.

Questions

1. In which organic reactions the amino needs to be protected? Why?
2. What are the characters of the reactions used to protect the functional groups? Explain with examples.

3.21.5 The multistep synthesis of 5,5-diphenylhydantion and *erythro*-1,2-diphenyl-1,2-ethandiol from benzaldehyde

Multi step synthesis route 6

The preparation of 5,5-diphenylhydantion and *erythro*-1,2-diphenyl-1,2-ethandiol

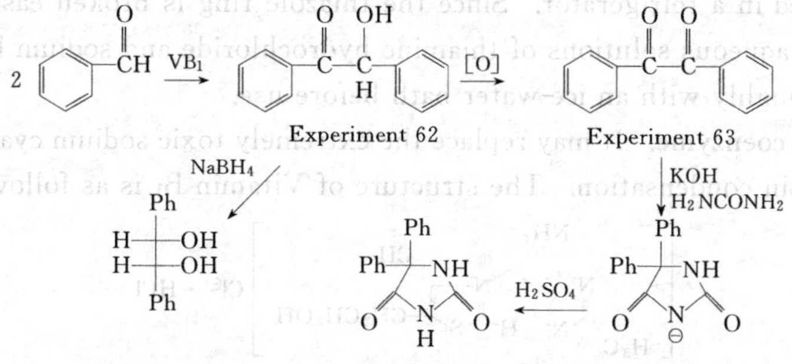

Experiment 62 Preparation of benzoin

In this experiment, a benzoin condensation of benzaldehyde is carried out with a biological coenzyme, thiamine hydrochloride, as the catalyst.

Reaction equation

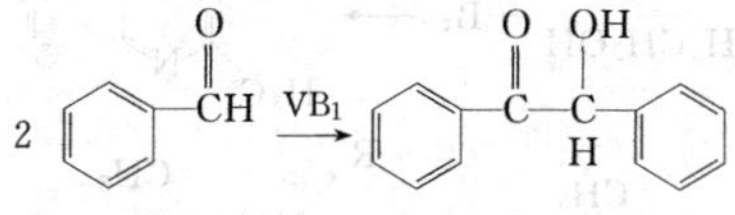

benzaldehyde → benzoin

Equipment Apparatus for: suction filtration, recrystallization.

Materials	
benzaldehyde (FW 106.13)	1.5 mL (1.56 g, 1.47 mmol)
vitamin B_1	0.30 g
ethanol (95%)	
sodium hydroxide solution (2.5 mol/L)	1.0 mL

Procedure

Dissolve 0.30 g of thiamine hydrochloride (Vitamin B1) in 1.0 mL of water in a 30 mL Erlenmeyer flask. When all of the VB_1 has dissolved, add 3.0 mL of 95% ethanol. Stopper the flask and cool the resulting solution with an ice-water bath, slowly add 1.0 mL of cold 2.5 mol/L sodium hydroxide to the flask, and make pH of the solution is about 10~11. Rapidly add 1.5 mL of benzaldehyde to the reaction mixture and sufficiently mix the solution. Stopper the flask and allow it to stand at room temperature for one day. At the end of to reaction period, the benzoin should have separated as fine while crystals. When the crystallization has completed, collect the crude product by suction filtration with a Büchner funnel and wash it with a small amount of ice-cold water. Press the crystals as dry as possible and spread them on a fresh filter paper to dry in air. Recrystallize the product from 95% ethanol. The weight of benzoin is 0.6 g, and the percentage yield is 38.5%. The pure product is a white needle-shape crystal, m. p. 134~136 ℃.

Notes

1. Vitamin B_1 (thiamine) exists in the form of thiamine hydrochloride. It is stable in the acidic condition, but it absorbs water easily, and it is a heat-sensitive reagent, the thiamine in aqueous solution is oxidated easily by oxygen in air. The rate of oxidation may be accelerated by light and some ions such as cupric ion, ironic ion and manganic ion. It should be stored in a refrigerator. Since the thiazole ring is broken easily in basic solution, both the aqueous solutions of thiamine hydrochloride and sodium hydroxide should be cooled thoroughly with an ice-water bath before use.

2. Vitamin B_1 is a coenzyme. It may replace the extremely toxic sodium cyanide, as the catalyst, in benzoin condensation. The structure of Vitamin B_1 is as follows:

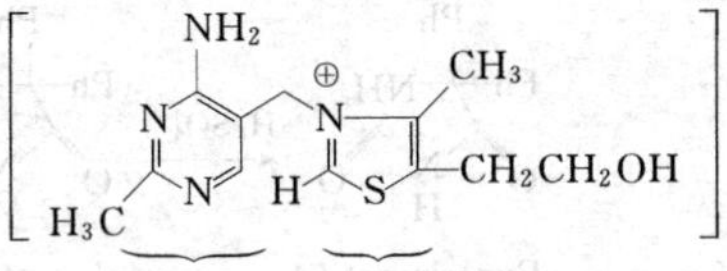

pyrimidine ring   thiazole ring

The proton on the thiazole ring component of thiamine is a relatively acidic proton. The proton is easily removed in the basic condition and a carbanion is produced. So these catalyze the formation of benzoin. The reaction mechanism is as follows:

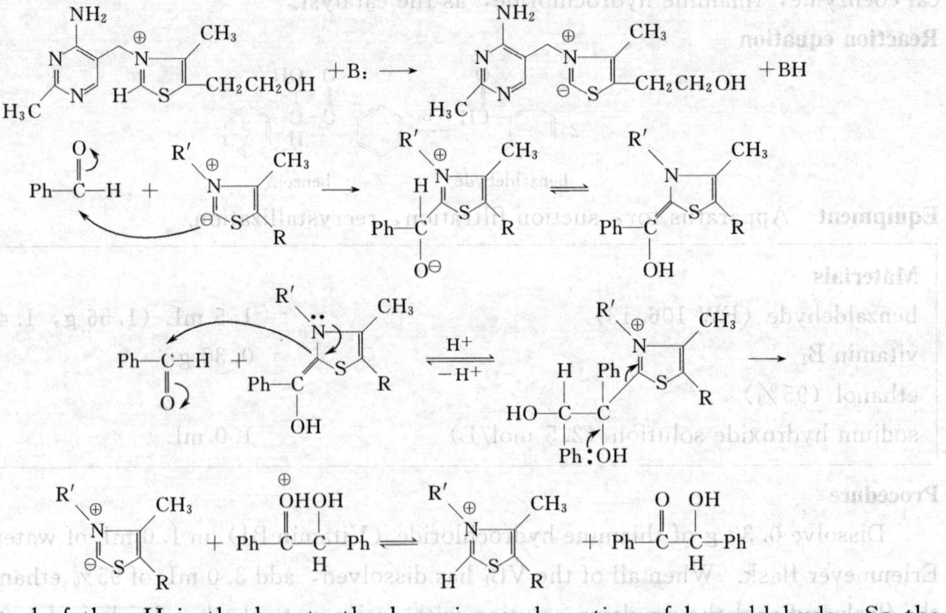

3. The control of the pH is the key to the benzoin condensation of benzaldehyde. So the benzaldehyde used for this experiment must be free of benzoic acid. The benzaldehyde must be redistilled before use.

4. Benzoin is a perfumery. The DL-type is a hexagon monoclinic rhombic crystal. Both D-

Chapter 3 Semimicro Synthetic Experiment of Typical Organic Compounds

type and L-type are needle-shape crystals.

Questions

1. How does Vitamin B_1 catalyze the reaction of benzoin condensation?
2. What is the function of the sodium hydroxide in this experiment? What is the theoretical amount of it?
3. Interpret the main absorbing peaks in the spectrum of benzoin.

Experiment 63   Preparation of benzil

α-Diketones are very useful in the preparation of cyclic compounds of great value. Benzil, an α-diketone, is prepared by the oxidation of an α-hydroxyketone, benzoin. This oxidation can easily be done with mild oxidizing agent. In this experiment, the oxidation is performed with ironic chloride hexahydrate.

Reaction equation

$$\underset{\text{benzoin}}{Ph-\underset{\underset{O}{\parallel}}{C}-\underset{\underset{OH}{|}}{CH}-Ph} \xrightarrow{[O]} \underset{\text{benzil}}{Ph-\underset{\underset{O}{\parallel}}{C}-\underset{\underset{O}{\parallel}}{C}-Ph}$$

Equipment   Apparatus for: reflux, suction filtration, recrystallization.

Materials	
benzoin (FW 212.3)	0.4 g (1.9 mmol)
ironic chloride hexahydrate (FW 270.30)	1.7 g (6.3 mmol)
ethanol	

Procedure

Place 2 mL of glacial acetic acid, 1 mL of water, 1.7 g of ironic chloride hexahydrate, and two boiling stones in a 30 mL round-bottomed flask fitted with a reflux condenser. Heat gently the mixture to boil and shake the mixture occasionally. Stop heating and cool the solution slightly. Add 0.4 g of benzoin to the flask and continue to heat. Reflux the mixture for 1 h. Then add 10 mL of water to the flask and heat the mixture again. After boil, cool the solution, yellow solid separates. Collect the crude product by suction filtration, and wash it three times with a small amount of cold water. It may be purified by recrystallization from 75% aqueous ethanol. The weight of the dry product is 0.3 g, the percentage yield is 75.2%.

Pure benzil is a yellow needle-shape crystal, m.p. 95 ℃.

Experiment 64   Preparation of 5,5-diphenylhydantion

Reaction equation

$$\underset{}{Ph-\underset{\underset{O}{\parallel}}{C}-\underset{\underset{O}{\parallel}}{C}-Ph} + \underset{}{H_2N-\underset{\underset{O}{\parallel}}{C}-NH_2} \xrightarrow{KOH} \underset{}{\overset{Ph}{\underset{}{\underset{}{\overset{Ph}{|}}}}\begin{matrix}\\\end{matrix}} \xrightarrow{H_2SO_4} \underset{}{}$$

Equipment   Motor stirrer; apparatus for: reflux, suction filtuation, recrystallization.

Materials	
benzil (FW 210.26)	0.30 g (1.43 mmol)
urea (FW 60.06)	0.18 g (3.0 mmol)

281

potassium hydroxide solution (15.2 mol/L)
sulfuric acid (6 mol/L)
ethanol (95%)

Procedure

To a 25 mL 3-neck flask provided with a motor stirrer and a reflux condenser, add 0.30 g of benzil, 0.18 g of urea, 3.0 mL of 95% ethanol, and 1.0 mL of 15.2 mol/L potassium hydroxide. Reflux the mixture with stirring in water bath for 2.5 h. Cool the reaction mixture and remove the sparingly insoluble solid by gravity filtration. Cool the filtrate further in an ice-water bath and acidify the filtrate with 6 mol/L sulfuric acid (pH≈3). The white powdered solid is obtained. Collect the product by suction filtration, and wash it thoroughly with water to remove the inorganic salt. The crude product may be recrystallized from 95% ethanol. The white needle-shape crystal is obtained. The weight is 0.2 g, the percentage yield is 55.6%, and the m.p. is 286~295 ℃.

Note

The sodium salt of 5,5-diphenylhydantion, dilantin, is an anticonvulsant used for the treatment of epilepsy. The synthetic process of 5,5-diphenylhydantion is as follows:

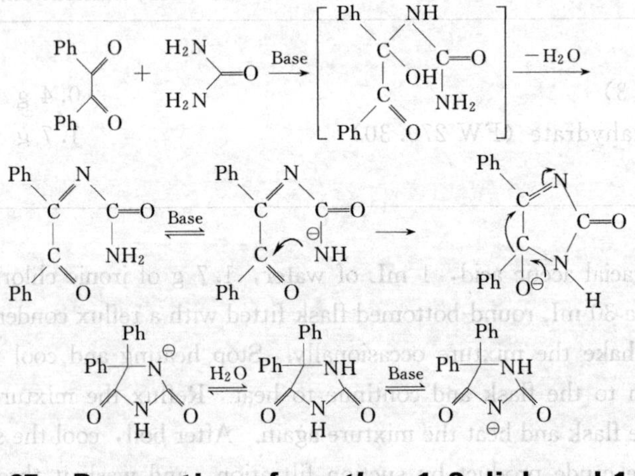

Experiment 65　Preparation of *erythro*-1,2-diphenyl-1,2-ethandiol

Reaction equation

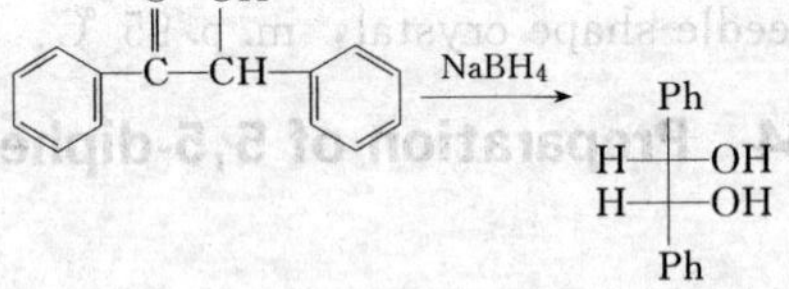

Equipment　Magnetic stirrer; apparatus for: suction filtration, recrystallization.

Materials	
benzoin (FW 212.3)	0.5 g (2.36 mmol)
sodium borohydride (FW 37.8)	0.05 g (1.32 mmol)
ethanol (FW 95%)	
petroleum ether (b.p. 60~90 ℃)	
hydrochloric acid (18%)	

Chapter 3 Semimicro Synthetic Experiment of Typical Organic Compounds

Procedure

Place a magnetic stirrer bar, 0.5 g of benzoin, and 5 mL of 95% ethanol in a 30 mL Erlenmeyer flask. While stirring, add 0.05 g of sodium borohydride to the mixture. Stir the mixture for another 15 minutes at room temperature. Cool the flask in an ice-water bath and decompose the excess sodium borohydride by adding 2.0 mL of water followed by careful and dropwise addition of 0.25 mL of 18% hydrochloric acid solution. Stir the mixture for a while. The white precipitation separates gradually. Stop the stir and continue to cool the flask until the product separates completely. Collect the white product by suction filtration, wash it with a small amount of cold water and allow the product to dry in air. The weight of the crude product is 0.4 g, the percentage yield is 79.1%. The crude product may be purified by recrystallization from acetone-petroleum ether (b.p. 60~90 ℃) or acetone-water. The melting point of the pure product is 138 ℃.

Note

Conceptually, reduction of the keto group of benzoin could give a mixture of the *erythro-* and *threo-*diols. However, the reduction of benzoin using sodium borohydride as reducing agent is a stereoselective reaction. The *erythro*-1,2-diphenyl-1,2-ethandiol is the major product. The structure of the diol can be confirmed by comparison of the melting point of the product with that of the authentic *erythro*- (m.p. 137 ℃) or *threo*-products (m.p. 119 ℃).

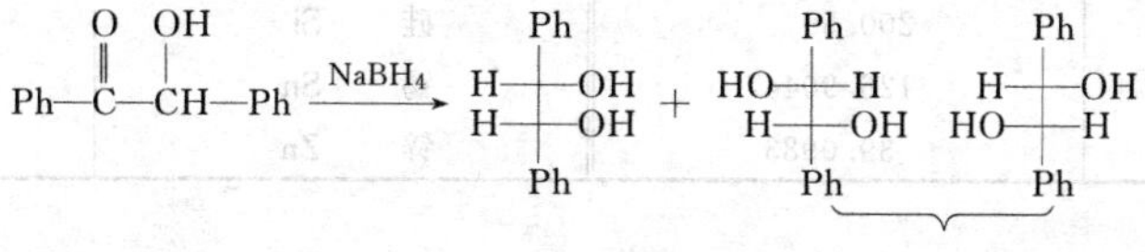

$$Ph-\overset{O}{\underset{\|}{C}}-\overset{OH}{\underset{|}{CH}}-Ph \xrightarrow{NaBH_4} \begin{array}{c}Ph\\H{-}{-}OH\\H{-}{-}OH\\Ph\end{array} + \begin{array}{c}Ph\\HO{-}{-}H\\H{-}{-}OH\\Ph\end{array} \quad \begin{array}{c}Ph\\H{-}{-}OH\\HO{-}{-}H\\Ph\end{array}$$

erythro   *threo*

Question

Please explain the reason that *erythro*-diol is a major product in the reduction of benzoin using sodium borohydride as reducing agent.

附 录

附录1 常用元素的相对原子质量（2004）

元素名称		相对原子质量	元素名称		相对原子质量
银	Ag	107.8682	镁	Mg	24.3050
铝	Al	26.981538	锰	Mn	54.938049
溴	Br	79.904	氮	N	14.0067
碳	C	12.0107	钠	Na	22.989770
钙	Ca	40.078	镍	Ni	58.6934
氯	Cl	35.453	氧	O	15.9994
铬	Cr	51.9961	磷	P	30.973761
铜	Cu	63.546	铅	Pb	207.2
氟	F	18.9984	钯	Pd	106.42
铁	Fe	55.845	铂	Pt	195.078
氢	H	1.00794	硫	S	32.065
汞	Hg	200.59	硅	Si	28.0855
碘	I	126.90447	锡	Sn	118.710
钾	K	39.0983	锌	Zn	65.409

附录2 常用有机溶剂的纯化

市售的有机溶剂有工业、化学纯、分析纯等各种规格。在有机合成中，通常根据反应特性来选择适宜规格的溶剂，以便使反应顺利进行而又不浪费试剂。但对某些反应来说，对溶剂纯度要求特别高，即使只有微量有机杂质和痕量水的存在，常常对反应速率和产率也会发生很大的影响，这就需要对溶剂进行纯化。

1. 无水乙醇

沸点78.3℃，折射率 n_D^{20} 1.3616，相对密度 d_4^{20} 0.7893。

普通乙醇含量为95%。与水形成恒沸溶液，不能用一般分馏法除去水分。初步脱水常以生石灰为脱水剂，这是因为：第一，生石灰来源方便；第二，生石灰或由它生成的氢氧化钙皆不溶于乙醇。操作方法：将600mL 95%乙醇置于1000mL圆底烧瓶内，加入100g左右新鲜煅烧的生石灰，放置过夜，然后在水浴中回流5～6h，再将乙醇蒸出。如此所得乙醇质量分数约为99.5%，相当于市售无水乙醇。若需要绝对无水乙醇，可用金属镁或金属钠将制得的无水乙醇或者分析纯的无水乙醇（含量不低于99.5%）进一步按下述方法处理。

方法1 取1000mL圆底烧瓶，安装回流冷凝管，在冷凝管上端附加一只氯化钙干燥管。瓶内放置2～3g干燥洁净的镁条与0.3g碘，加入30mL 99.5%乙醇，水浴加热至碘粒完全消失（如果不起反应，可再加入数小粒碘），然后继续加热，待镁完全溶解后，加入500mL 99.5%乙醇和几粒沸石，继续加热回流1h，蒸出乙醇。弃去先蒸出的10mL，其后蒸出的收集于干燥洁净的瓶内储存。如此所得乙醇纯度可超过99.95%。此方法脱水是按下列反应进行的。

$$Mg + 2C_2H_5OH \longrightarrow H_2 + Mg(OC_2H_5)_2$$

$$Mg(OC_2H_5)_2 + 2H_2O \longrightarrow Mg(OH)_2 + 2C_2H_5OH$$

由于无水乙醇具有非常强的吸湿性，故在操作过程中必须防止吸入水汽，所用仪器需事先置于烘箱内干燥。

方法 2 可采用金属钠除去乙醇中含有的微量水分。金属钠与金属镁的作用是相似的，但是单用金属钠并不能达到完全去除乙醇中含有的水分的目的，因为这一反应有如下平衡。

$$C_2H_5ONa + H_2O \rightleftharpoons NaOH + C_2H_5OH$$

若要使平衡向右移动，可以加过量的金属钠，增加乙醇钠的生成量。但这样做，造成了乙醇的浪费。因此，通常的办法是加入高沸点的酯，如邻苯二甲酸二乙酯或琥珀酸乙酯，通过皂化反应除去反应中生成的氢氧化钠。这样制得的乙醇，只要能严格防潮，含水质量分数可以低于 0.01%。

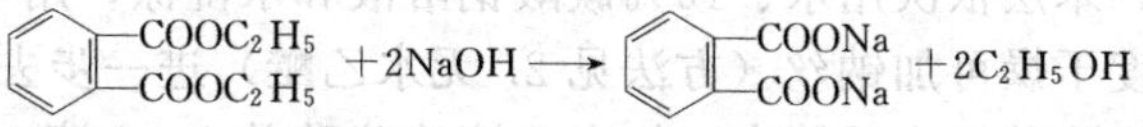

$$\text{邻苯二甲酸二乙酯} + 2NaOH \longrightarrow \text{邻苯二甲酸二钠} + 2C_2H_5OH$$

操作方法：取 1000mL 圆底烧瓶加入 500mL 99.5% 乙醇，安装回流冷凝管和干燥管，加入 3.5g 金属钠，待其完全作用后，再加入 12.5g 琥珀酸乙酯或 14g 邻苯二甲酸二乙酯，回流 2h，然后蒸出乙醇。弃去先蒸出的 10mL，其后的收集于干燥洁净的瓶内储存。

测定乙醇中含有的微量水分，可加入乙醇铝的苯溶液，若有大量的白色沉淀生成，证明乙醇中含有的水的质量分数超过 0.05%。此法还可测定甲醇中含 0.1%、乙醚中含 0.005% 及醋酸乙酯中含 0.1% 的水分。

2. 无水乙醚

沸点 34.6℃，折射率 n_D^{20} 1.3527，相对密度 d_4^{15} 0.7193。

普通乙醚中常含有水和乙醇。在储存乙醚期间，由于与空气接触和光的照射，还可能产生二乙基过氧化物 $(C_2H_5)_2O_2$。这些杂质的存在，对于一些要求用无水乙醚作溶剂的实验（如 Grignard 试剂的合成）是不适合的，特别是有过氧化物存在时，还有发生爆炸的危险。

纯化乙醚可选择下述方法。

① 在 1000mL 分液漏斗中加入 500mL 的普通乙醚，再加入 50mL 10% 新配制的亚硫酸氢钠溶液；或加入 10mL 硫酸亚铁溶液和 100mL 水充分振摇（若乙醚中不含过氧化物，则可省去这步操作）。然后分出醚层，用饱和食盐溶液洗涤两次，再用无水氯化钙干燥数天，间歇振摇，过滤，蒸馏。将蒸出的乙醚放在干燥的磨口试剂瓶中，压入金属钠丝干燥。如果乙醚干燥不够，当压入钠丝时，即会产生大量气泡。遇到这种情况，暂时先用装有氯化钙干燥管的软木塞塞住，放置 24h 后，过滤到另一干燥试剂瓶中，再压入金属钠丝，至不再产生气泡，钠丝表面保持光泽，即可盖上磨口玻璃塞备用。

硫酸亚铁溶液的制备：在 100mL 水中，慢慢加入 6mL 浓硫酸，再加入 60g 硫酸亚铁溶解即得。此溶液必须使用时配制，放置过久易氧化变质。

② 经无水氯化钙干燥后的乙醚，也可用 4A 型分子筛干燥，所得绝对无水乙醚能直接用于格氏反应。

为了防止发生事故，对于一般条件下保存的或储存过久的乙醚，除已鉴定不含过氧化物的以外，蒸馏时，都不要全部蒸干。

3. 甲醇

沸点 64.96℃，折射率 n_D^{20} 1.3288，相对密度 d_4^{20} 0.7914。

通常所用的甲醇含水质量分数不超过 0.5%～1%。由于甲醇和水不能形成共沸混合物，因此可通过高效的精馏柱将少量水除去。精制甲醇含有 0.02%的丙酮和 0.1%的水，一般已可应用。如需制得无水甲醇，可用金属镁（方法见 1. 无水乙醇）。甲醇有毒，处理时应避免吸入其蒸气。

4. 无水、无噻吩苯

沸点 80.1℃，折射率 n_D^{20} 1.5011，相对密度 d_4^{20} 0.87865。

分析纯的苯通常可供直接使用，若需要无水苯则可直接用无水氯化钙干燥过夜，过滤后压入钠丝（方法见 2. 无水乙醚）。普通苯中含有少量的水（可达 0.02%），由煤焦油加工得来的苯还含有少量噻吩（沸点 84℃），不能用分馏或分步结晶等方法分离除去。为制得无水、无噻吩的苯可采用下列方法精制。

在分液漏斗内将普通苯及相当于苯体积 15%的浓硫酸一起摇荡，摇荡后将混合物静置，弃去底层的酸液，再加入新的浓硫酸，这样重复操作直至酸层呈现无色或淡黄色，且检验无噻吩为止。分去酸层，苯层依次用水、10%碳酸钠溶液和水洗涤，用氯化钙干燥，蒸馏收集 80℃的馏分。若需高度干燥可加钠丝（方法见 2. 无水乙醚）进一步去水。

噻吩的检验：取 5 滴苯于小试管中，加入 5 滴浓硫酸及 1～2 滴 1% α, β-吲哚醌-浓硫酸溶液，振荡片刻。如呈墨绿色或蓝色，表示有噻吩存在。

5. 丙酮

沸点 56.2℃，折射率 n_D^{20} 1.3588，相对密度 d_4^{20} 0.7899。

普通丙酮中往往含有少量水及甲醇、乙醛等还原性杂质，可用下列方法精制。

① 于 1000mL 丙酮中加入 5g 高锰酸钾回流，以除去还原性杂质。若高锰酸钾紫色很快消失，需要加入少量高锰酸钾继续回流，直至紫色不再消失为止。蒸出丙酮，用无水碳酸钾或无水硫酸钙干燥后，过滤，蒸馏，收集 55～56.5℃的馏分。

② 于 1000mL 丙酮中加入 40mL 10%硝酸银溶液及 35mL 0.1mol/L 氢氧化钠溶液，振荡 10min，除去还原性杂质。过滤，滤液用无水硫酸钙干燥后，蒸馏，收集 55～56.5℃的馏分。

6. 乙酸乙酯

沸点 77.06℃，折射率 n_D^{20} 1.3723，相对密度 d_4^{20} 0.9003。

乙酸乙酯沸点在 76～77℃部分的质量分数达 99%时，已可应用。普通乙酸乙酯含量为 95%～98%，含有少量水、乙醇及醋酸，可用下列方法精制。

于 1000mL 乙酸乙酯中加入 100mL 醋酸酐、10 滴浓硫酸，加热回流 4h，除去乙醇及水等杂质，然后进行分馏。馏出液用 20～30g 无水碳酸钾振荡，再蒸馏。最后产物的沸点为 77℃，纯度达 99.7%。

7. 二硫化碳

沸点 46.25℃，折射率 n_D^{20} 1.6319，相对密度 d_4^{20} 1.2632。

二硫化碳是有毒的化合物（有使血液和神经组织中毒的作用），又具有高度的挥发性和易燃性，所以在使用时必须注意，避免接触其蒸气。一般有机合成实验中对二硫化碳纯度要求不高，在普通二硫化碳中加入少量磨碎的无水氯化钙，干燥数小时，然后在水浴上（温度 55～65℃）蒸馏收集。

如需要制备较纯的二硫化碳，则需将试剂级的二硫化碳用质量分数为 0.5%的高锰酸钾水溶液洗涤 3 次，除去硫化氢，再用汞不断振荡除硫。最后用 2.5%硫酸汞溶液洗涤，除去所有恶臭（剩余的 H_2S），再经氯化钙干燥，蒸馏收集。其纯化过程的反应式如下。

$$3H_2S + 2KMnO_4 \longrightarrow 2MnO_2\downarrow + 3S\downarrow + 2H_2O + 2KOH$$
$$Hg + S \longrightarrow HgS\downarrow$$
$$HgSO_4 + H_2S \longrightarrow HgS\downarrow + H_2SO_4$$

8. 氯仿

沸点 61.7℃，折射率 n_D^{20} 1.4459，相对密度 d_4^{20} 1.4832。

普通用的氯仿含有质量分数为 1% 的乙醇，这是为了防止氯仿分解为有毒的光气，作为稳定剂加进去的。为了除去乙醇，可以将氯仿用其体积一半的水在分液漏斗中振荡数次，然后分出下层氯仿，用无水氯化钙干燥数小时后蒸馏。

另一种精制方法是将氯仿与少量浓硫酸一起振荡两三次。每 1000mL 氯仿，用浓硫酸 50mL 洗涤。分去酸层以后的氯仿用水洗涤，干燥，然后蒸馏。除去乙醇的无水氯仿应保存于棕色瓶子里，并且不要见光，以免分解。

9. 二氯甲烷

沸点 40℃，折射率 n_D^{20} 1.4242，相对密度 d_4^{20} 1.3266。

使用二氯甲烷比氯仿安全，因此常用它来代替氯仿作为比水密度大的萃取溶剂，普通的二氯甲烷一般都能直接作为萃取剂使用。如需纯化，可用 5% 碳酸钠溶液洗涤，再用水洗涤，然后用无水氯化钙干燥，蒸馏收集 40~41℃ 的馏分。

10. 石油醚

石油醚为轻质石油产品，是低相对分子质量的烃类（主要是戊烷和己烷）的混合物。其沸程为 30~150℃，收集的温度区间一般为 30℃ 左右，如有 30~60℃、60~90℃、90~120℃ 等沸程规格的石油醚。石油醚中含有少量不饱和烃，沸点与烷烃相近，用蒸馏法无法分离，必要时可用浓硫酸和高锰酸钾把它除去。通常将石油醚用其体积 1/10 的浓硫酸洗涤两三次，再用 10% 硫酸加入高锰酸钾配成的饱和溶液洗涤，直至水层中的紫色不再消失为止。然后再用水洗，经无水氯化钙干燥后蒸馏。如需要绝对干燥的石油醚则加入钠丝（方法见 2. 无水乙醚）除水。

11. 吡啶

沸点 115.5℃，折射率 n_D^{20} 1.5095，相对密度 d_4^{20} 0.9819。

分析纯的吡啶含有少量水分，但已可供一般应用。如要制得无水吡啶，可与粒状氢氧化钾或氢氧化钠一同回流，然后隔绝潮气蒸出备用。干燥的吡啶吸水性很强，保存时应将容器口用石蜡封好。

12. N,N-二甲基甲酰胺（DMF）

沸点 149~156℃，折射率 n_D^{20} 1.4305，相对密度 d_4^{20} 0.9487。

N,N-二甲基甲酰胺含有少量的水分。在常压蒸馏时会分解，产生二甲胺与一氧化碳。若有酸或碱存在时，分解加快，所以在加入固体氢氧化钾或氢氧化钠室温放置数小时后，即有部分分解。因此，最好用硫酸钙、硫酸镁、氧化钡、硅胶或分子筛干燥，然后减压蒸馏，收集 76℃/4.8kPa（36mmHg）以下的馏分。如其中含水较多时，可加入其体积 1/10 的苯，在常压及 80℃ 以下蒸去水和苯，然后用硫酸镁或氧化钡干燥，再进行减压蒸馏。

N,N-二甲基甲酰胺中如有游离胺存在，可用 2,4-二硝基氟苯产生颜色来检查。

13. 四氢呋喃（THF）

沸点 67℃（64.5℃），折射率 n_D^{20} 1.4050，相对密度 d_4^{20} 0.8892。

四氢呋喃是具有乙醚气味的无色透明液体，常在 Grignard 反应和氢化锂铝的还原中用来代替乙醚作为溶剂。市售的四氢呋喃常含有少量水及过氧化物。如要制得无水四氢呋喃可与氢化锂铝在隔绝潮气下回流（通常 1000mL 约需 2~4g 氢化锂铝）除去其中的水和过氧化物，然后在常压下蒸馏，收集66℃的馏分。精制后的液体应在氮气中保存，如需较久放置，应加质量分数为0.025%的2,6-二叔丁基-4-甲基苯酚作为抗氧化剂。处理四氢呋喃时，应先少量进行试验，确定只有少量水和过氧化物，作用不过于猛烈时，方可进行。

四氢呋喃中的过氧化物可用酸化的碘化钾溶液来检验。如过氧化物很多，应另行处理为宜。

14. 二甲亚砜（DMSO）

沸点189℃，折射率n_D^{20} 1.4770，相对密度d_4^{20} 1.1014。

二甲亚砜是能与水互溶的高极性的非质子溶剂，因而广泛用作有机反应和光谱分析中的试剂。它易吸潮，常压蒸馏时还会分解。若要制备无水二甲亚砜，可以用活性氧化铝、氧化钡或硫酸钙干燥过夜。然后滤去干燥剂，在减压下蒸馏收集75~76℃/1.6kPa（12mmHg）的馏分，放入分子筛储存待用。

附录3 常用酸碱溶液的质量分数、相对密度和溶解度

盐　酸

质量分数/%	相对密度	$S(HCl)/g \cdot (100mLH_2O)^{-1}$	质量分数/%	相对密度	$S(HCl)/g \cdot (100mLH_2O)^{-1}$
1	1.0032	1.003	22	1.1083	24.38
2	1.0082	2.006	24	1.1187	26.85
4	1.0181	4.007	26	1.1290	29.35
6	1.0279	6.167	28	1.1392	31.90
8	1.0376	8.301	30	1.1492	34.48
10	1.0474	10.47	32	1.1593	37.10
12	1.0574	12.69	34	1.1691	39.75
14	1.0675	14.95	36	1.1789	42.44
16	1.0776	17.24	38	1.1885	45.16
18	1.0878	19.58	40	1.1980	47.92
20	1.0980	21.96			

硫　酸

质量分数/%	相对密度	$S(H_2SO_4)/g \cdot (100mLH_2O)^{-1}$	质量分数/%	相对密度	$S(H_2SO_4)/g \cdot (100mLH_2O)^{-1}$
1	1.0051	1.005	70	1.6105	112.7
2	1.0118	2.024	80	1.7272	138.2
3	1.0184	3.055	90	1.8144	163.3
4	1.0250	4.100	91	1.8195	165.6
5	1.0317	5.159	92	1.8240	167.8
10	1.0661	10.66	93	1.8279	170.2
15	1.1020	16.53	94	1.8312	172.1
20	1.1394	22.79	95	1.8337	174.2
25	1.1783	29.46	96	1.8355	176.2
30	1.2185	36.56	97	1.8364	178.1
40	1.3028	52.11	98	1.8361	179.9
50	1.3951	69.76	99	1.8342	181.6
60	1.4983	89.90	100	1.8305	183.1

硝 酸

质量分数/%	相对密度	$S(HNO_3)/g \cdot (100mLH_2O)^{-1}$	质量分数/%	相对密度	$S(HNO_3)/g \cdot (100mLH_2O)^{-1}$
1	1.0036	1.004	65	1.3913	90.43
2	1.0091	2.018	70	1.4134	98.94
3	1.0146	3.044	75	1.4337	107.5
4	1.0201	4.080	80	1.4521	116.2
5	1.0256	5.128	85	1.4686	124.8
10	1.0543	10.54	90	1.4826	133.4
15	1.0842	16.26	91	1.4850	135.1
20	1.1150	22.30	92	1.4873	136.8
25	1.1469	28.67	93	1.4892	138.5
30	1.1800	35.40	94	1.4912	140.2
35	1.2140	42.49	95	1.4932	141.9
40	1.2463	49.85	96	1.4952	143.5
45	1.2783	57.52	97	1.4974	145.2
50	1.3100	65.50	98	1.5008	147.1
55	1.3393	73.66	99	1.5056	149.1
60	1.3667	82.00	100	1.5129	151.3

氢 氧 化 钾

质量分数/%	相对密度	$S(KOH)/g \cdot (100mLH_2O)^{-1}$	质量分数/%	相对密度	$S(KOH)/g \cdot (100mLH_2O)^{-1}$
1	1.0083	1.008	28	1.2695	35.55
2	1.0175	2.035	30	1.2905	38.72
4	1.0359	4.144	32	1.3117	41.97
6	1.0554	6.326	34	1.3331	45.33
8	1.0730	8.584	36	1.3549	48.78
10	1.0918	10.92	38	1.3769	52.32
12	1.1108	13.33	40	1.3991	55.96
14	1.1299	15.82	42	1.4215	59.70
16	1.1493	19.70	44	1.4443	63.55
18	1.1588	21.04	46	1.4673	67.50
20	1.1884	23.77	48	1.4907	71.55
22	1.208	26.58	50	1.5143	75.72
24	1.2285	29.48	52	1.5382	79.99
26	1.2489	32.47			

氢 氧 化 钠

质量分数/%	相对密度	$S(NaOH)/g \cdot (100mLH_2O)^{-1}$	质量分数/%	相对密度	$S(NaOH)/g \cdot (100mLH_2O)^{-1}$
1	1.0095	1.010	26	1.2848	33.40
5	1.0538	5.269	30	1.3279	39.84
10	1.1089	11.09	35	1.3798	48.31
16	1.1751	18.80	40	1.4300	57.20
20	1.2191	24.38	50	1.5253	76.27

碳 酸 钠

质量分数/%	相对密度	$S(Na_2CO_3)/g \cdot (100mLH_2O)^{-1}$	质量分数/%	相对密度	$S(Na_2CO_3)/g \cdot (100mLH_2O)^{-1}$
1	1.0086	1.009	12	1.1244	13.49
2	1.0190	2.038	14	1.1463	16.05
4	1.0398	4.159	16	1.1682	13.50
6	1.0606	6.364	18	1.1905	21.33
8	1.0816	8.653	20	1.2132	24.26
10	1.1029	11.03			

氨 水

质量分数/%	相对密度	$S(NH_3)/g \cdot (100mLH_2O)^{-1}$	质量分数/%	相对密度	$S(NH_3)/g \cdot (100mLH_2O)^{-1}$
1	0.9939	9.94	16	0.9362	149.8
2	0.9895	19.79	18	0.9295	167.3
4	0.9811	39.24	20	0.9229	184.6
6	0.9730	58.38	22	0.9164	201.6
8	0.9651	77.21	24	0.9101	218.4
10	0.9575	95.75	26	0.9040	235.0
12	0.9501	114.0	28	0.8980	251.4
14	0.9430	132.0	30	0.8920	267.6

附录4 水的饱和蒸气压

温度/℃	蒸气压/Pa	温度/℃	蒸气压/Pa	温度/℃	蒸气压/Pa	温度/℃	蒸气压/Pa
1	6.57×10^2	26	3.36×10^3	51	1.29×10^4	76	4.02×10^4
2	7.06×10^2	27	3.56×10^3	52	1.36×10^4	77	4.19×10^4
3	7.58×10^2	28	3.78×10^3	53	1.43×10^4	78	4.36×10^4
4	8.13×10^2	29	4.0×10^3	54	1.49×10^4	79	4.55×10^4
5	8.72×10^2	30	4.24×10^3	55	1.57×10^4	80	4.73×10^4
6	9.35×10^2	31	4.49×10^3	56	1.65×10^4	81	4.93×10^4
7	1.0×10^3	32	4.75×10^3	57	1.73×10^4	82	5.13×10^4
8	1.07×10^3	33	5.03×10^3	58	1.81×10^4	83	5.34×10^4
9	1.15×10^3	34	5.32×10^3	59	1.9×10^4	84	5.56×10^4
10	1.23×10^3	35	5.62×10^3	60	1.99×10^4	85	5.78×10^4
11	1.31×10^3	36	5.94×10^3	61	2.08×10^4	86	6.01×10^4
12	1.4×10^3	37	6.23×10^3	62	2.18×10^4	87	6.25×10^4
13	1.5×10^3	38	6.62×10^3	63	2.28×10^4	88	6.49×10^4
14	1.6×10^3	39	6.99×10^3	64	2.39×10^4	89	6.75×10^4
15	1.7×10^3	40	7.37×10^3	65	2.49×10^4	90	7.0×10^4
16	1.81×10^3	41	7.78×10^3	66	2.61×10^4	91	7.28×10^4
17	1.94×10^3	42	8.2×10^3	67	2.73×10^4	92	7.56×10^4
18	2.06×10^3	43	8.64×10^3	68	2.86×10^4	93	7.85×10^4
19	2.2×10^3	44	9.09×10^3	69	2.98×10^4	94	8.14×10^4
20	2.34×10^3	45	9.58×10^3	70	3.12×10^4	95	8.45×10^4
21	2.49×10^3	46	1.01×10^4	71	3.25×10^4	96	8.77×10^4
22	2.64×10^3	47	1.06×10^4	72	3.39×10^4	97	9.09×10^4
23	2.81×10^3	48	1.12×10^4	73	3.54×10^4	98	9.42×10^4
24	2.98×10^3	49	1.17×10^4	74	3.69×10^4	99	9.77×10^4
25	3.17×10^3	50	1.23×10^4	75	3.85×10^4	100	1.013×10^5

附录5 部分共沸混合物的性质

二元共沸混合物的性质

混合物的组分[①]	760mmHg[②]时的沸点/℃		质量分数/%	
	纯组分	共沸物	第一组分	第二组分
水~	100			
甲苯	110.8	84.1	19.6	81.4
苯	80.2	69.3	8.9	91.1
乙酸乙酯	77.1	70.4	8.2	91.8
正丁酸丁酯	125	90.2	26.7	73.3
异丁酸丁酯	117.2	87.5	19.5	80.5

续表

混合物的组分①	760mmHg②时的沸点/℃		质量分数/%	
	纯组分	共沸物	第一组分	第二组分
苯甲酸乙酯	212.4	99.4	84.0	16.0
2-戊酮	102.25	82.9	13.5	86.5
乙醇	78.4	78.1	4.5	95.5
正丁醇	117.8	92.4	38	62
异丁醇	108.0	90.0	33.2	66.8
仲丁醇	99.5	88.5	32.1	67.9
叔丁醇	82.8	79.9	11.7	88.3
苄醇	205.2	99.9	91	9
烯丙醇	97.0	88.2	27.1	72.9
甲酸	100.8	107.2(最高)	22.5	77.5
硝酸	86.0	120.5(最高)	32	68
氢碘酸	−34	127(最高)	43	57
氢溴酸	−67	126(最高)	52.5	47.5
氢氯酸	−84	110(最高)	79.76	20.24
乙醚	34.5	34.2	1.3	98.7
丁醛	75.7	68	6	94
三聚乙醛	115	91.4	30	70
乙酸乙酯	77.1			
二硫化碳	46.3	46.1	7.3	92.7
己烷	69			
苯	80.2	68.8	95	5
氯仿	61.2	60.8	28	72
丙酮	56.5			
二硫化碳	46.3	39.2	34	66
异丙醚	69.0	54.2	61	39
氯仿	61.2	65.5	20	80
四氯化碳	76.8			
乙酸乙酯	77.1	74.8	57	43
环己烷	80.8			
苯	80.2	77.8	45	55

① 下标有"~"符号者为第一组分。
② 760mmHg=101.325kPa。

三元共沸混合物的性质

第 一 组 分		第 二 组 分		第 三 组 分		沸点/℃
名称	质量分数/%	名称	质量分数/%	名称	质量分数/%	
水	7.8	乙醇	9.0	乙酸乙酯	83.2	70.0
水	4.3	乙醇	9.7	四氯化碳	86.0	61.8
水	7.4	乙醇	18.5	苯	74.1	64.9
水	7	乙醇	17	环己烷	76	62.1
水	3.5	乙醇	4.0	氯仿	92.5	55.5
水	7.5	异丙醇	18.7	苯	73.8	66.5
水	0.81	二硫化碳	75.21	丙酮	23.98	38.0

附录6 常用酸碱的相对分子质量及浓度

化合物	相对分子质量	相对密度	质量分数/%	物质的量浓度/mol·L^{-1}
HCl	36.5	1.18	37	12
HNO_3	63.0	1.41	70	16
H_2SO_4	98.1	1.84	98	18
H_3PO_4	98.0	1.69	85	14.7
HCOOH	46.0	1.20	90	23.7
CH_3COOH	60.0	1.06	99.7	17.5
NH_4OH	35.0	0.90	29	7.4

附录7 实验室仪器（Laboratory Equipment）

1. Glass equipment with standard-taper ground glass joints

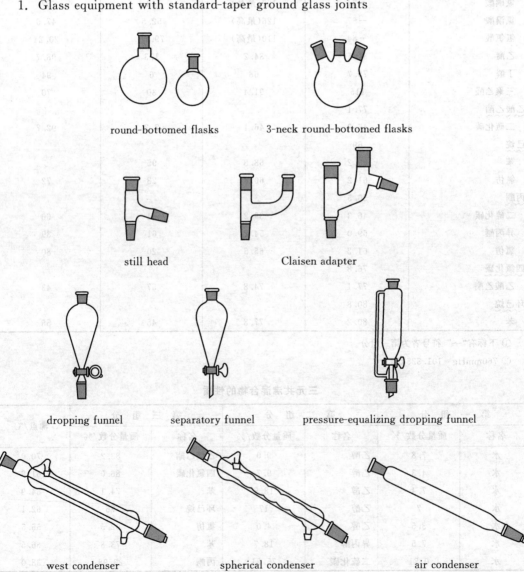

round-bottomed flasks   3-neck round-bottomed flasks

still head   Claisen adapter

dropping funnel   separatory funnel   pressure-equalizing dropping funnel

west condenser   spherical condenser   air condenser

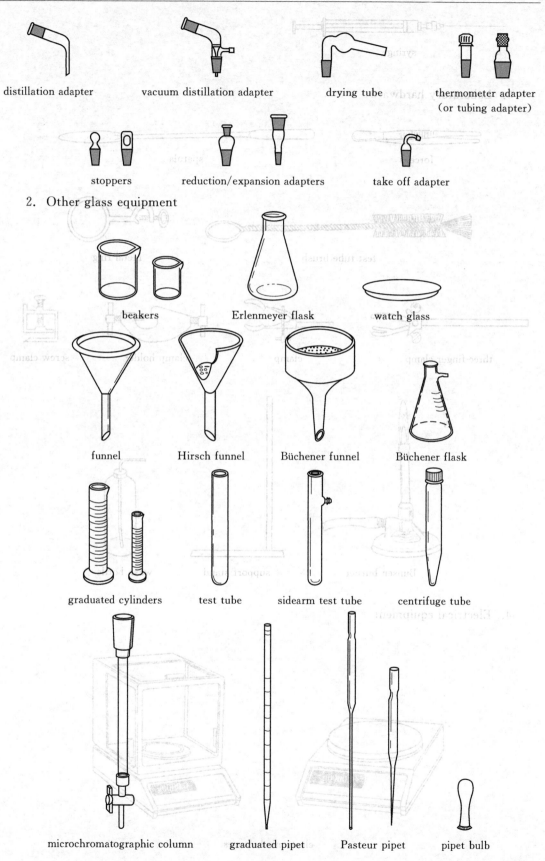

3. Laboratory hardware

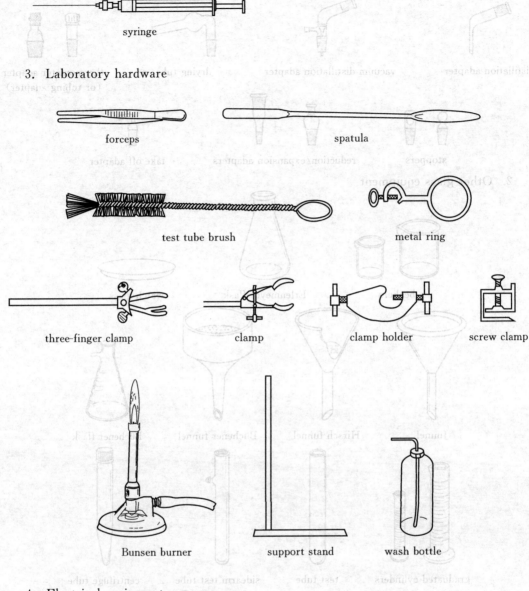

4. Electrical equipment

electronic balances

pneumatic conveyer dryer

heating mantle

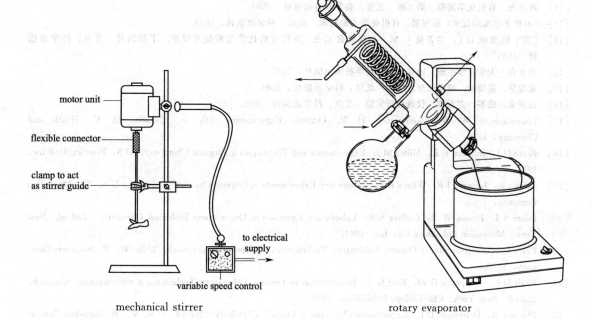

mechanical stirrer

rotary evaporator

参 考 文 献

[1] 曾昭琼. 有机化学实验. 第3版. 北京：高等教育出版社，2000.
[2] 李兆陇，阴金香，林天舒. 有机化学实验. 北京：清华大学出版社，2001.
[3] 关烨第，葛树丰，李翠娟，田桂玲. 小量-半微量有机化学实验. 北京：北京大学出版社，1999.
[4] 焦家俊. 有机化学实验. 上海：上海交通大学出版社，2000.
[5] 兰州大学、复旦大学有机化学教研室. 有机化学实验. 第2版. 北京：高等教育出版社，1994.
[6] 周科衍，高占先. 有机化学实验. 第3版. 北京：高等教育出版社，1996.
[7] 谷珉珉，贾韵仪，姚子鹏. 有机化学实验. 上海：复旦大学出版社，1991.
[8] 傅春玲. 有机化学实验. 杭州：浙江大学出版社，2000.
[9] 奚关根，赵长宏，高建宝. 有机化学实验. 上海：华东理工大学出版社，1999.
[10] 周宁怀，王德琳. 微型有机化学实验. 北京：科学出版社，1999.
[11] 谷亨杰. 有机化学实验. 第2版. 北京：高等教育出版社，2002.
[12] 《有机化学实验技术》编写组. 有机化学实验技术. 北京：科学出版社，1978.
[13] [美] 帕维亚 D L，兰普曼 G M，小克里兹 G S. 现代有机化学实验技术导论. 丁新腾译. 北京：科学出版社，1985.
[14] 印永嘉. 大学化学手册. 山东：山东科学技术出版社，1985.
[15] 常建华，董绮功. 波谱原理及解析. 北京：科学出版社，2001.
[16] 张剑荣，戚苓，方惠群. 仪器分析实验. 北京：科学出版社，2002.
[17] Linstromberg W W, Baumgarten H E. Organic Experiments. 5th ed. Canada：D. C. Heath and Company，1983.
[18] Pasto D J, Johnson C R, Miller M J. Experiments and Techniques in Organic Chemistry. U S：Prentice-Hall Inc，1992.
[19] Adaws R, Johnson J R. Wilcox C F. Laboratory Experiments in Organic Chemistry. 5th ed. U S：The Macmillan Company，1963.
[20] Baum S J. Bowen W R, Poulter S R. Laboratory Exercises in Organic and Biological Chemistry. 2nd ed. New York：Macmillan Publishing Co，Inc，1981.
[21] Pavia D L. Introduction to Organic Laboratory Techniques a contemporary approach. U S：W. B. Saunders Company，1976.
[22] Pavia D L, Lampman G M, Kriz G S. Introduction to Organic Laboratory Techniques a contemporary approach. 2nd ed. New York：CBS College Publishing，1982.
[23] Moore J A, Dalrymple D L. Experimental Methods in Organic Chemistry. 2nd ed. U S：W. B. Saunders Company，1976.
[24] Wilcox C F. Experimental Organic Chemistry Theory and Practice. U S：Macmillan Publishing Company，a division of Macmillan，Inc，1984.
[25] Landgrebe J A. Theory and Practice in the Organic Laboratory. 3rd ed. Canada：D. C. Heath and Company，1982.
[26] Moore J A. Dalrymple D L, Rodig O R. Experimental Methods in Organic Chemistry. 3rd ed. U S：CBS College Publishing，1982.
[27] Harwood L M, Moody C J. Experimental organic Chemistry Principles and Practice. U S：Blackwell Scientific Publications Editorial offices，1989.
[28] Lehman J W. Operational Organic Chemistry A Laboratory Course. U S：Allyn and Bacon，Inc，1981.
[29] Pavia D L, Lampman G M, Kriz G S, Engel R G. Introduction to Organic Laboratory Techniques a small scale approach. U S：Harcourt Brace & Company，1998.
[30] Mcmurry J. Fundamentals of Organic Chemistry. China：Thomson Learning Asia and China Machine Press under the Authorization of Thomson Learning，2004.
[31] Wingrove A S, Caret R L. Organic Chemistry. U S：Alan S. Wingrove and Robert L. Caret，1981.
[32] Jonathan C, Nick G, Stuart W, Peter W. Organic Chemistry. New York：Oxford University Press, 2001.